Soft Computing and Machine Learning

This reference text covers the theory and applications of soft computing and machine learning and presents readers with the intelligent fuzzy and neutrosophic rules that require situations where classical modeling approaches cannot be utilized, such as when there is incomplete, unclear, or imprecise information at hand or inadequate data. It further illustrates topics such as image processing, and power system analysis.

This book:

- Discusses soft computing techniques including fuzzy Logic, rough sets, neutrosophic sets, neural networks, generative adversarial networks, and evolutionary computation
- Examines novel and contemporary advances in the fields of soft computing, fuzzy computing, neutrosophic computing, and machine learning systems, as well as their applications in real life
- Serves as a comprehensive reference for applying machine learning and neutrosophic sets in real-world applications such as smart cities, healthcare, and the Internet of Things
- Covers topics such as image processing, bioinformatics, natural language processing, supply chain management, and cybernetics
- Illustrates classification of neutrosophic machine learning, neutrosophic reinforcement learning, and applications of neutrosophic machine learning in emerging industries

The text is written for senior undergraduate students, graduate students, and academic researchers in the fields of electrical engineering, electronics and communications engineering, computer science and engineering, and information technology.

Computational Methods for Industrial Applications

Series Editor: Bharat Bhushan

In today's world IoT platforms and processes in conjunction with the disruptive blockchain technology and path breaking AI algorithms lay out a sparking and stimulating foundation for sustaining smarter systems. Further computational intelligence (CI) has gained enormous interests from various quarters in order to solve numerous real-world problems and enable intelligent behavior in changing and complex environment. This book series focuses on varied computational methods incorporated within the system with the help of artificial intelligence, learning methods, analytical reasoning and sense making in big data. Aimed at graduate students, academic researchers and professionals, the proposed series will cover the most efficient and innovative technological solutions for industrial applications and sustainable smart societies in order to alter green power management, effect of carbon emissions, air quality metrics, industrial pollution levels, biodiversity and ecology.

Blockchain for Industry 4.0: Emergence, Challenges, and Opportunities
Anoop V.S, Asharaf S, Justin Goldston and Samson Williams

Intelligent Systems and Machine Learning for Industry: Advancements, Challenges and Practices
P.R Anisha, C. Kishor Kumar Reddy, Nguyen Gia Nhu, Megha Bhushan, Ashok Kumar, Marlia Mohd Hanafiah

Sustainability in Industry 5.0
Theory and Applications
Edited by C Kishor Kumar Reddy, P R Anisha, Samiya Khan, Marlia Mohd Hanafiah, Lavanya Pamulaparty and R Madana Mohana

Soft Computing and Machine Learning
A Fuzzy and Neutrosophic View of Reality
Edited by Mohd Anas Wajid, Aasim Zafar, Mohammad Saif Wajid, Akib Mohi Ud Din Khanday and Pronaya Bhattacharya

Soft Computing and Machine Learning

A Fuzzy and Neutrosophic View of Reality

Edited by
Mohd Anas Wajid
Aasim Zafar
Mohammad Saif Wajid
Akib Mohi Ud Din Khanday
Pronaya Bhattacharya

CRC Press
Taylor & Francis Group
Boca Raton London New York

CRC Press is an imprint of the
Taylor & Francis Group, an **informa** business

First edition published 2025
by CRC Press
2385 NW Executive Center Drive, Suite 320, Boca Raton FL 33431

and by CRC Press
4 Park Square, Milton Park, Abingdon, Oxon, OX14 4RN

CRC Press is an imprint of Taylor & Francis Group, LLC

ISBN: 978-1-032-74632-6 (hbk)
ISBN: 978-1-032-99795-7 (pbk)
ISBN: 978-1-003-60605-5 (ebk)

DOI: 10.1201/9781003606055

Typeset in Sabon
by SPi Technologies India Pvt Ltd (Straive)

Contents

About the Editors

Mohd Anas Wajid is a Postdoctoral Research Associate in Data Science at the Institute for the Future of Education, TEC de Monterrey, Mexico. He is also Assistant Professor at Galgotias University, Greater Noida, India, in the Department of Computer Science and Application. He received his PhD degree in Computer Science from Aligarh Muslim University, India. He was awarded with the MITACS-SICI Globalink Research Award by Mitacs in collaboration with HRD ministry, Government of India to do a project at the University of Athabasca, Edmonton, Alberta, Canada. ACM India Council named him an ACM India Anveshan Setu Fellow, and he received a fellowship to conduct a part of his research at IIIT-Delhi. He was also a recipient of Maulana Azad National Fellowship (SRF) from UGC, Government of India. He qualified various prestigious national exams such as UGC-NET and GATE multiple times. His contribution to Neutrosophic research earned him a Diploma from the Neutrosophic Science International Association (NSIA), University of New Mexico, United States (USA). He has several research papers to his credit in refereed journals and conferences of international repute such as *Journal of Cloud Computing*, *Nature Scientific Reports*, *Journal of Computational Intelligence & Neuroscience*, and *Neutrosophic Sets & Systems*, etc. He has co-authored one book and two patents. He has keen interest in Soft Computing, Machine Learning, Data Science, Information Retrieval, and Neutrosophy. He has academic as well as industrial experience.

Prof. Aasim Zafar is a Senior Professor at Computer Science Department, Aligarh Muslim University, Aligarh, India. He holds a master's in computer science and applications and obtained the degree of PhD in Computer Science from Aligarh Muslim University, Aligarh, India. His research areas and special interests include Mobile Ad hoc and Sensor Networks, Software Engineering, Information Retrieval, E-Systems, e-Security, Virtual Learning Environment, Neuro-Fuzzy and Soft Computing, Knowledge Management Systems and Web Mining. His areas of teaching interest include Computer Networks, Network Security, Software Engineering, E-Systems, Database Management Systems and

Computer Programming. He has presented many papers in National and International Conferences and published various research papers in journals of international repute.

Mohammad Saif Wajid received bachelor's degree in computer science and engineering from Dr. A.P.J. Abdul Kalam Technical University, India, and Master's Degree from BBD University, India. He received his Full-Time PhD degree from Tecnologico de Monterrey, Monterrey Campus, Mexico. He also worked as a Visiting Research Scholar at the University of Texas, San Antonio, USA. For the past eight years, he has been working as an Assistant Professor in Computer Science and Engineering and worked as a Software Engineer for a year in India. His research jumps around the potential applications of Digital Twin, Knowledge Graph, Machine Learning, Computer Vision, NLP, and Neutrosophy, touching areas such as Violence Detection, Healthcare Systems, Dense Captioning of images/videos, and DTWIN for Smart cities. He is passionate about digital learning, real-time and asynchronous, and an early adopter of emerging software and hardware solutions to facilitate teaching and research. His other favored tool is LaTeX, which he likes to use for all academic writings and presentations. He has published over 25 research papers in international journals, conferences, and one book. He has supervised eight master's thesis.

Akib Mohi Ud Din Khanday has completed his PhD from Department of Computer Sciences, Baba Ghulam Shah Badshah University, Rajouri, India. He has received his master's degree in information technology from Islamic University of Science and Technology, Jammu and Kashmir, India. The author has qualified UGC-NET (December 2020) in Computer Science and Applications. He has worked as an Assistant Professor in PG Department of Information Technology, Cluster University Srinagar for the year 2022. He is currently associated with United Arab Emirates University Al Ain, UAE as a post-doctoral researcher. His research interests are Computational Social Sciences, Computations for Social Good, NLP and Machine/Deep Learning. He has authored many research articles in the reputed journals indexed in SCOPUS and Web of Science. He has presented his work in many international conferences. His research has been cited by many researchers around the Globe. He has served as a reviewer in reputed journals over the years like Scientific Reports, Journal of Social Science and Humanities, IEEE Access, Processes, etc. He has also served as the TPC member of various reputed International Conferences.

Pronaya Bhattacharya is currently employed as an Associate Professor in the Department of Computer Science and Engineering, Amity School of Engineering and Technology and Research and Innovation Cell, Amity University, Kolkata, West Bengal, India. He completed his PhD in Optical

Networks from Dr. A. P. J Abdul Kalam Technical University in 2021. He has over ten years of teaching experience and 2 years of Industrial experience. He has authored or co-authored more than 110 research papers in leading SCI/SCIE journals and top core IEEE COMSOC A* conferences. Some of his top-notch findings are published in reputed SCI journals. He has an H-index of 22 and an i10-index of 46 as per Google Scholar. His research interests include healthcare analytics, optical switching and networking, federated learning, blockchain, and the IoT. Four of his works on COVID-19 have been added by the World Health Organization (WHO) in its research database for the technical novelty of proposed solutions which orients the societal aspect. He has been granted three international patents in the fields of healthcare, blockchain, and network communications. He has been appointed in the capacity of a Keynote Speaker, Technical Committee Member, Program Chair, and the Session Chair across the globe. He was awarded nine best paper awards in Springer ICRIC-2019, IEEE-ICIEM-2021, IEEE-ECAI-2021, Springer COMS2-2021, and IEEE-ICIEM-2022. He has three internationally granted patents to his credit. He is a Reviewer of 25 reputed SCI journals, such as *IEEE Internet of Things Journal*, *IEEE Transactions on Industrial Informatics*, *IEEE Transactions of Vehicular Technology*, *IEEE Journal of Biomedical and Health Informatics*, *IEEE Access*, *IEEE Network*, *ETT*, *IJCS*, *MTAP*, *OSN*, *WPC*, and others.

Contributors

Afa
Information and Technology Group
Indian Institute of Management (IIM)
Kolkata, India

Agrawal Atul
Department of Computer Science and Engineering
ITS Engineering College, (AKTU)
Greater Noida, India

Alam Mahfooz
Department of Computer Science
Aligarh Muslim University
Aligarh, India

Baniya Pashupati
Department of Computer Science and Engineering
ITS Engineering College, (AKTU)
Greater Noida, India

Bhushan Bharat
Department of Computer Science and Engineering
School of Engineering and Technology
Sharda University
Greater Noida, India

Dass Gurcharan
Manohar Memorial PG College
Fatehabad, India

Debnath Bijoy Krishna
Department of Applied Sciences
School of Engineering
Tezpur University
Tezpur, India

Dowlatshahi Mohammad Bagher
Department of Computer Engineering
Faculty of Engineering
Lorestan University
Khorramabad, Iran

Enam Marghoob
Department of Commerce
Aligarh Muslim University
Aligarh, India

Faisal Syed Mohd
Department of CSE
Koneru Lakshmaiah Education Foundation
Vijayawada, India

Goswami Shivani
Motilal Nehru National Institute of Technology
Prayagraj, India

Hashemi Amin
Department of Computer Engineering
Faculty of Engineering
Lorestan University
Khorramabad, Iran

Hassan Shabbir
Department of Commerce
Aligarh Muslim University
Aligarh, India

Ishrat Mohammad
Department of CSE
Koneru Lakshmaiah Education Foundation
Vijayawada, India

Jain Vishal
Department of Computer Science and Engineering
School of Engineering and Technology
Sharda University
Greater Noida, India

Kaunert Christian
Dublin City University
Dublin, Ireland
University of South Wales
Newport, United Kingdom

Madas Michael
School of Information Sciences
Department of Applied Informatics
University of Macedonia
Thessaloniki, Greece

Miri Mohsen
Department of Computer Engineering
Faculty of Engineering
Lorestan University
Khorramabad, Iran

Mohammad Umair Rizwan Khan
Aligarh Muslim University
Aligarh, India

Mohammed Al Farsi Monia
Department of Information Technology
University of Technology & Applied Science
Mussanah, Oman

Pandey Shraiyash
Department of Computing and Information Technology
Purdue University
West Lafayette, Indiana

Paraskevas Antonios
School of Information Sciences
Department of Applied Informatics
University of Macedonia
Thessaloniki, Greece

R Khan Mohd Ovais
Sapienza University Rome
Rome, Italy

R Narmadhagnanam
Ramanujan Research Centre
P.G. & Research Department of Mathematics
Government Arts College
Kumbakonam, India

Sahoo Brajamohan
Department of Applied Sciences
School of Engineering
Tezpur University
Tezpur, India

Samuel A. Edward
Ramanujan Research Centre
P.G. & Research Department of Mathematics
Government Arts College
Kumbakonam, India

Sheikhi Akram
Electrical Engineering Department
Lorestan University
Khorramabad, Iran

Singh Anil Kumar
Motilal Nehru National Institute of Technology
Prayagraj, India

Singh Bhupinder
Sharda University
Greater Noida, India

Singh Vikram
Computer Science & Engineering
Chaudhary Devi Lal University
Sirsa, India

Tahir Abdullah
Université de Franche Comte
Besançon, France

Vojka Ornel
Université de Franche Comte
Besançon, France

Wajid Mohd Anas
Institute for the Future of Education
Tecnologico de Monterrey
Monterrey, Mexico

Wasim Khan
Department of CSE
Koneru Lakshmaiah Education Foundation
Vijayawada, India

Chapter 1

Enhancing the performance of power amplifiers by leveraging on fuzzy logic techniques

Akram Sheikhi and Mohammad Bagher Dowlatshahi

1.1 INTRODUCTION

Fuzzy logic is employed in power amplifiers (PAs) to enhance their performance, particularly in scenarios where traditional control methods may struggle to handle the complexity of nonlinear and imprecise systems. Fuzzy logic provides a mathematical framework that can model and control systems with uncertainty, imprecision, and varying conditions. The chapter begins by introducing PAs and fuzzy logic as a computational paradigm inspired by human reasoning and decision-making. It addresses the need to handle uncertainty and imprecision in PA design. Fuzzy logic techniques are used to optimize impedance matching in PAs. This is crucial for maximizing power transfer efficiency. Also, fuzzy logic techniques are utilized to enhance the efficiency of PAs. This is a key consideration in PA design to reduce power consumption and heat dissipation. Moreover, fuzzy logic is employed for controlling linearity in PAs. Linearity is essential for ensuring accurate amplification of the input signal without distortion. The chapter reviews different fuzzy logic techniques and adaptive neuro-fuzzy inference systems. These are applied specifically to PAs to improve linearity and efficiency. And introduces a novel approach called Continuous Fuzzy Logic Mode Technique (CFLMT) to address bandwidth limitations in continuous-mode-based Doherty power amplifiers (DPAs). This technique combines a continuous-mode technique and a K-means unsupervised learning clustering algorithm, leading to a significant increase in fractional bandwidth.

1.1.1 Introduction to PA

A switch-mode power amplifier (SMPA) is a type of electronic amplifier that uses switching devices, such as transistors, to regulate the flow of electrical energy. Unlike traditional linear amplifiers, which operate in their active region to amplify signals continuously, SMPAs utilize a switching process to efficiently control the power delivered to the load. This design offers advantages in terms of power efficiency, size, and weight, making switch-mode PAs particularly suitable for various applications, including audio

DOI: 10.1201/9781003606055-1

amplification, radio frequency (RF) communication, and power supply systems. The fundamental principle behind switch-mode PAs involves rapidly switching the input signal between fully on and fully off states. This on-off switching generates a series of pulses that can be controlled to replicate the characteristics of the input signal. The amplified output is then reconstructed by a low-pass filter, which smoothens the pulsed waveform, resulting in a faithful reproduction of the original signal with minimal distortion. Key features and benefits of switch-mode PAs include high efficiency, reduced heat dissipation, and the ability to handle a wide range of input voltages. These amplifiers are widely employed in applications where energy efficiency and compact design are crucial considerations. In summary, switch-mode PAs offer an innovative approach to amplification by leveraging switching technology, providing advantages in terms of efficiency and size, making them well-suited for various electronic applications.

The design of PAs is crucial in wireless communication, radar, etc. One of the key challenges in amplifier design is finding the right balance between efficiency and linearity. Efficiency refers to how effectively the amplifier converts input power into output power, while linearity ensures signal integrity, especially in applications like wireless communication where high linearity is essential. Class-A, -B, and -C PAs operate based on specific bias points and conduction angles. Class-A operates in the active region of the transistor, providing equal amplitudes for both DC and fundamental components of the current waveform. However, this results in a drain efficiency of 50%. Class-B and Class-C operate with reduced conduction angles, leading to decreased fundamental components and increased efficiency. The choice of amplifier class depends on the application's requirements, whether it prioritizes, output power, efficiency, gain, linearity, and bandwidth. In practical scenarios, the design of a PA involves meeting specific requirements related to linearity, efficiency, gain, bandwidth, and output power, often necessitating a combination of these factors. Achieving high linearity and efficiency simultaneously is challenging because the relationship between input and output power in PAs is complex. The output power in the linear region ideally contains only the fundamental frequency to ensure a fixed gain. However, operating the PA in the compression region can enhance efficiency but at the expense of degraded linearity. The challenge for designers lies in optimizing the PA's performance to meet the specific demands of the application, ensuring that the trade-offs between efficiency and linearity are carefully managed. This delicate balance is crucial for the overall functionality and energy consumption of systems employing PAs. Figure 1.1 shows the relationship between the output power and the corresponding input power. In the linear region, the output power ideally contains only the fundamental frequency of the input signal. The gain of the amplifier is fixed within this linear range. The linear operation ensures that the output signal faithfully reproduces the input signal without introducing significant distortion or harmonics. This is crucial in applications where signal integrity is a top

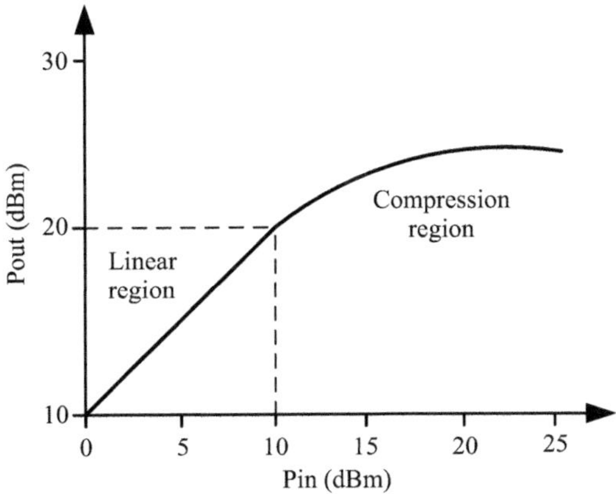

Figure 1.1 Linear and compression region of PAs.

priority, such as in high-quality audio or sensitive communication systems. When the PA operates in the compression region, it means the amplifier is being driven to its maximum output power capacity and efficiency is improved. However, operating in the compression region comes at a cost: the linearity of the amplifier is degraded. Linearity degradation means that the output signal starts to deviate from the ideal replica of the input signal. This deviation can result in signal distortion and the generation of harmonics, which are frequencies that are integer multiples of the input signal frequency. In applications where high-efficiency PAs are crucial, such as RF transmission, this trade-off between efficiency and linearity might be acceptable. However, in applications where signal fidelity is paramount, such as in high-quality audio systems, maintaining linearity is essential even at the expense of efficiency.

In PA design, particularly in applications where constant envelope modulation techniques like Binary Phase Shift Keying (BPSK) and Frequency Shift Keying (FSK) are used, there is often a delicate balance between linearity and efficiency. Constant envelope modulation means that the amplitude of the transmitted signal remains constant, making these techniques more tolerant to amplitude distortion. In such cases, designers can sacrifice some linearity for improved efficiency and higher output power. One way to enhance efficiency is by using overdriven PAs. Overdriven amplifiers are designed to operate in the compression region where the amplifier is driven to its maximum output power capacity. This approach increases efficiency but typically results in some distortion in the output signal. To manage this distortion, appropriate harmonic terminations are used to achieve efficient current and voltage waveforms. For instance, Class-F amplifiers represent one approach to achieving high efficiency. They transform the sine voltage waveform to a square waveform at the output of the active device. This transformation helps in reducing the overlap of the switch waveforms, minimizing power dissipation as heat, and enhancing overall efficiency. This chapter provides valuable insights for designers aiming for the design of PAs for specific applications, making informed decisions regarding the trade-offs between linearity and efficiency based on bias point and harmonic control techniques.

1.1.2 Bias classes of power amplifier

In the 1930s, amplifiers operating with 100% and 50% duty ratios were referred to as Class-A and Class-B PAs, respectively [1]. The operation of these amplifiers can be analyzed by considering an active device as an ideal current or voltage-controlled current source.

1.1.2.1 Class-A power amplifier

In Class-A amplifiers, the amplifying device conducts throughout the entire input cycle. This means the conduction angle θ is 360 degrees, representing 100% of the input cycle, Figure 1.2. Class-A amplifiers have the advantage of linearity but are highly inefficient usually around 25–30% efficiency as they are in the "on" state all the time.

1.1.2.2 Class-B power amplifier

As can be seen from Figure 1.3, Class-B amplifiers conduct for 180 degrees of the input cycle, [2, 3]. Class-B amplifiers are more efficient than Class-A amplifiers typically around 78.5% theoretical maximum efficiency but suffer from crossover distortion, where the transition between the on and off states distorts the output waveform. To improve efficiency, especially for high-power applications, variations of Class-B amplifiers are used.

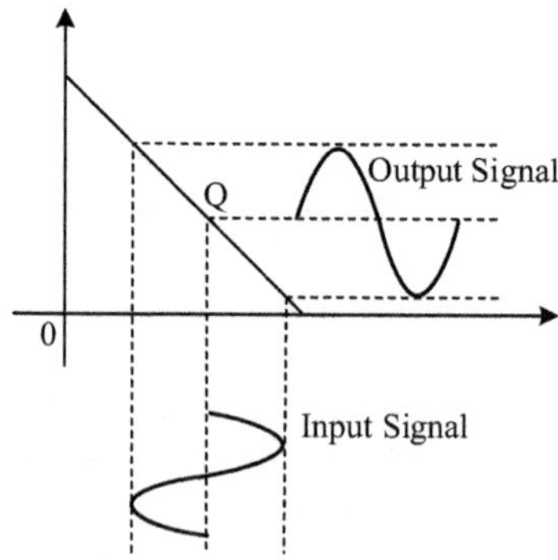

Figure 1.2 Waveforms in Class-A operation.

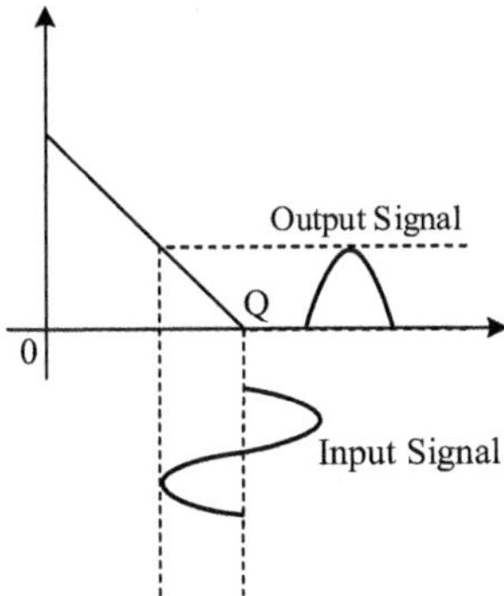

Figure 1.3 Waveforms in a Class-B operation.

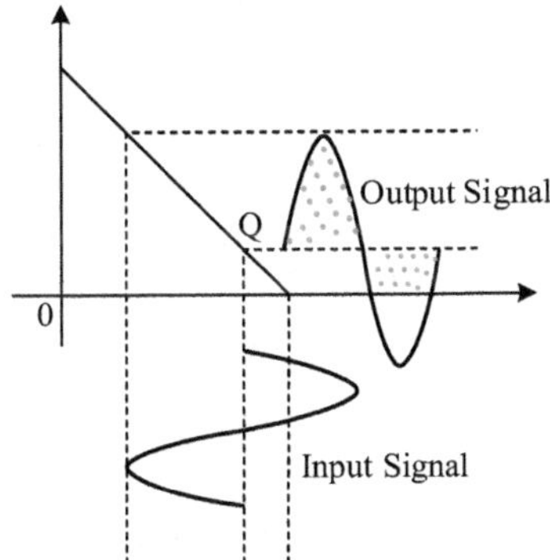

Figure 1.4 Waveforms in a Class-AB operation.

1.1.2.3 Class-AB power amplifier

Class-AB amplifiers operate between Class-A and Class-B, conducting slightly more than half of the input cycle. This reduces crossover distortion and increases efficiency compared to pure Class-B amplifiers. The input and output waveforms in Class AB are shown in Figure 1.4.

1.1.2.4 Class-C power amplifier

Class-C amplifiers conduct for less than half of the input cycle, usually around 120–150 degrees, Figure 1.5. They are highly efficient more than 80% but are suitable only for applications where high efficiency is more critical than waveform fidelity such as in RF applications, [2, 3].

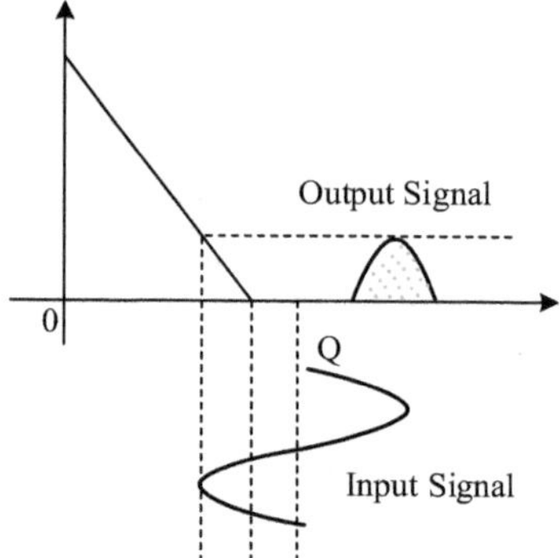

Figure 1.5 Waveforms in a Class-C operation.

1.1.3 Classes of amplifiers based on a finite number of Harmonics

PAs can produce harmonic components in addition to the fundamental frequency of the input signal. Harmonics are multiples of the fundamental frequency and can be present in both the voltage and current waveforms of the amplifier. In Class-AB, -B, or -C operation, the drain current consists of harmonics. These classes of operation are common in PAs and have specific conduction angles, leading to different harmonic content in the output. Class-F and Class-E operation modes involve specific configurations where certain harmonic components are present either in the drain voltage or drain current. These modes are engineered to achieve higher efficiency by utilizing specific harmonics.

The efficiency of the PA is influenced by the number of harmonic components utilized, rather than the class of operation. Proper adjustment of waveforms and the fundamental-frequency load reactance allows for achieving maximum efficiency. This means that different classes of operation can be optimized to achieve similar efficiency levels by adjusting the harmonic components and load parameters. In the realm of wireless communication and active radar applications, PAs play a pivotal role by amplifying various RF input signals to provide ample transmitting power to antennas. Similarly, in the medical field, RF power sources are indispensable for applications like induction heating, plasma generation, RF-driven lighting, and imaging processes. Each unique application demands a specific set of requirements from the PA. Consequently, designers are tasked with meticulously evaluating the crucial performance factors tailored to each application. This involves the judicious selection or invention of suitable components and techniques, ranging from transistors to the overall architecture. The bespoke nature of this process renders PAs among the costliest components within a given system.

In switching mode operation, transistors function as switches, necessitating appropriate output harmonic termination. However, the linearity of switching mode amplifiers is compromised due to the deep saturation region they are driven into. As a result, these amplifiers find common use in RF power generation systems within the Industrial, Scientific, and Medical (ISM) sectors, as well as in communication systems modulated with constant-envelope signals. Designing switching-mode PAs entails a multitude of considerations, including selecting the right operating class, transistors, output power, gain, efficiency, operating frequency, bandwidth, stability, and linearity. This section delves into various types of switching-mode PAs. Achieving high efficiency in PAs involves employing biharmonic or polyharmonic manipulation through single or multiresonant circuits tuned to even or odd harmonics. The discussion in this chapter provides a concise overview of PA types, including Class-F [4], inverse Class-F [5], Class-E [6], inverse Class-E [7], and mixed mode amplifiers such as Class-DE [8], Class-E/F [9, 10], and Class-F/E [11, 12] amplifiers, Figure 1.6. The exploration

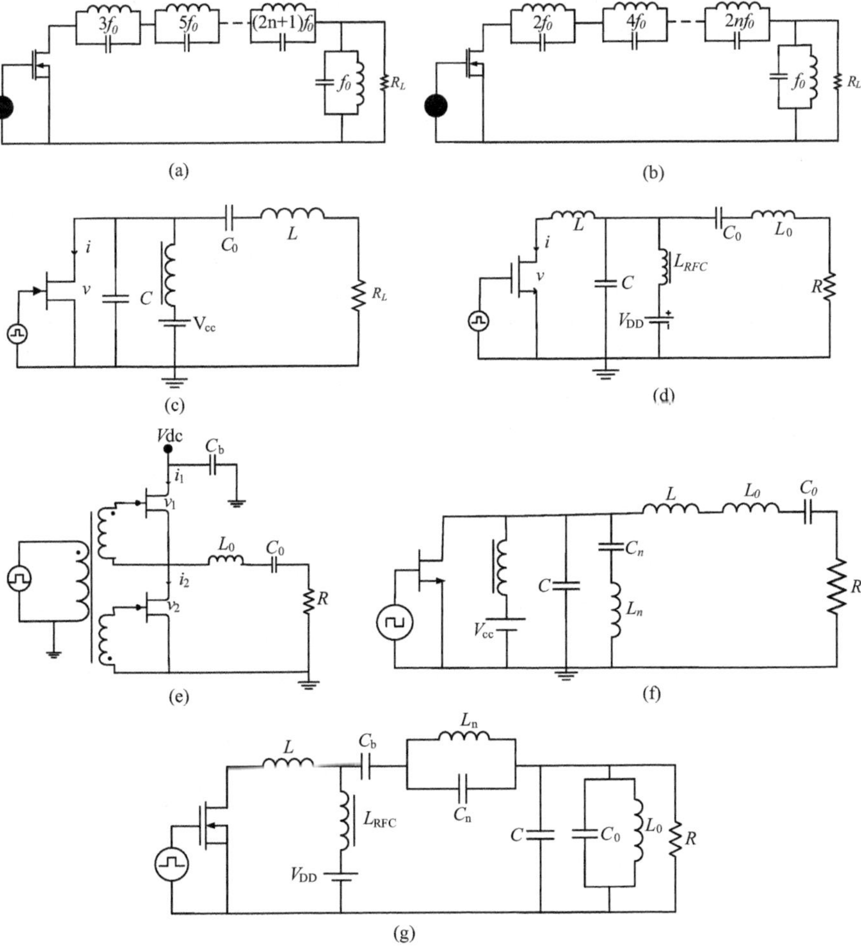

Figure 1.6 The structure of the (a) Class-F, and (b) inverse Class-F, (c) Class-E, (d) inverse Class-E, (e) Class-DE, (f) Class-E/F, (g) and Class-F/E.

commences with simpler approaches like Class-F and inverse Class-F amplifiers, gradually introducing more complex amplifier designs.

1.1.4 Introduction to fuzzy logic

Fuzzy logic is a computational paradigm inspired by human reasoning and decision-making processes that involve uncertainty, imprecision, and vagueness. Developed by Lotfizadeh in the 1960s, fuzzy logic provides a framework for dealing with situations where traditional binary (true or false) logic may be insufficient or overly rigid. This paradigm recognizes the nuanced and ambiguous nature of many real-world problems, allowing for more flexible and human-like reasoning. At the core of fuzzy logic is the concept of "fuzziness" or "degree of membership," where instead of clear-cut distinctions between true and false, values exist on a spectrum between 0 and 1. This spectrum enables the representation of uncertainty in a manner that mirrors human cognitive processes. In traditional logic, an object or condition is either a member or non-member of a set, expressed as 1 or 0. In contrast, fuzzy logic introduces the idea that an element can belong to a set to a certain degree, expressed as a value between 0 and 1. The fundamental components of fuzzy logic include fuzzy rules, linguistic variables, an inference engine, and fuzzy sets. Linguistic variables are terms used to describe imprecise or vague concepts, like "high" or "low." Fuzzy sets define the degree of membership of an element about a linguistic variable. Fuzzy rules are conditional statements that relate these fuzzy sets, and the inference engine processes these rules to make decisions or derive conclusions. Fuzzy logic finds applications in a wide range of fields, including control systems, artificial intelligence, decision-making, pattern recognition, and various engineering disciplines. Its ability to model and handle uncertainty has proven particularly valuable in situations where precise mathematical models may be challenging to define. Overall, fuzzy logic provides a powerful tool for capturing the richness and complexity of real-world problems that go beyond the binary constraints of traditional logic, [13–15].

Fuzzy logic techniques in PA design involve leveraging the principles of fuzzy logic, a mathematical framework that handles uncertainty and imprecision, to enhance the efficiency and performance of PAs. In the context of PAs, fuzzy logic techniques are employed to optimize the design process, improve system robustness, and adapt to varying operating conditions. The application of fuzzy logic techniques in PA design provides a flexible and adaptive framework for addressing challenges such as impedance matching, efficiency optimization, linearization, and robustness in the face of changing operating conditions. The incorporation of fuzzy logic contributes to the creation of high-performance PAs capable of meeting the demands of modern communication systems. Here are key aspects of applying fuzzy logic techniques in PA design:

- Impedance Matching Optimization: Fuzzy logic can be used to optimize the impedance-matching network in PAs. By considering various operating conditions and input parameters, fuzzy logic controllers can adjust the matching network to ensure optimal power transfer.

- Adaptive Biasing: Fuzzy logic controllers can dynamically adjust the biasing of PA stages based on the input signal characteristics and environmental conditions. This adaptability improves the overall efficiency and linearity of the amplifier.
- Efficiency Enhancement: Fuzzy logic techniques can be applied to control the operating point of the PA, aiming to achieve the highest efficiency while maintaining acceptable linearity. This involves dynamically adjusting parameters such as biasing and supply voltage.
- Linearization and Distortion Control: Fuzzy logic controllers can be implemented for linearization purposes, mitigating nonlinearity in the amplifier's response. By adjusting the bias points and gain based on the input signal, fuzzy logic can help minimize distortion and enhance linearity.
- Load Modulation Compensation: Fuzzy logic can be utilized to compensate for variations in load conditions, ensuring that the PA adapts to changes in the load impedance. This adaptive approach enhances the robustness of the amplifier across different operating scenarios.
- Clustering Algorithms for Design Space Exploration: Fuzzy logic, in combination with clustering algorithms such as K-means, can be employed for efficient exploration of the amplifier design space. This aids in identifying optimal configurations and impedance values for different amplifier stages.

Following this introduction to the application of fuzzy logic techniques in PA design, the paper will delve into specific areas of enhancement. The next section will focus on the optimization aspects, including impedance matching, adaptive biasing, efficiency enhancement, linearity control, and load modulation compensation using fuzzy logic. Subsequently, a review of various fuzzy logic techniques such as interval type-2 fuzzy inference engines, self-organizing fuzzy neural networks, and adaptive neuro-fuzzy inference systems applied to PAs will be presented. The chapter will then introduce the innovative CFLMT for addressing bandwidth limitations in continuous-mode-based DPAs. Following this, the chapter will discuss improvements in dynamic response, compensation of nonlinear distortion, and linearization of RF PAs through the application of fuzzy logic techniques. Each section will provide insights into how fuzzy logic contributes to these enhancements, offering a comprehensive view of the role of fuzzy logic in advancing PA performance.

1.2 VARIOUS FUZZY LOGIC TECHNIQUES IN PAs

Various fuzzy logic techniques have been explored in the field of RF and microwave research to address linearization challenges and minimize power consumption in PAs, Figure 1.7. These techniques include the interval type-2 fuzzy inference engine [16], self-organizing fuzzy neural networks [17],

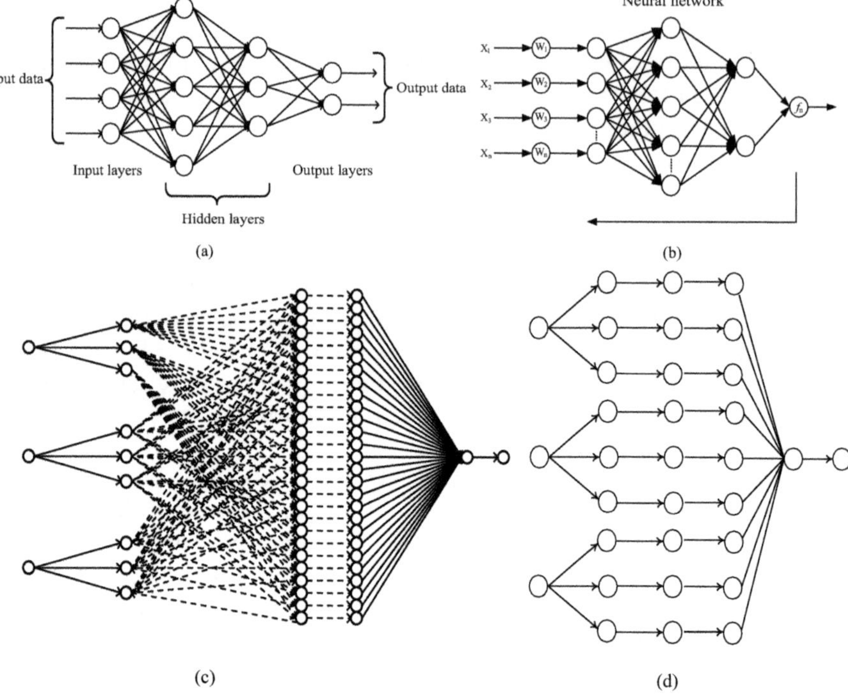

Figure 1.7 (a) Multilayer perceptron, (b) RBF, (c) ANFIS architecture, and (d) MANFIS architecture.

radial-basis function neural networks (RBFNNs) [18], adaptive neuro-fuzzy inference system (ANFIS) [19], modified adaptive neuro-fuzzy inference system (MANFIS) [20], and other techniques [21–26]. Researchers in this domain have successfully employed these methods to significantly enhance the linearity performance of reported amplifiers.

In [18], an RBFNN was employed to capture the dynamic nonlinear characteristics of a third-generation RF PA, utilizing the signal's envelope in the modeling process. This approach demands less training effort compared to a model relying on *IQ* data. The identification and validation stages involved sampled input and output signals, incorporating noise-like signals with bandwidths of 4 and 20 MHz. A comparative analysis was conducted with a parallel Hammerstein (PH) model, revealing similar performance between the two models in the absence of memory. In the case of the 4-MHz signal, the RBFNN and PH models exhibited superior in-band and out-of-band performance, respectively.

1.2.1 Various fuzzy logic techniques for bandwidth extension in PAs

Despite the notable improvements achieved through the application of fuzzy logic techniques, an aspect that remains unexplored is the consideration of the fractional bandwidth (FBW) performance of the amplifiers. In the reported studies, the PAs were primarily designed to operate at single frequency points. This focus can be attributed to the underlying motivation of these designs, which primarily revolves around augmenting the linearity performance in amplifiers. Future investigations could potentially delve into optimizing both linearity and fractional bandwidth to achieve a more comprehensive understanding of PA behavior and performance. Numerous techniques and architectural approaches grounded in the Doherty operating principle have been put forth to address bandwidth extension, as documented in a range of studies. Among these methodologies, notable examples include the real frequency, Bayesian optimization, and the continuous-mode techniques [28–34]. For instance, Sun and Jansen [28] introduced a design for two-way symmetrical Doherty PAs (TW-SDPAs) utilizing the real frequency technique. However, the recorded fractional bandwidth (FBW) in their implementation fell short, registering at less than 30%. Similarly, [29] also applied Bayesian optimization in their TW-SDPA design, but the measured FBW yielded results below 47%. In [30–34], the reported bandwidths remained below 52%, despite the positive impact of the continuous-mode technique on the efficiency. The authors [35–37] present TW-SDPAs without employing the continuous-mode technique in their designs. These endeavors collectively highlight the ongoing pursuit of achieving optimal bandwidth extension in Doherty-based architectures.

In response to the bandwidth limitations observed in TW-SDPAs shown in Figure 1.8, the authors [27] introduce a novel approach termed CFLMT. The flowchart of the proposed CFLMT is shown in Figure 1.9. The proposed technique integrates a continuous-mode technique and K-means unsupervised learning clustering algorithm within a fuzzy logic system environment, Figure 1.10. This innovative combination aims to effectively extend the bandwidth of TW-SDPAs. The CFLMT methodology directly addresses the bandwidth challenge by solving PA design parameters in sub-clustered regions. This results in the automatic determination of optimal impedances required by the carrier and peaking sub-amplifiers to operate efficiently at saturation and output-power-back-off (OPBO) levels. While the primary focus is on resolving the bandwidth issue in continuous-mode-based TW-SDPAs, the article also considers crucial performance indicators such as output power, efficiency, and gain. To validate the effectiveness of the proposed technique, a 1.2–2.4-GHz TW-SDPA with an impressive 66.7% fractional bandwidth (FBW) was designed, implemented, and measured. A comparative analysis of the FBW performance reported by Chen et al. [30–32] and Shi et al. [33, 34] reveals a notable success for the proposed TW-SDPA. The results showcase a remarkable 15.5%–29.7% increase in FBW, underscoring the efficacy of the CFLMT in addressing and surpassing the bandwidth limitations previously observed in the proposed TW-SDPAs. The results of the proposed DPA in [33] are shown in Figures 1.11 and 1.12.

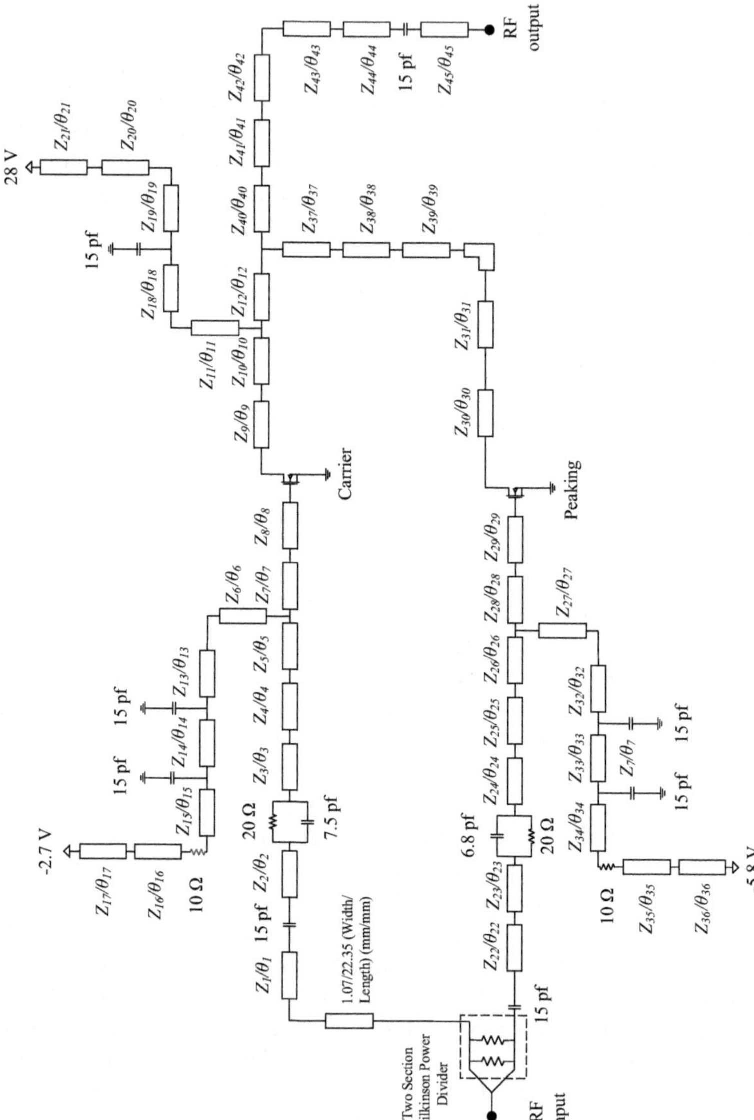

Figure 1.8 Complete schematic of the proposed TW-SDPA in [27].

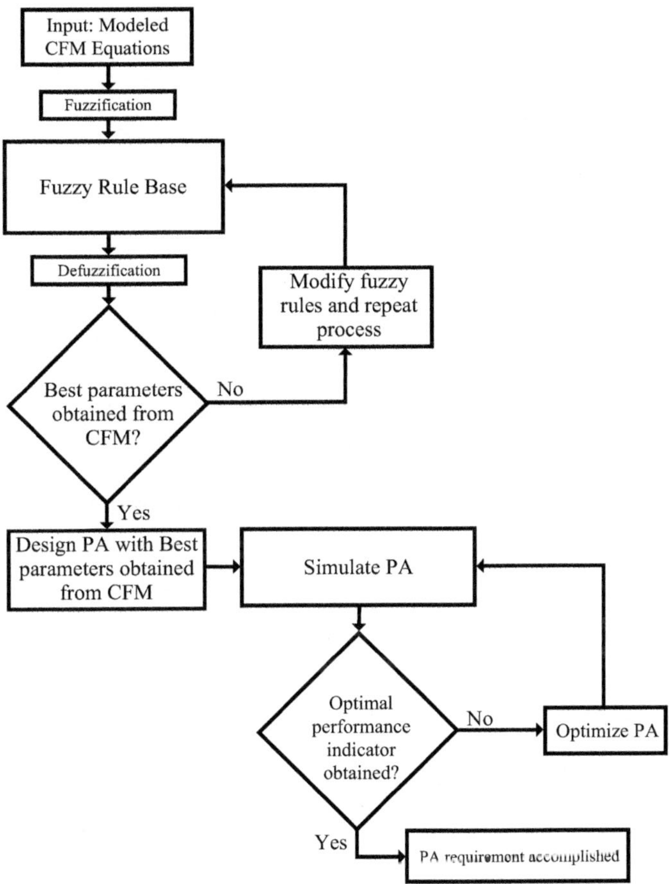

Figure 1.9 The proposed CFLMT in [27].

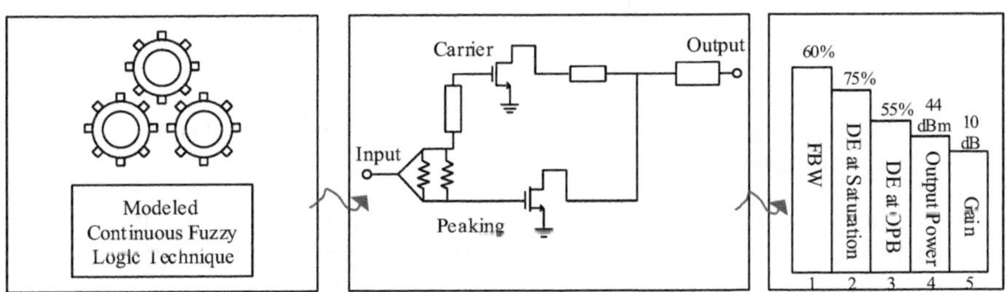

Figure 1.10 The proposed TW-SDPA design steps [27].

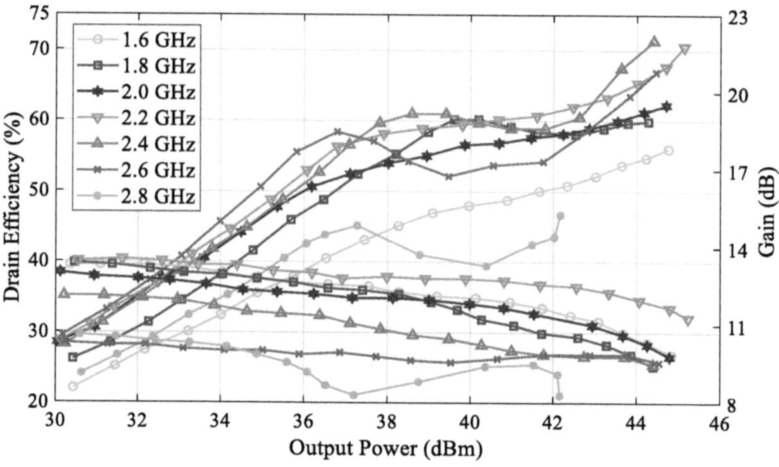

Figure 1.11 The measured efficiencies and gains of the proposed DPA [33].

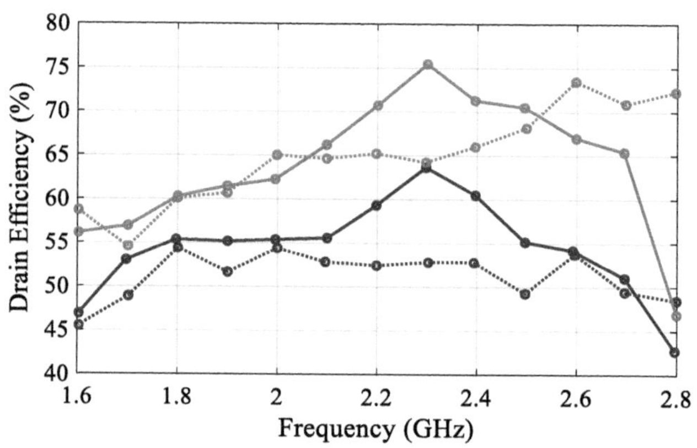

Figure 1.12 The measured and simulated drain efficiencies at 6-dB OBO and maximum power levels of the proposed DPA [33].

1.2.2 Fuzzy logic techniques for improved dynamic response in PAs

The authors in [18] introduce an innovative framework for dynamically modeling the behavior of RF PAs utilizing an ANFIS-based Hammerstein model followed by a Finite Impulse Response (FIR) filter. For parameter identification in the ANFIS, a hybrid learning algorithm is employed, while the FIR filter parameters are estimated using a straightforward least-squares method. In [19], a novel framework for dynamically modeling the behavior of PAs by employing an ANFIS-based Hammerstein model has been presented. To validate the model, the input and output signals of the PA were sampled during the identification and validation phases on a test bench. The experimental results demonstrate the accuracy of the proposed model in providing a precise approximation to characterize wideband RF PAs. This application of MANFIS in modeling RF PAs with memory effects has been explored by [20]. Both simulated and measured results in the time and frequency domains indicate that the MANFIS model exhibits better accuracy, faster convergence, and lower computational complexity compared to the ANFIS model. The MANFIS model shows lower normalized mean squared errors than those observed in the RBF neural network and real-valued time delay neural network. In a notable application, the MANFIS model proves successful in a digital pre-distortion system, delivering an impressive improvement of over 10 dB in adjacent channel leakage ratio (ACLR) for three-carrier wideband code division multiple access (WCDMA) signals. The proposed configuration is shown in Figure 1.13.

RF Heating stands out as a prominent application of PAs, where the key parameters defining their performance are stability and efficiency. Among various amplifiers, Class-F_3 PAs emerge as an optimal choice, particularly in the interface between rectifiers and induction heating systems. Theoretically, Class-F amplifiers can achieve an efficiency of 88.4%, a figure that can be maximized up to 100% with the implementation of harmonic tuning in the amplifier. To reduce power dissipation, and enhance operating frequency, M. Thian proposed two novel amplifiers. By incorporating an Electromagnetic Interference (EMI) filter at the input of the PA, the total harmonic distortions (THD) have been minimized. The integration of this filter not only

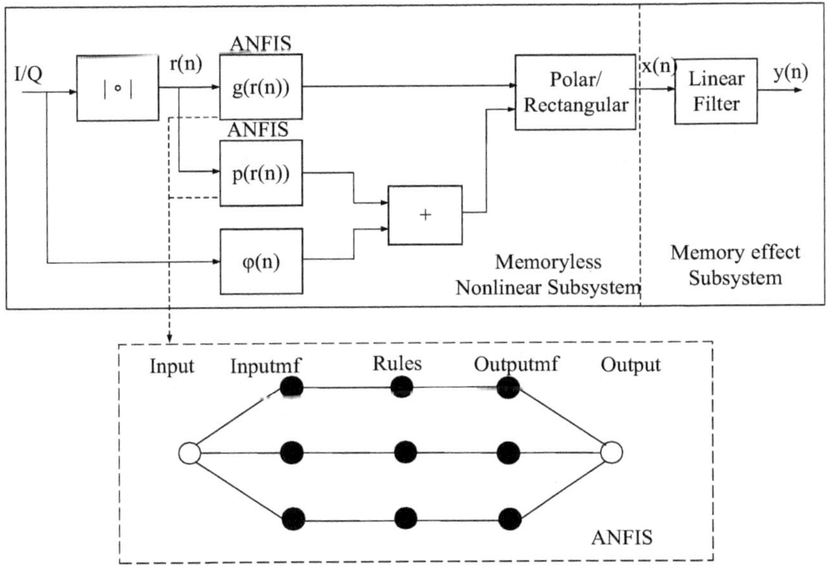

Figure 1.13 The proposed TW-SDPA design steps in [20].

drastically reduces THD but also allows for a more streamlined amplifier system, utilizing a single power switch [38]. Notably, this approach contrasts with existing Induction Heating (IH) systems, which commonly rely on Class-D/E amplifiers. Arumugam introduced a Class-E inverter-based IH system, while many IH systems employ frequency-varying methods such as pulse frequency and amplitude modulation to obtain and control output power [39–42]. The characteristics of Class-F and inverse Class-F amplifiers were compared through numerical analysis by Junghwan Moon [43]. Eswaran proposed new design methods to improve the efficiency of PAs, and Carrubba introduced a new mode of Class-F amplifier to maintain constant power output over a wide range of first and second harmonics [44, 45].

1.2.3 Fuzzy logic techniques for compensation of nonlinear distortion in PAs

The contemporary landscape of mobile communication systems demands radio transceivers capable of supporting high data rates and throughput. A significant trend in the design of wireless transmitters involves incorporating enhanced functionalities through digital signal processing (DSP) schemes. However, the modulated signals of these wireless systems exhibit a higher peak-to-average ratio (PAR) and broader bandwidth, necessitating PAs with exceptional linearity [46]. The conventional assumption of linearity in wireless communication systems poses challenges to their efficiency due to the nonlinear properties inherent in the front-end components and semiconductor devices within PAs [47]. While one simple method to achieve high linearity is to back off the power level, this comes at the cost of significantly reduced efficiency. To address this trade-off, various linearization techniques have been developed, including feedforward methods [48], feedback methods [49], analog pre-distortion [50], and digital pre-distortion (DPD) schemes [51]. The presented methodology in [51] relies on an innovative concept bolstered by the advancements in digital receiver technology. This approach incorporates an envelope detector to capture envelope variations, which are subsequently digitized to serve as indices for a lookup table integrated into the distorting generator block, as illustrated in Figure 1.14.

Among these approaches, DPD has emerged as a promising method, thanks to its high accuracy, flexibility, and advancements in digital-to-analog converters and digital signal processors. In a DPD system, the behavioral modeling and inverse modeling of PAs play pivotal roles. Numerous schemes for behavioral modeling have been explored in the literature, encompassing memory or memoryless models such as the Volterra series [52], Look-Up Tables (LUT) [53], Wiener [54], polynomial [55], neural network [56], and neuro-fuzzy [57]. These diverse modeling techniques cater to the complex and dynamic nature of PAs in modern wireless communication systems, enabling the optimization of linearity and efficiency simultaneously.

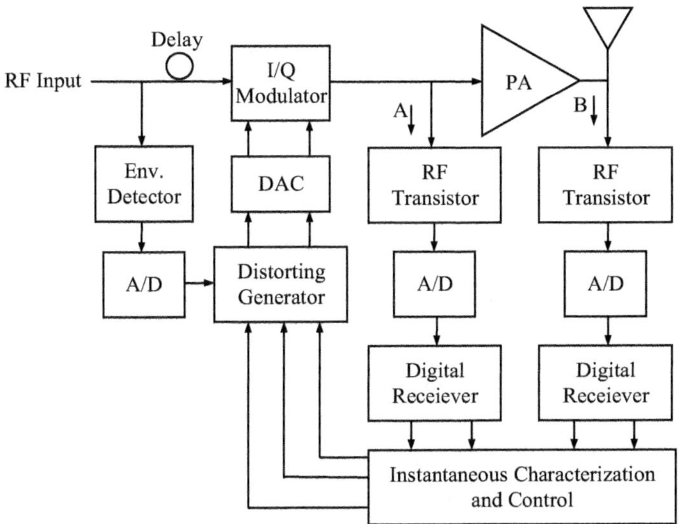

Figure 1.14 The block diagram of the linearizer [51].

The authors [57] introduce a pioneering pre-distorter for High PAs (HPA) designed for Orthogonal Frequency Division Multiplexing (OFDM) signals, utilizing ANFIS. The proposed technique is comprehensively examined using models of both solid-state PAs and traveling wave tube amplifiers, encompassing memoryless and memory-equipped scenarios. Upon training, the ANFIS effectively linearizes the HPA response, resulting in a signal that closely mirrors the original. Their approach consistently achieves an average Error Vector Magnitude of 10^{-6}, thereby minimizing Bit Error Rate (BER) degradation to a level surpassing alternative methods in the literature. Additionally, the proposed scheme reduces the overall complexity of the system. The configuration of the neuro-fuzzy predistorter during the operational phase is depicted in Figure 1.15.

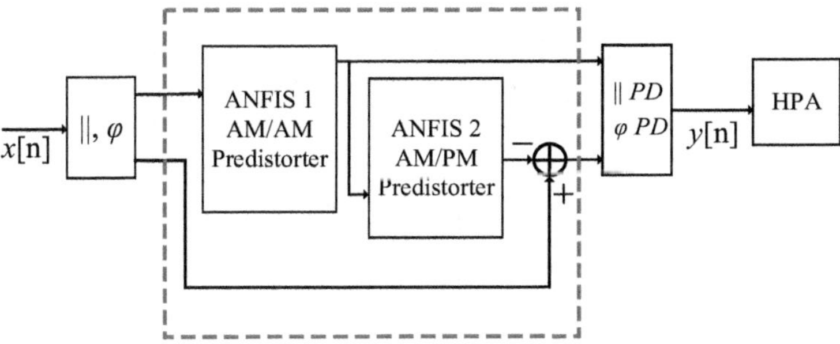

Figure 1.15 The implemented ANFIS architecture in [57].

In [58], the highly nonlinear MIMIX CFH2162-P3 PA is modeled by an adaptive fuzzy logic system (AFLS). Additionally, a pre-distorter unit has been developed by an AFLS-based inverse model. For both the PA's forward and inverse dynamics excellent modeling performance was achieved by driving an LTE 1.4 MHz 64 QAM signal at 880 MHz as the center frequency. A comparative analysis between AFLS and neural networks demonstrating AFLS as an effective baseband model suitable for inverse modeling of PAs for serving as an efficient digital pre-distorter with a distortion suppression of 84.2%, a significant improvement over the 48.4% obtained using the neural network-based equivalent scheme.

1.2.4 Fuzzy logic techniques for RF PA linearization

In contemporary communication systems, the widespread use of Quadrature Amplitude Modulation (QAM) and Quadrature Phase Shift Keying (QPSK), in combination with nonlinear RF PAs, poses challenges related to intermodulation distortion (IMD). This issue becomes particularly critical in applications like base stations and mobile phones. While linear PAs, for example, Class-A in Figure 1.16, could be employed to mitigate adjacent channel interference (ACI) caused by nonconstant envelope signals, this approach is highly inefficient, especially for portable devices like mobile phones, where power efficiency is a crucial factor due to limited battery life. Linearization techniques are essential to balance the spectral efficiencies and power of the transmitter without introducing significant distortions. The primary goal of linearization is imparting a more robust saturation characteristic not increasing the saturated output power capability. This enables the PA to be linearized up to a higher input power level before reaching saturation, effectively reducing unwanted IMD products [59]. While traditional solutions, like using a band-pass filter after the PA, exist to reduce unwanted IMD products, they lack flexibility and add unnecessary insertion loss to the system. A promising linearization technique for PAs is Adaptive Digital Predistortion (ADP) due to its adaptability, stability, and correction accuracy. In ADP, different techniques represent the predistorter (PD), falling into conventional and advanced categories. Conventional techniques, such as LUT and polynomial representations, are common. Polynomial PDs are straightforward to apply but are limited by the order of the function

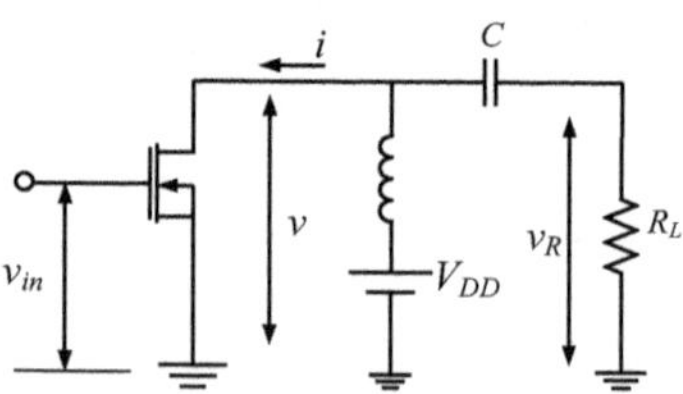

Figure 1.16 The structure of the Class-A PA.

and may involve complex fast Fourier transform (FFT) or post-distortion. Additionally, LUT PDs can suffer from nonlinearity or quantization noise. Advanced techniques involve the use of neural networks and fuzzy methods, offering more sophisticated and flexible approaches to PD representation. Polynomial PDs, while easy to apply, may face limitations in performance, while LUT PDs could encounter challenges with nonlinearity and convergence issues in iterative algorithms [60–64].

Neural network PDs [62–64] excel in modeling the inverse characteristics of PA responses but the number of hidden neurons in most of them increases with the desired improvement, posing scalability challenges. Extracting structured knowledge from the weights or system configuration in MLPs is also a complex task. The reported fuzzy system PD in [65] utilizes complex fuzzy basis function networks (CFBFNs), which limits its correction capabilities based on polynomial order. To address these challenges, this paper explores a hybrid neuro-fuzzy technique that amalgamates the strengths of both fuzzy systems and neural networks. Specifically, the technique employed is known as the ANFIS, where the tuning parameters are updated using least square estimate (LSE) and gradient descent [24, 66–68]. An innovative digital predistortion technique utilizing an ANFIS has been presented in [66]. The proposed method utilizes real-time input and output signals from a nonlinear PA to train the ANFIS, enabling it to approximate the inverse functions of the PA. The parameters of the Fuzzy Inference System (FIS) constructed by the ANFIS are fine-tuned using a combination of least squares and backpropagation algorithms. Simulation results demonstrate that this novel technique enhances the linearity of a WCDMA signal by an additional 4 dBc compared to a conventional LUT approach. Furthermore, the proposed technique exhibits the capability to adapt to instantaneous variations in the PA response over time, addressing a crucial aspect often overlooked by researchers in this field. The authors [68] introduce an adaptive digital pre-distortion technique for linearizing RF PAs, leveraging a combined approach of neural networks and fuzzy systems. The proposed method integrates both neural network and fuzzy logic components to create a sophisticated adaptive system for predistortion of the input signals to counteract nonlinearities in RF PAs, thus improving overall linearity. Measurement results demonstrate that the ANFIS PD effectively linearizes both narrowband and WCDMA systems. Notably, the ANFIS PD showcases adaptability to variations in PA responses, highlighting its feasibility in adjusting to environmental changes. This paper provides an in-depth analysis of ANFIS theory applied to distortion, presenting simulated and measured results that extend previous findings [24, 66–68]. Below, we summarize the advantages of the ANFIS technique. Figure 1.17(a) illustrates a proposed predistortion linearizer in [24]. The linearizer is designed to address two common nonlinear distortions in a PA such as amplitude and phase distortion (AM/AM and AM/PM). In this approach, distinct PDs are employed to correct AM/AM and AM/PM distortions separately. The PDs are

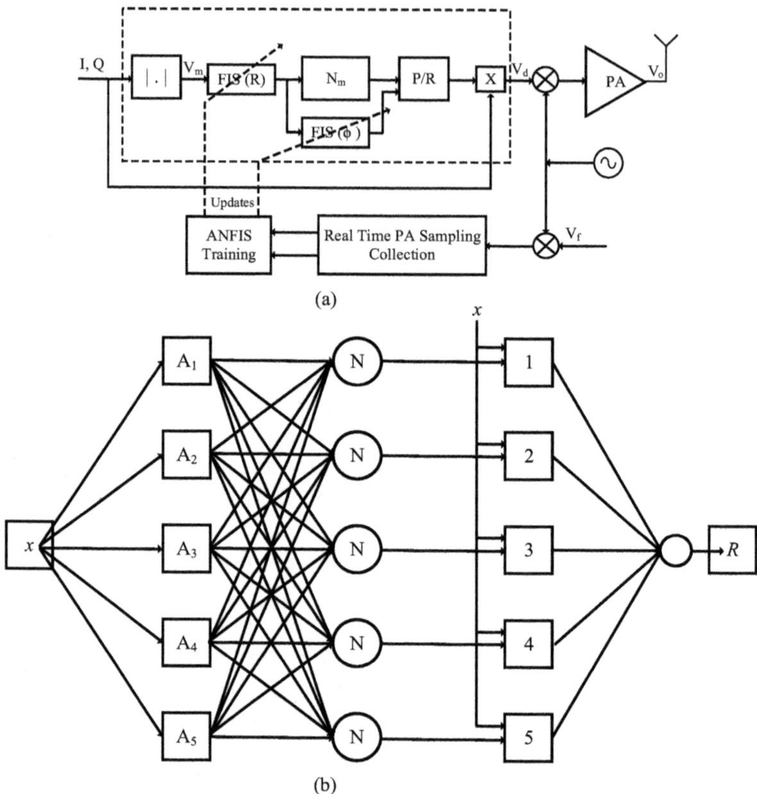

Figure 1.17 (a) The structure, and (b) Equivalent of the proposed ANFIS architecture, [24].

represented using a first-order Sugeno FIS. Each correction is defined by a set of five rules, denoted as $i = 1, 2, \ldots, 5$.

Amplitude: If x is A_i, then $R = G_i\, x + H_i$
Phase: If R is B_i, then $\Phi = J_i R + K_i$

Where

- A_i represents the fuzzy set associated with the input voltage x.
- B_i represents the fuzzy set associated with R.
- R is the input to the predistorter.
- Φ is the phase function of the predistorter.
- G_i, H_i, J_i, and K_i are the consequent parameters of the FIS that define the predistortion corrections for the specific fuzzy rules.

In this context, the FIS is utilizing fuzzy sets A_i and B_i to make decisions about the amplitude and phase corrections (R and Φ) based on the input voltage x and the associated fuzzy sets. The parameters G_i, H_i, J_i, and

K_i are then adjusted accordingly to implement the desired predistortion corrections based on the fuzzy rules. This structure allows the predistortion linearizer to adapt and correct for both amplitude and phase distortions in the PA. Figure 1.17(b) shows the equivalent ANFIS architecture for the PD amplitude FIS (R). Note that PD phase FIS (Φ) has a similar architecture.

Linearizing nonlinear RF PAs becomes crucial when deploying spectrally efficient modulation signals in mobile communications. One promising technique for this task is ADP. In [68] a novel approach combining neural networks and fuzzy systems is introduced for RF PA linearization, termed Adaptive ANFIS. This hybrid method offers several advantages: real-time processing of PA signals, offline adaptation, effective IMD suppression, absence of convergence issues, and flexible tuning. Experimental findings indicate an enhancement of approximately 12 dB for a distorted WCDMA signal, with a -60 dBm noise floor. The adaptability of the ANFIS approach to instantaneous variations in PA response shows its capacity to adjust to environmental changes. Further tests reveal that the tuning parameters from training can be significantly reduced by more than half for moderately nonlinear PAs, without substantial degradation in distortion suppression capability. This reduction in parameters translates to a substantial reduction in the DSP processing burden, offering considerable efficiency gains.

1.3 CONCLUSION

In conclusion, the exploration of various fuzzy logic techniques in the context of RF PAs has revealed significant advancements in addressing challenges such as linearization, bandwidth extension, improved dynamic response, compensation of nonlinear distortion, and overall PA linearization. Researchers have investigated diverse fuzzy logic methodologies, including interval type-2 fuzzy inference engines, radial-basis function neural networks, self-organizing fuzzy neural networks, adaptive neuro-fuzzy inference systems, and modified adaptive neuro-fuzzy inference systems, among others. The first section highlights the utilization of fuzzy logic techniques for addressing linearization challenges in PAs, with a focus on different neural network models such as RBFNN and ANFIS-based Hammerstein models. Additionally, the need for exploring fractional bandwidth (FBW) performance in PA designs is identified as an avenue for future research. The second section delves into techniques for bandwidth extension in continuous-mode-based TW-SDPAs, introducing the CFLMT. This innovative approach combines a modeled continuous-mode technique and a modeled K-means unsupervised learning clustering algorithm within a fuzzy logic system to automatically determine optimal impedances, resulting in a substantial increase in fractional bandwidth. The third section explores fuzzy logic techniques for improved dynamic response in PAs, emphasizing the use of ANFIS-based Hammerstein models and MANFIS for modeling

RF PAs with memory effects. The application of these models showcases commendable accuracy in modeling, with MANFIS demonstrating superiority in certain scenarios. The fourth section discusses the application of fuzzy logic techniques in RF heating, particularly in the context of Class-F_3 PAs. Various designs and configurations are presented, including those incorporating EMI filters to reduce total harmonic distortions, offering potential improvements in efficiency for RF heating applications. The fifth section addresses the need for linearization techniques to achieve high linearity in wireless communication systems, focusing on DPD schemes. The integration of envelope detectors, lookup tables, and advanced techniques like ANFIS is discussed, with a particular emphasis on achieving high linearity while maintaining efficiency. The final section provides an overview of fuzzy logic techniques for RF PA linearization, emphasizing the challenges posed by nonconstant envelope signals in contemporary communication systems. The paper explores the use of adaptive DPD techniques, with a hybrid neuro-fuzzy approach, specifically the ANFIS, proving effective in linearizing RF PAs. The proposed ANFIS technique exhibits adaptability, stability, and correction accuracy, making it a promising method for addressing intermodulation distortion challenges. In summary, the comprehensive exploration of fuzzy logic techniques in RF PAs demonstrates their versatility and effectiveness in addressing various challenges across different aspects of PA design and performance. The presented methodologies open avenues for further research and innovation in the field of RF and microwave engineering.

REFERENCES

1. Barton, L.E., High audio power from relatively small tubes. *Proceedings of the Institute of Radio Engineers*, 1931. 19(7): p. 1131–1149.
2. Fay, C., The operation of vacuum tubes as Class B and Class C amplifiers. *Bell System Technical Journal*, 1932. 11(1): p. 28–52.
3. Terman, F.E. and J.H. Ferns, The calculation of class C amplifier and harmonic generator performance of screen-grid and similar tubes. *Proceedings of the Institute of Radio Engineers*, 1934. 22(3): p. 359–373.
4. Hayati, M., A. Sheikhi, and A. Grebennikov, Class-F power amplifier with high power added efficiency using bowtie-shaped harmonic control circuit. *IEEE Microwave and Wireless Components Letters*, 2015. 25(2): p. 133–135.
5. Grebennikov, A. and F.H. Raab, History of class-F and inverse class-F techniques: Developments in high-efficiency power amplification from the 1910s to the 1980s. *IEEE Microwave Magazine*, 2018. 19(7): p. 99–115.
6. Sheikhi, A., M. Thian, and M. Vafaee, Broadband parallel-circuit class-E amplifier with second harmonic control circuit. *IEEE Transactions on Circuits and Systems II: Express Briefs*, 2018. 66(6): p. 928–932.
7. Sheikhi, A., H. Hemesi, and A. Grebennikov, Employing inverse Class-E power amplifier series output filter in parallel Doherty power amplifier. *Microwave and Optical Technology Letters*, 2023. 65(2): p. 425–433.

8. Sheikhi, A., et al., Effect of shunt capacitances on performance of current-mode class-de power amplifier at any active time. *IEEE Transactions on Power Electronics*, 2017. 33(8): p. 7038–7045.
9. Grebennikov, A., High-efficiency class E/F lumped and transmission-line power amplifiers. *IEEE Transactions on Microwave Theory and Techniques*, 2011. 59(6): p. 1579–1588.
10. Sheikhi, A., M. Hayati, and A. Grebennikov, A design methodology of class--E/F 3 power amplifier considering linear external and nonlinear drain–source capacitance. *IEEE Transactions on Microwave Theory and Techniques*, 2016. 65(2): p. 548–554.
11. Sheikhi, A. and H. Hemesi, Analysis and design of the novel class-F/E power amplifier with series output filter. *IEEE Transactions on Circuits and Systems II: Express Briefs*, 2021. 69(3): p. 779–783.
12. Sheikhi, A., M. Hayati, and A. Grebennikov, High-Efficiency Class-$\ mbox {E}^{-1} $ and Class-F/E Power Amplifiers at Any Duty Ratio. *IEEE Transactions on Industrial Electronics*, 2015. 63(2): p. 840–848.
13. Dowlatshahi, M., Application of fuzzy logic in detecting irregular heartbeat using ECG. *Fuzzy Systems and its Applications*, 2020. 3(1): p. 33–61.
14. Hashemi, A. and M. Dowlatshahi, Fuzzy Multi-Feature Decision Making and Its Applications. *Fuzzy Systems and its Applications*, 2020. 3(1): p. 181–235.
15. Joodaki, M., M.B. Dowlatshahi, and N.Z. Joodaki, An ensemble feature selection algorithm based on PageRank centrality and fuzzy logic. *Knowledge-Based Systems*, 2021. 233: p. 107538.
16. de Souza, G.A.F., R.B. dos Santos, and L. de Abreu Faria, Low-power current-mode interval type-2 fuzzy inference engine circuit. *IEEE Transactions on Circuits and Systems I: Regular Papers*, 2019. 66(7): p. 2639–2650.
17. Han, H., et al., An efficient second-order algorithm for self-organizing fuzzy neural networks. *IEEE Transactions on Cybernetics*, 2017. 49(1): p. 14–26.
18. Isaksson, M., D. Wisell, and D. Ronnow, Wide-band dynamic modeling of power amplifiers using radial-basis function neural networks. *IEEE Transactions on Microwave Theory and Techniques*, 2005. 53(11): p. 3422–3428.
19. Zhai, J., et al., Dynamic behavioral modeling of power amplifiers using ANFIS-based Hammerstein. *IEEE Microwave and Wireless Components Letters*, 2008. 18(10): p. 704–706.
20. Zhai, J., et al., The dynamic behavioral model of RF power amplifiers with the modified ANFIS. *IEEE Transactions on Microwave Theory and Techniques*, 2008. 57(1): p. 27–35.
21. Zhai, J., et al., Behavioral modeling of power amplifiers with dynamic fuzzy neural networks. *IEEE Microwave and Wireless Components Letters*, 2010. 20(9): p. 528–530.
22. Jain, S. Analysis & simulation of electrical parameters of fuzzy system using different types of current amplifiers with different membership functions for AkT pathway. in *2017 Fourth International Conference on Image Information Processing (ICIIP)*. 2017. IEEE.
23. Limin, J., *Fuzzy-neural tool for topology extraction of RF and microwave transistors*. 2005, University of Ottawa (Canada).
24. Lee, K.C. and P. Gardner, Adaptive neuro-fuzzy inference system (ANFIS) digital predistorter for RF power amplifier linearization. *IEEE Transactions on Vehicular Technology*, 2006. 55(1): p. 43–51.

25. Nuñez-Perez, J., et al. FPGA realization of RF-PA models with memory effects based on ANFIS. In *2016 IEEE International Autumn Meeting on Power, Electronics and Computing (ROPEC)*. 2016. IEEE.
26. Rezaei, M.J., A.A. Shahraki, and S.B. Shokouhi, A review of intelligent predistortion methods for the linearization of RF power amplifiers. In *2013 International Conference on Computer Applications Technology (ICCAT)*. 2013. IEEE.
27. Naah, G. and R. Giofrè, Empowering the bandwidth of continuous-mode symmetrical Doherty amplifiers by leveraging on fuzzy logic techniques. *IEEE Transactions on Microwave Theory and Techniques*, 2019. 68(7): p. 3134–3147.
28. Sun, G. and R.H. Jansen, Broadband Doherty power amplifier via real frequency technique. *IEEE Transactions on Microwave Theory and Techniques*, 2011. 60(1): p. 99–111.
29. Chen, P., et al., Multiobjective Bayesian optimization for active load modulation in a broadband 20-W GaN Doherty power amplifier design. *IEEE Transactions on Microwave Theory and Techniques*, 2016. 65(3): p. 860–871.
30. Chen, X., et al. A 200 watt broadband continuous-mode Doherty power amplifier for base-station applications. In *2017 IEEE MTT-S International Microwave Symposium (IMS)*. 2017. IEEE.
31. Chen, X., et al., A broadband Doherty power amplifier based on continuous-mode technology. *IEEE Transactions on Microwave Theory and Techniques*, 2016. 64(12): p. 4505–4517.
32. Gan, D., et al., Design of continuous mode Doherty power amplifiers using generalized output combiner. *AEU-International Journal of Electronics and Communications*, 2020. 116: p. 153069.
33. Shi, W., et al., Broadband continuous-mode Doherty power amplifiers with noninfinity peaking impedance. *IEEE Transactions on Microwave Theory and Techniques*, 2017. 66(2): p. 1034–1046.
34. Shi, W., S. He, and Q.-A. Liu, Deign of broadband highly efficienct Doherty power amplifiers by using series of continuous modes. In *2016 IEEE MTT-S International Microwave Symposium (IMS)*. 2016. IEEE.
35. Giofrè, R., et al., A closed-form design technique for ultra-wideband Doherty power amplifiers. *IEEE Transactions on Microwave Theory and Techniques*, 2014. 62(12): p. 3414–3424.
36. Yang, Z., et al., Bandwidth extension of Doherty power amplifier using complex combining load with noninfinity peaking impedance. *IEEE Transactions on Microwave Theory and Techniques*, 2018. 67(2): p. 765–777.
37. Rubio, J.J.M., et al., Design of an 87% fractional bandwidth Doherty power amplifier supported by a simplified bandwidth estimation method. *IEEE Transactions on Microwave Theory and Techniques*, 2017. 66(3): p. 1319–1327.
38. Wajid, M. A., & Zafar, A. (2022). Neutrosophic Image Segmentation: An Approach for the Treatment of Uncertainty in Multimodal Information Systems. *International Journal of Neutrosophic Science (IJNS)*, 19(1).
39. Arumugam, S. and S. Ramareddy, Experimental Studies on Class E Inverter based Induction Heater. *International Journal of Computer Theory and Engineering*, 2011. 3(2): p. 311.

40. Matysik, J.T., A new method of integration control with instantaneous current monitoring for class D series-resonant converter. *IEEE Transactions on Industrial Electronics*, 2006. 53(5): p. 1564–1576.
41. Wajid, M. S., Terashima-Marin, H., Najafirad, P., Pablos, S. E. C., & Wajid, M. A. (2024). DTwin-TEC: An AI-based TEC district digital twin and emulating security events by leveraging knowledge graph. *Journal of Open Innovation: Technology, Market, and Complexity*, 10(2), 100297.
42. Acero, J., et al., Modeling of planar spiral inductors between two multilayer media for induction heating applications. *IEEE Transactions on Magnetics*, 2006. 42(11): p. 3719–3729.
43. Moon, J., et al., Behaviors of Class-F and Class-${\hbox {F}}^{-1} $ Amplifiers. *IEEE Transactions on Microwave Theory and Techniques*, 2012. 60(6): p. 1937–1951.
44. Eswaran, U., H. Ramiah, and J. Kanesan, Power amplifier design methodologies for next generation wireless communications. *IETE Technical Review*, 2014. 31(3): p. 241–248.
45. Carrubba, V., et al. The continuous class-F mode power amplifier. In *The 5th European Microwave Integrated Circuits Conference*. 2010. IEEE.
46. Kenington, P.B., *High linearity RF amplifier design*. 2000: Artech House, Inc.
47. Wajid, M. S., & Wajid, M. A. (2021). The importance of indeterminate and unknown factors in nourishing crime: a case study of South Africa using neutrosophy. *Neutrosophic Sets and Systems*, 41(2021), 15.
48. Braithwaite, R.N. and A. Khanifar. High efficiency feedforward power amplifier using a nonlinear error amplifier and offset alignment control. in *2013 IEEE MTT-S International Microwave Symposium Digest (MTT)*. 2013. IEEE.
49. Kim, Y., et al. Linearization of 1.85 GHz amplifier using feedback predistortion loop. In *1998 IEEE MTT-S International Microwave Symposium Digest (Cat. No. 98CH36192)*. 1998. IEEE.
50. Yi, J., et al., Analog predistortion linearizer for high-power RF amplifiers. *IEEE Transactions on Microwave Theory and Techniques*, 2000. 48(12): p. 2709–2713.
51. Jeckeln, E.G., F.M. Ghannouchi, and M.A. Sawan, A new adaptive predistortion technique using software-defined radio and DSP technologies suitable for base station 3G power amplifiers. *IEEE Transactions on Microwave Theory and Techniques*, 2004. 52(9): p. 2139–2147.
52. Zhu, A., M. Wren, and T.J. Brazil, An efficient Volterra-based behavioral model for wideband RF power amplifiers. in *IEEE MTT-S International Microwave Symposium Digest, 2003*. 2003. IEEE.
53. Jardin, P. and G. Baudoin, Filter lookup table method for power amplifier linearization. *IEEE Transactions on Vehicular Technology*, 2007. 56(3): p. 1076–1087.
54. Ku, H., M.D. McKinley, and J.S. Kenney, Quantifying memory effects in RF power amplifiers. *IEEE Transactions on Microwave Theory and Techniques*, 2002. 50(12): p. 2843–2849.
55. Morgan, D.R., et al., A generalized memory polynomial model for digital predistortion of RF power amplifiers. *IEEE Transactions on Signal Processing*, 2006. 54(10): p. 3852–3860.

56. Boumaiza, S. and F. Mkadem, Wideband RF power amplifier predistortion using real-valued time-delay neural networks. in *2009 European Microwave Conference (EuMC)*. 2009. IEEE.
57. Jiménez, V.P.G., et al., High power amplifier pre-distorter based on neural-fuzzy systems for OFDM signals. *IEEE Transactions on Broadcasting*, 2010. 57(1): p. 149–158.
58. Vaskovic, M., V.S. Kodogiannis, and D. Budimir, An adaptive fuzzy logic system for the compensation of nonlinear distortion in wireless power amplifiers. *Neural Computing and Applications*, 2018. 30: p. 2539–2554.
59. Cripps, S.C., *RF power amplifiers for wireless communications*. Vol. 250. 2006: Artech house Norwood, MA.
60. Stapleton, S. and J. Cavers, A new technique for adaptation of linearizing predistorters. in *[1991 Proceedings] 41st IEEE Vehicular Technology Conference*. 1991. IEEE.
61. Wright, A.S. and W.G. Durtler, Experimental performance of an adaptive digital linearized power amplifier (for cellular telephony). *IEEE transactions on Vehicular Technology*, 1992. 41(4): p. 395–400.
62. Bernardini, A. and S. De Fina, Application of neural waveform predistortion to experimental TWT data. in *[1991 Proceedings] 6th Mediterranean Electrotechnical Conference*. 1991. IEEE.
63. Benvenuto, N., F. Piazza, and A. Uncini, A neural network approach to data predistortion with memory in digital radio systems. in *Proceedings of ICC'93-IEEE International Conference on Communications*. 1993. IEEE.
64. Watkins, B.E., R. North, and M. Tummala, Neural network based adaptive predistortion for the linearization of nonlinear RF amplifiers. in *Proceedings of MILCOM'95*. 1995. IEEE.
65. Li, Y. and P.-H. Yang, Data predistortion with adaptive fuzzy systems. In *IEEE SMC'99 Conference Proceedings. 1999 IEEE International Conference on Systems, Man, and Cybernetics (Cat. No. 99CH37028)*. 1999. IEEE.
66. Lee, K.C. and P. Gardner, A novel digital predistorter technique using an adaptive neuro-fuzzy inference system. *IEEE communications letters*, 2003. 7(2): p. 55–57.
67. Lee, K. and P. Gardner, Neuro-fuzzy approach to adaptive digital predistortion. *Electronics Letters*, 2004. 40(3): p. 1.
68. Lee, K. and P. Gardner, A combined neural network and fuzzy systems based adaptive digital predistortion for RF power amplifier linearization. In *The 2004 47th Midwest Symposium on Circuits and Systems, 2004. MWSCAS'04*. 2004. IEEE.

Chapter 2

Neutrosophic cognitive maps

Theoretical and mathematical formulations, literature review, and applications

Antonios Paraskevas and Michael Madas

2.1 INTRODUCTION

Classical logic, with its binary nature of true or false propositions, excels in providing precise representations of information in problem-solving contexts. However, in practice, many real-world scenarios involve uncertainties, ambiguities, and incomplete information, where classical logic's strict dichotomy may fail to capture the nuances and complexities of the situation, leading to limited or inaccurate conclusions. This is because a less rigorous and more intuitive approach to knowledge representation is needed. Additionally, there is a desire to reduce the amount of information about a topic, such as the number of symbols and phrases used to describe it. These issues have led to the development of knowledge representation methods and have occupied philosophers and scientists since ancient times.

The Greek philosopher Porphyry (234–305 BC) commenting on Aristotle's *Categories* designed what we could characterize as the first semantic network. He presented a hierarchical classification system for organizing concepts and categories, building upon Aristotle's framework laid out in the Categories. Porphyry's classification system aimed to categorize different types of entities and their relationships in a structured manner. Porphyry's model can be seen as a precursor to modern semantic networks, which represent knowledge in terms of nodes and edges. While Porphyry's system was primarily textual rather than graphical, it laid the groundwork for later developments in semantic networks and knowledge representation.

The need to represent related concepts led to the development of the mathematical entity called graph which is used in many different research fields. With the help of graphs, it is possible to visualize complex physical situations that are dependent on a number of concepts and require a significant number of logical procedures. If the concepts are represented as points in space, and connected in pairs, a graph is formed. Graph theory, which was established with the work of Euler in 18th century, deals with mathematical structures used to model pairwise relations between objects. Graphs consist of vertices (or nodes) representing the objects and edges (or links)

DOI: 10.1201/9781003606055-2

representing the relations between them. These relationships can take many different forms, including social ties in networks, road networks, chemical structures, and communication systems. Graph theory has found several applications over time in domains such as computer science, operations research, social sciences, and network analysis. Graph theory, due to its variety and adaptability, has become a vital tool in modern mathematics and applications.

Charles Pierce developed existential graphs which were the first semantic network specification using modern logic in 1909. With semantic networks, an attempt is made to organize and represent the expert's knowledge in a way analogous to that which humans use to represent their knowledge [1]. These networks are modeled with nodes and links-arcs. Nodes represent various events that can be physical objects (John, sofa bed, cat), concepts (numbers, mountain), states and descriptions (is, has). On the other hand, links describe the relationships that exist between events. One of the notable aspects of existential graphs is their ability to capture not only the structure of individual propositions but also the relationships between multiple propositions. This makes them particularly useful for analyzing complex logical arguments and reasoning processes. While existential graphs were not widely adopted during Peirce's lifetime, they have gained renewed interest in recent years, especially in fields such as artificial intelligence and cognitive science. Researchers have explored the potential of existential graphs as a framework for representing and reasoning with knowledge in intelligent systems.

A schema is an organized collection of concepts that can be used to describe events, sequences of events, situations, connections, and objects. Schemas were one of the earliest hypotheses proposed by researchers in their attempt to explain how complex information is organized in human memory, with the aim of understanding human behavior. The theory of schemas is based on a series of philosophical theories, starting with those of E. Kant (1724–1804), followed by J. Piaget (1896–1980), and culminating in more recent proposals in the late 1970s and early 1980s [2]. These proposals argue that there are molecular phenomena (such as events, scenes in space, and discourse structure) in the psychological world.

Driven by an interest in human memory and language processing, Ross Quillian (1961) began working on semantic networks within the field of artificial intelligence. A highly influential paper, written by Marvin Minsky (1975), introduced a version of semantic networks known as frames. In this context, a frame is a representation of an object or class, including its properties and relationships to other objects and classes [3]. Frames are made up of slots that represent characteristics or qualities of objects or concepts, and their hierarchical organization allows for the inheritance of properties and relationships. Default values manage missing information, but procedural attachments allow you to define behaviors. This paradigm enables the effective representation of complex domains, aiding applications in natural language comprehension, expert systems, and cognitive modeling, among others, so altering the landscape of knowledge representation in AI.

In the meantime, interdisciplinary research studies were proposed in an effort to understand the mental representations of spatial information and relationships constructed in people's minds. In the 1940s, Edward Tolman, an American psychologist, introduced the idea of cognitive maps in his studies on rats' behavior in mazes [4]. Tolman's insights contributed to the cognitive revolution in psychology by pointing the way toward comprehending the mind's internal representations. Conceptual graphs were proposed in 1976 by Sowa [5] and serve as a knowledge representation language rooted in modern linguistics, psychology, and philosophy. Additionally, this formalism is supported by suitable data structures and manipulation techniques. According to Sowa, perception involves the brain's construction of an active model that represents and interprets sensory stimuli. This model consists of two parts: the sensory part, encompassing perception elements corresponding to sensory stimuli, and an abstract part known as the conceptual graph. The conceptual graph describes how these individual perception elements combine to form perception in its broadest sense.

At around the same time, Axelrod's work contributed to understanding how individuals strategize and make decisions, which can relate to the formation of cognitive maps in social and decision-making processes [6]. Cognitive maps are primarily representations of a decision maker or expert's views and opinions about his or her own subjective world, rather than objective truth. What is considered in cognitive maps is whether the causal consequences are positive (+) or negative (–), and the relative strength of those causal relationships is ignored. Cognitive maps make documentary coding easier by creating symbolic representations of expert documents. In general, cognitive maps are often considered too limiting for the construction of knowledge-based systems.

In response to the aforesaid issue, Kosko made significant contributions to the use of fuzzy logic in cognitive research, including the development and application of fuzzy cognitive mapping (FCMs). FCMs are used to model and simulate complex systems with numerous variables and interactions, as well as to offer a framework for capturing knowledge, perspectives, and interactions between system components [7]. FCMs are fuzzy signed diagrams with feedback [8]. They connect causal events, actors, values, objectives, and trends in a fuzzy feedback dynamical system. In this sense, we may argue that FCMs create virtual worlds that alter over time. An FCM lists the fuzzy rules or causal flow channels that connect occurrences. Nodes represent fuzzy collections or events that occur to varying degrees. These nodes are used as causal ideas to depict concepts or processes with multiple parameters, which is how professionals create FCM causal images of the virtual world. However, FCMs have several disadvantages. According to [9], these can be summarized as follows: (i) FCM models lack temporal delay in node interactions, (ii) edge weight is linear, (iii) they cannot describe logical operators between incoming nodes, and (iv) FCMs cannot depict multi-meaning (gray) environments. The final limitation of FCMs arises from the constraints of fuzzy theory, which solely focuses on membership grades and does not adequately capture perception when the connections between concepts are unclear.

To address the latter, Smarandache [10] proposed neutrosophy theory as a way for dealing with the idea of neutrality. Neutrosophy proposes a continuum-power spectrum of neutralities between ideas <A> and <anti-A>. As a result, Smarandache developed neutrosophic logic, because fuzzy logic is assumed to be incapable of displaying indeterminacy by itself. According to the definition given in [11], *neutrosophic logic* is a particular form of logic that encompasses fuzzy logic, paraconsistent logic, intuitionistic logic, and others. It has three components: the degree of membership (T), indeterminacy (I), and falsehood (F). Neutrosophic logic introduces a formalism where each statement is assigned a percentage of truth in subset T, a percentage of indeterminacy in subset I, and a percentage of falsehood in subset F. In cases of unsupervised data analysis, there are instances where no relationships between ideas can be established. Neutrosophic cognitive maps (NCMs) were introduced in [12] as a mathematical tool to extend the capabilities of FCMs, by handling additional layers of uncertainty and indeterminacy in the representation of relationships between variables. In this sense, NCMs utilize neutrosophic logic, which is a well-suited logic to deal with the uncertainty often present in human judgment and opinion.

In general, an NCM is created by combining existing experience and information about a system. This can be accomplished by assembling a team of human experts to describe the system's structure and behavior under various scenarios. NCMs are a simple method for determining which factors should be changed and how. They find applications in scenarios where uncertainty is prevalent, such as decision-making in highly ambiguous or dynamic environments, modeling systems with incomplete data, or analyzing complex networks where relationships are not clearly defined. Additionally, NCMs can represent dynamic systems through vector-matrix operations [13, 14]. Knowledge engineers have recently focused their efforts on the topic of strategic decision-making in dynamic environments. Strategic concerns are becoming more complicated, unstructured, and quantitative [15]. As a result, NCMs have caught the interest of researchers and are regarded as a beneficial approach in a wide range of scientific domains, from knowledge modeling to decision-making.

Despite its promise, no detailed literature review study on NCMs has been conducted yet. This book chapter aims to address this gap by providing an analysis of the topic, offering scholars a solid foundation for future research efforts. Our work will provide the theoretical and mathematical formulations of NCMs, as well as a literature review based on the current level of knowledge on the issue, which has attracted the interest of scholars, mainly over the past decade.

Moreover, we will suggest and examine, for the first time in the related literature, a conceptual framework that can be viewed as an extension to NCMs, addressing and representing the concept of time and how this can be dealt with when modeling a system utilizing NCMs. Introducing NCMs with time representation could contribute to the existing literature in several ways, offering novel perspectives and addressing limitations of traditional models, such as improved handling of temporal dynamics and enhanced modeling of uncertainty and ambiguity. Our understanding of the world reveals that it is constantly evolving and changing. Consequently, processes, whether natural or technological, are dynamic, and abstract concepts must embrace change in order to be useful. In a dynamic system, the time needed before the effects of a cause are felt can vary. We have acknowledged earlier in this section that traditional NCMs do not consider or model the dynamics of time. Therefore, the idea of evolving representations over time is crucial, and incorporating the idea of time into NCMs is considered an important research contribution. One of the primary goals of this chapter, within this framework, is to assess the ability of NCMs to deal with the ontological and epistemological issues of situation analysis by proposing a conceptual framework that will allow the decision-maker to achieve a certain level of awareness of the situation in order to make informed decisions.

In this study, apart from the provision of the theoretical formulation of NCMs and the extended literature review on the current state of knowledge of NCMs over the last decade, we propose an innovative conceptual framework aimed at extending NCMs to incorporate the representation of time, thus addressing a significant gap in the existing literature and offering novel perspectives on modeling dynamic systems. The major research contributions of the current chapter could be summarized in the following:

- **Systematic literature review in NCMs:** Our chapter attempts to shed light on the subject by conducting a systematic literature review (SLR) with the purpose of identifying and classifying the prominent categories in the context of NCMs.
- **Conceptual framework for time representation in NCMs:** The chapter introduces a novel conceptual framework that extends NCMs to incorporate the representation of time. This addresses a significant gap in the existing literature where traditional NCMs do not adequately model temporal dynamics.
- **Enhanced modeling of dynamic systems:** By incorporating time into NCMs, the framework facilitates a better understanding and modeling of dynamic systems. This is crucial because many real-world processes, whether natural or technological, are dynamic and constantly evolving.

- **Improved handling of temporal dynamics**: The proposed framework offers improved handling of temporal dynamics within NCMs. It acknowledges that the time needed for the effects of a cause to be felt can vary, and thus, evolving representations over time are essential for accurate modeling.
- **Enhanced representation of uncertainty and ambiguity**: NCMs with time representation offer enhanced modeling of uncertainty and ambiguity. This is valuable because real-world systems often involve uncertainty, and traditional models may struggle to adequately capture this aspect.

The remainder of the chapter is as follows: Section 2.2 delves into the theoretical underpinnings and mathematical formulations essential for comprehending the modeling and construction processes involved in developing a NCM. This section aims to provide readers with the necessary background knowledge and tools required to engage effectively with the subsequent discussions on NCMs.

Following this, Section 2.3 offers a comprehensive literature review of previous studies that have utilized NCMs, highlighting the diverse purposes and application areas in which they have been employed. By synthesizing existing research findings, this section aims to elucidate the breadth and depth of NCM applications across various domains, providing valuable insights into their versatility and efficacy.

In Section 2.4, we introduce a novel extension to NCMs known as Neutrosophic Time Cognitive Maps (NTCMs). Building upon the foundational concepts of NCMs, NTCMs enable the analysis and reasoning of causal relationships among factors across discrete time intervals. By incorporating temporal dynamics into the cognitive mapping framework, NTCMs offer enhanced capabilities for modeling and predicting changes in causalities over time, thereby facilitating more nuanced and dynamic decision-making processes.

Demonstrating the utility of NTCMs, this section showcases our conceptual formalism for constructing and analyzing these temporal cognitive maps, illustrating their potential applications in various real-world contexts.

Lastly, the chapter concludes with Section 2.5, where we offer concluding remarks and outline potential directions for future research in the field of neutrosophic cognitive mapping. By summarizing key findings and insights garnered throughout the chapter, we reflect on the significance of NCMs and NTCMs as powerful tools for understanding complex systems and guiding decision-making processes. Additionally, we propose avenues for further exploration and refinement of these methodologies, identifying areas ripe for future investigation and innovation.

2.2 THEORETICAL AND MATHEMATICAL FORMULATIONS OF NCMS

2.2.1 Basic definitions

DEFINITION [16]. A graph G is a structure made up of a finite collection of objects known as vertices, and a set of pairings of vertices known as edges. An edge's endpoints are two adjacent vertices that the edge connects.

DEFINITION [16]. A digraph consists of a finite collection of objects known as vertices and a finite set of directed edges, or arcs, which are ordered pairings of vertices.

According to the definitions presented above, a digraph is identical to a graph except that each edge is assigned a direction--one vertex is designated as a start and the other as a finish. It is important to note that, unlike graphs, a digraph can have two arcs with identical endpoints if they are oriented in opposite directions. A directed multigraph is an object that can contain several arcs.

DEFINITION [17]. Assume G is a graph with n vertices labeled 1, 2, 3,...n. The adjacency matrix A(G) of G is a $n \times n$ matrix in which the item in row i and column j represents the number of edges connecting vertices i and j.

DEFINITION [17]. Assume D is a digraph with n vertices marked 1, 2, 3,...n. D's adjacency matrix A(D) is the number of arcs between vertex i and vertex j.

An adjacency matrix allows us to represent each graph or digraph with a rectangular array of numbers known as a matrix. These matrices are suitable for computational procedures and are often the simplest way to convey information. There are several types of matrices that can be used to describe a given graph or digraph. In this chapter, we will focus on the most basic type, the adjacency matrix.

DEFINITION [17]. A walk of length k in a graph consists of k edges of the form uv, vw, wx, and yz. This walk, indicated by $uvwx...yz$, is also known as a walk between a and z.

DEFINITION [17]. A cycle is a closed walk with unique edges and intermediate vertices.

DEFINITION [12]. A neutrosophic graph is one in which at least one edge is an indeterminacy, shown by dotted lines.

Example Figure 2.1 shows a neutrosophic graph:

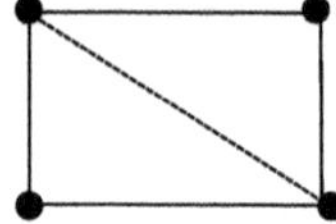

Figure 2.1 A simple neutrosophic graph.

> **DEFINITION** [12]. A neutrosophic directed graph (or digraph) is a directed graph with at least one edge that is indeterminate.

> **DEFINITION** [12]. An NCM is a neutrosophic directed graph with nodes representing policies, occurrences, and so on, and edges expressing causality or indeterminacy. It expresses a causal relationship between ideas.

NCMs are collections of ideas (linguistic terms) represented by nodes. Arrows with weights in the set $\{-1, 0, 1, I\}$ represent the relationship between ideas. The weights indicate the intensity of the causal influence.

> **DEFINITION** [12]. Let $C_1, C_2 \ldots C_n$ be the nodes of an NCM. Let the neutrosophic matrix N(A) be defined as $N(A) = (w_{ij})$ where w_{ij} is the weight of the directed edge $C_i C_j$, where $w_{ij} \in \{0,1,-1,I\}$. We call N(A) to be the adjacency matrix of the NCM.

Let us examine in more detail the definition of an NCM and its adjacency matrix with the following example.

Example Consider the following simple NCM.

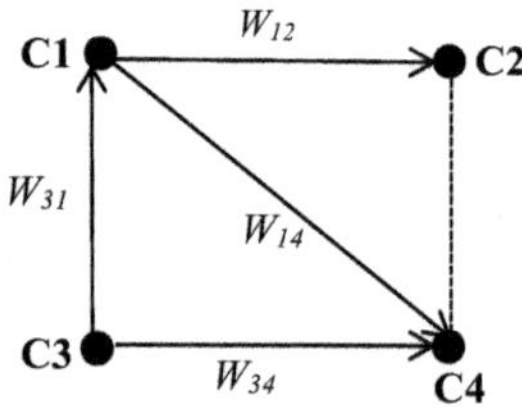

Figure 2.2 A simple neutrosophic cognitive map.

Figure 2.2 shows concept variables represented as nodes, such as C1, C2, C3, and C4. Causal variables always display concept variables at the origin of the arrow, while effect variables represent concept variables at the terminal points of the arrow. In the above example C1 is said to impact C2 denoted as C1→C2. Causality between concepts allows degrees of causality, so the weights of the interconnections can range in the set {−1, 0, 1, I}. So every edge in the NCM (e.g. W_{14}, W_{34}) is weighted with a number in the set {−1, 0, 1, I}. The map's node structure and linkages hold existing information about the system's behavior. Conceptual relationships may be divided into four types: (i) express positive causation between ideas ($W_{ij}>0$), (ii) express negative causality ($W_{ij}<0$), (iii) no causality or relationship ($W_{ij}=0$), and (iv) the relation or causality of Ci on Cj is indeterminate, i.e. impact is unknown or cannot be identified ($W_{ij} = I$). The sign of W_{ij} determines whether the link between ideas C_i and C_j is direct or inverse. The direction of causation determines whether idea C_i causes concept C_j or vice versa.

The NCM shown in Figure 2.2 has the following adjacency matrix:

$$\mathbf{N(A)} = \begin{bmatrix} 0 & W_{12} & 0 & W_{14} \\ 0 & 0 & 0 & I \\ W_{31} & 0 & 0 & W_{34} \\ 0 & I & 0 & 0 \end{bmatrix}$$

Figure 2.3 Adjacency matrix of NCM.

We should note that in the adjacency matrix N(A) shown in Figure 2.3, the i_{th} row lists the connection strength of the edges e_{ik} directed out from causal concept C_i. The i_{th} column lists the edges e_{ki} directed into C_i.

> DEFINITION [12]. An NCM with cycles is considered to have feedback. When the NCM contains feedback or when the causal linkages run through a cycle in a recursive form, it is referred to as a dynamical system.

2.2.2 Basic methodologies

In this subsection, we will briefly describe the dynamic and static analysis of an NCM which are used as the basic methodologies in order to describe qualitatively and quantitatively, respectively, the causalities among the concepts represented in the map. These methodologies have found great applicability in modeling complex systems when using NCMs and this is confirmed in the next section where related literature review on the application areas of NCMs is conducted.

2.2.2.1 Dynamic analysis of NCM

Assume an NCM consists of n nodes. The $n \times n$ adjacency matrix N(A) denotes the causality of the node interactions. Each element w_{ij} of the matrix N(A) represents the weight w_{ij} between the nodes C_i and C_j. Then we claim that the basic operation that defines its representation at each time step is given by the following equation:

$$A^{\kappa} = \mathcal{F}\left(A^{\kappa-1} + A^{\kappa-1} \times \mathcal{W}\right) \tag{2.1}$$

where A^{κ} is the column matrix with values of concepts at iteration step k and $\mathcal{F}$ is the threshold function. The latter could be a simple function with a threshold value T which could result in binary values, such as:

$$\text{Aj} = \begin{cases} O\, if\ activation_j\ \leq T \\ 1 if\ activation_j\ \geq T \\ I\ if\ A_j\ is\ not\ an\ integer \end{cases} \tag{2.2}$$

The FCM model of any system is free to interact and acquires initial concept values based on real-world data. Changes in the value of one or more concepts also lead to interactions. These interactions will continue until the model meets one of the following conditions:

1. The model reaches an equilibrium fixed point, where the output values stabilize at specific numerical values.
2. The model exhibits limit cycle behavior, with concept values decreasing in a numerical loop.
3. The model exhibits chaotic behavior, with concept values reaching a wide range of numerical values in a non-deterministic and unexpected manner.

In most circumstances, we state that an NCM settles or converges to a fixed point or limit cycle, as specified in conditions (1) and (2) above. The output equilibrium responds to a causal what-if question: What if C(0) occurs? Next, we provide explicit definitions for the aforementioned NCM states.

DEFINITION [12]. Let $\overrightarrow{C1C2},\ \overrightarrow{C2C3},\ldots,\ \overrightarrow{C_{n-1}C_n}$ be a cycle when C_i is turned on, and if causation flows along the edges of a cycle and triggers C_i again, we say the dynamical system swings round and round. This is true for any node C_i, for i = 1, 2,..., n. The *hidden pattern* is the *equilibrium* state for this dynamical system.

DEFINITION [12]. A fixed point is a dynamical system's equilibrium state defined by a unique state vector. Consider the NCM with nodes C_1, C_2, ..., C_n. For example, let us start the dynamical system by activating C_1. Assuming the NCM stabilizes with both C_1 and C_n enabled, the state vector will stay (1, 0,..., 1). The fixed point is a state vector with neutrosophic features (1, 0,..., 1).

DEFINITION [12]. The NCM's limit cycle occurs when the neutrosophic state vector follows the pattern A1 → A2 → .... → A_i → A1.

NCMs retain when the NCM dynamical system achieves equilibrium. Simple NCM inference rules for matrix-vector multiplication [19]. The state vectors C_n cycle through the NCM adjacency matrix N(A) as follows: $C_1 \rightarrow$ N(A) → C_2 → N(A) → C_3... Equation 2.1 describes how the system nonlinearly changes each node's weighted input. Simple threshold NCMs rapidly converge on stable limit cycles or fixed points. These limit cycles reveal hidden patterns within the causal chain of the NCM.

2.2.3 Static analysis of NCM

Kosko [18, 19] defines centrality as a measure of node significance in an FCM using fuzzy causal algebra. Similarly, the centrality of a term (C_j) in an NCM is determined by adding the number of ideas that directly or indirectly cause C_j to the number of words that cause C_j. These numbers are calculated based on the total number of nodes on the pathways that flow into and out of C_j. The following are the numbers and strengths of nodes in an NCM.

$$\text{Concept centrality}(Ci) = \text{IN}(Ci) + \text{OUT}(Ci) \tag{2.3}$$

where IN(C_i) is the sum of weights of causal connections linking nodes C_j, i ≠ j, to C_i, whereas OUT(C_i) is the sum of all of the weights of causal links connecting node C_i to all nodes C_i, i ≠ j. The suggested framework employs the measurements mentioned above, which are based on the absolute values of the neutrosophic adjacency matrix [20]:

- *Outdegree* is the sum of the row elements in the neutrosophic adjacency matrix and represents the strength of the variable's outgoing connections (c_{ij}).

$$od(vi) = \sum_{i=1}^{n} c_{ij} \tag{2.4}$$

- *Indegree* is the sum of the column components in the neutrosophic adjacency matrix and it represents the strength of the relationships (c_{ij}) that emerge from the variable.

$$id(vi) = \sum_{i=1}^{n} c_{ji} \tag{2.5}$$

- *Total centrality* is the sum of the *indegree* and the *outdegree* of the variable.

$$td(vi) = id(vi) + od(vi) \tag{2.6}$$

2.3 LITERATURE REVIEW

In this part, we will present a complete summary of current research on NCMs and their applications. NCMs are widely used in many areas, including agriculture, social sciences, medical, business, management, decision-making, education, psychology, and environmental research, among others. Researchers have enhanced NCM models by merging them with other decision-making techniques such as PESTEL, genetic algorithms, neutrosophic logic, static analysis, and other approaches to build hybrid NCM models for decision-making.

For our purpose, we used the Scopus research database in order to identify studies related to NCMs that were published in journals, conferences, or books during the last decade (i.e. from 2013 to 2023) and were written in English. In Figure 2.4, 78 research studies accomplished in each one of the most common application domains, such as mathematics, computer science, engineering, and others during the past decade are depicted.

At this point, it is important to note that the results shown in Figure 2.4 have been further elaborated by the authors. This is because there is an overlap observed in research domains regarding the number of included studies retrieved from Scopus. For example, when we used Scopus to search for NCM articles in the field of mathematics, which is a broad subject with various applications, we found that some studies could also be categorized

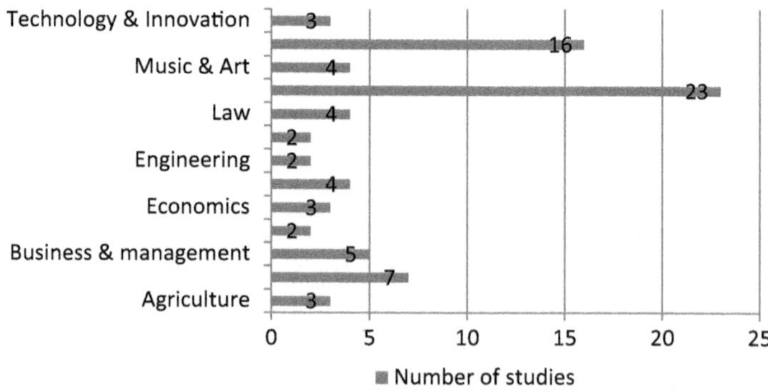

Figure 2.4 Number of NCM studies per research domain during the past decade.

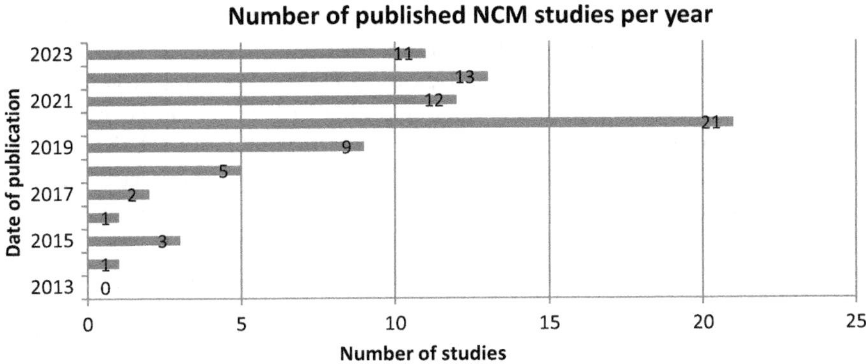

Figure 2.5 Number of NCM publications per year the past decade.

under other domains. To obtain more precise outcomes, additional human logic and effort are required to calibrate the findings of Scopus and achieve a more accurate classification.

Figure 2.5 depicts the distribution of papers based on their publication year. The great majority of the selected publications (about 73%) were published between 2020 and 2023, demonstrating a dramatic rise in this specific study topic in recent years. Observing the broad interest in literature on NCMs applied in various research fields, one can predict that the number of articles will continue to grow in the future.

It is important to mention that it was difficult to cover all of the innovative and useful applications performed by NCMs and their expansions in this book chapter. Instead, our goal was to identify some of the most often stated applications in the literature and present current research objectives for NCMs. Therefore, for the sake of brevity, in the following subsections, we will review the most representative studies applied in the major research disciplines based on the conducted literature review.

2.3.1 Medicine

Over the last decade, FCMs have found significant applications in medical diagnostics and decision assistance. FCMs use fuzzy logic to describe complicated interactions between different aspects in medical disorders, allowing for a more nuanced portrayal of uncertainty and ambiguity in diagnostic processes. By merging expert knowledge and patient data, FCM-based solutions can help healthcare practitioners make better judgments, improve accuracy, and perhaps reduce mistakes in diagnosis and treatment planning. Using dynamic analysis as a methodology of study, researchers in [21] conducted a comparative study between FCMs and NCMs to analyze the parameters responsible for variations in COVID-19. They comment on the results obtained without claiming which model gives better results, although it is acknowledged that the indeterminacy entailed in NCMs sheds light on parameters that are not so obvious in FCMs. In a similar vein, Zafar and Wajid [22] utilize NCMs to objectively illustrate how indeterminate and unknown components contribute to the spread of COVID-19 throughout India. The findings not only emphasize the importance of determinate factors (government actions) in controlling the spread of this lethal virus, but also shed light on the critical role of indeterminate and uncertain factors such as poverty, negligence, ineffective healthcare, age, immunity, unscientific religious practices, and illiteracy.

Next, we mention the research work of [23] where NCMs are used to determine the factors that will facilitate decision-making to establish an accurate diagnosis in somatoform disorders which are characterized by a multitude of symptoms that, due to their similarities to functional neurological illnesses, sometimes confound professionals when establishing a diagnosis. In the field of decision-making, the research article of Al-Subhi et al. [24] propose a novel model based on the NCM that combines diagnosis, therapy, and prognosis processes to improve clinical decision-making in pregnant women with cardiovascular illnesses.

The pandemic period motivated the work of [25] who utilizing FCMs and NCMs, quantitatively analyzed the COVID-19 like symptoms, spreading method, and precaution method. Both strategies rely on expert judgment and the selection of fuzzy sets and neutrosophic sets were employed for their observations since the medical area contains ambiguity and indeterminacy. One of the first studies observed in related literature is that of Kumar et al. [26] used NCMs with genetic algorithms to handle scenarios when judgments are not clearly distinct in the diagnosis of the most probable condition by increasing the number of factor concepts.

The causes of cancer in modern civilization are highlighted in the study of [27]. Neutrosophy method was introduced to address the idea of indeterminacy in choice with positive and negative influencing elements. Connections and importance among the attributes are discussed based on the expert's idea. Cancer dangers are caused by thirteen reasons mentioned in the article. In the research of Shakil et al. [28] NCMs are employed to investigate the elements that might contribute to health decline. The approach used not only demonstrates how NCMMs may be utilized, but it also suggests strategies for the general public to identify and regulate health-related aspects.

In their study, Reyes Salgado et al. [29] attempt to shed light on the causes of autism spectrum disorders. With a sample of 42 cases diagnosed with any of the autism subtypes, they seek to determine the key factors to which the patients were exposed in order to identify potential commonalities among disease-causing variables. The data were analyzed using NCMs, leading to the conclusion that preterm birth and a family history of autism were the most significant contributing factors.

2.3.2 Sociology

In sociology, NCMs may be used to represent the complex and frequently ambiguous interactions between diverse social elements such as cultural norms, socioeconomic position, political beliefs, and demographic data. NCMs facilitate a more thorough understanding of social phenomena and dynamics by allowing the portrayal of unclear and contradictory opinions within a social environment.

In this context, writers in [30] use NCMs to simulate the causes and repercussions of crimes against the homeless, as well as how institutions may better prepare to confront the issue. Similarly, the authors of [31] intend to perform a cause-and-effect research of crimes perpetrated against the homeless, enhance their services by providing information, standards, and processes for handling the situation, and assist the homeless in participating in the Good Living Plan. They used a hermeneutical approach to build their study and created a cause-and-effect tree with NCMs.

Dimitri et al. [32] examine health concerns such as stress, depression, anxiety, and addiction and their connection to the global success of technology or any other industry. To achieve their goals, they use NCMs to analyze the daily difficulties individuals face while neglecting their personal and social lives. In the same research framework, scholars in [33] present an NCM for violence analysis based on triangular neutrosophic values. Their proposed approach allows experts to express their preferences while considering varying degrees of truth, indeterminacy, and falsity in the underlying map links.

The custody of children and adolescents is a phenomenon that occurs daily in many countries worldwide. For this reason, del Pozo Franco et al. [34] study this situation in Ecuador using NCMs. They conclude that government decision-making should be aided by outlining ways to offset the negative consequences of custody not envisioned under parental responsibility and to preserve the rights of childhood and adolescence.

Wajid et al. [35] contribute to identifying mathematically the elements, whether certain (known) or uncertain (indeterminate and unknown), that lead to criminal activity in humans. This assists policymakers in taking the necessary actions to reduce crime in South Africa. The study also demonstrates the superiority of NCMs over FCMs by modeling the problem using FCMs.

Researchers in [36] examine the migration problem in Ecuador and study the economic impact of Venezuelan emigration to the city of Santo Domingo for citizens and migrants using the NCM approach. Huera Castro et al. [37] examine the many elements that influence child custody and the development of childhood and adolescence using NCMs. They claim that by prioritizing the incidence variables, government decision-making will be facilitated, detailing methods to reduce the negative consequences of custody not conceived under parental responsibility and to defend the rights of children and adolescents.

Recognizing the benefits of NCMs in terms of interpretability, scalability, information aggregation, dynamism, and the ability to depict feedback and indeterminacy linkages, academics in [38] investigate the critical problem of femicide in Ecuador. They propose providing the government with a tool that will enable them to focus on critical initiatives aimed at reducing and/or eliminating femicide. The purpose is to use NCMs to assess and recommend suitable actions based on cognitive variables, personal living situations, the victim's physical status, the strength of laws and counseling, profession, and ethnicity-environment.

2.3.3 Applied mathematics

NCMs provide a novel approach to applied mathematics by combining neutrosophic logic with cognitive mapping techniques. In applied mathematics, NCMs can be used to represent complicated systems and processes that are indeterminate, imprecise, or uncertain. They provide a framework for expressing and analyzing dynamic connections among system components, taking into account both deterministic and indeterminate elements.

An NCM is a mathematical model used to describe and analyze complex systems, particularly those involving decision-making and cognitive processes. In their article [39], the authors provide an overview of neutrosophy and its mathematical developments over the past two decades. This resource serves to educate readers on neutrosophy as an extension of dialectics, along with its various applications in mathematical algebra and calculus.

Taking a more algorithmic approach, Chithra et al. [40] use a Dynamic NCM with an Improved Cuckoo Search Algorithm to obtain the gene expression profile that distinguishes individuals affected by Rheumatoid Arthritis. Their work involves four major processes: data preparation, feature selection, prediction, and classification.

Researchers in [41] focus on developing a nature-inspired unsupervised learning model with the goal of managing indeterminacy and ambiguity while clustering datasets for Alzheimer's disease identification. This paper suggests a neutrosophic clustering method powered by the intelligence of a chaotic crow search algorithm. The performance results demonstrate the usefulness of this approach in dealing with noisy data and indeterminacy, leading to more accurate diagnosis of Alzheimer's disease.

Martin et al. [42] attempt to fill the literature gap where NCMs are predominantly discussed under expert-based methods, with less emphasis on automated methods such as Non-linear Hebbian (NLH). To bridge this disparity, this paper presents an anxiety predictive model based on the NCM-NLH approach.

2.3.4 Business, management, and accounting

In business management, NCMs may be used to capture the interaction of numerous aspects such as market dynamics, customer behavior, competitive landscape, and internal organizational structures. They let managers evaluate the effect of various tactics, predict market trends, and make educated decisions in dynamic contexts where traditional models may overlook ambiguity and imprecision.

Due to the above, NCMs have gained popularity in related field for their use in product development, analysis, and decision-making. Several interesting applications are worth mentioning. In a study by Lara et al. [43], a real case study of a retail company is examined to investigate the interrelationship among financial performance indicators in the company's accounting process. NCMs are used to establish a more realistic model for assessing cause-effect relationships, allowing for the identification of indirect aspects of accounting variables such as "customer satisfaction."

Another research [44] uses NCMs to analyze the investment process, taking into account elements such as economic return on investment, recovery time, resources (financial, human, and time), production capacity, execution duration, workload, market demand, and staff training. The study investigates the interconnectivity and indeterminacy of these characteristics.

To measure the mood swings of women in employment, scholars in [45] employ a new Triangular Combined Overlap NCMs to rank factors that affect their behavior at work under uncertain situations. Commercial ties resulting from economic growth can be classified as formal or informal based on the nature of their interactions. Cortes et al. [46] propose a solution to assess the impact of informal commerce using a multi-expert multi-criteria approach. Modeling causal linkages with NCMs and the Weighted Power Mean operator are recommended methods. The study of [47] focuses on the analysis of the municipal government of Babahoyo's collection concerns on municipal market "4 de Mayo" taxes. The Ishikawa Diagram of causes and effects is obtained for this purpose. The NCM approach is also used to identify the variables with the greatest effect on the problem. The study concludes that the three variables requiring greater attention are "lack of collection policies," "lack of procedures," and "late payment of accounts receivable," in that order.

The reported inadequacy of NCMs in quantifying the level of uncertainty in decision-making problems, where decisions are viewed as a series of interconnected decisions made in sequence, inspired Al-Subhi et al. [48] to introduce a new NCM for multistage sequential decision-making problems (MS-TrNCM) based on triangular neutrosophic numbers. This proposed model represents all connections on the map as triangular neutrosophic numbers, enabling decision-makers to convey their preferences while considering the levels of truth, uncertainty, and falsity.

The goal of researchers in [49] is to examine the impact of technological innovation on Ecuador's GDP by employing an NCM that identifies the components directly influencing technological innovation. The PESTEL framework is used to determine the political, economic, social, technical, environmental, and legal aspects influencing technological innovation in Ecuador's GDP. A quantitative study based on static analysis and neutrosophic numbers is performed to facilitate the application of the proposal.

2.3.5 Education

NCMs show potential in educational applications, providing a diverse framework for modeling and analyzing complex educational systems and processes. In education, NCMs may be used to depict the complex interactions between teaching techniques, curriculum design, student learning styles, educational technology, and socioeconomic backgrounds.

The introduction of digital learning portals has expanded learners' access to information and challenged the idea that teachers are the sole source of knowledge. In this context, Martin and Broumi [50] examine how changing a teacher's role to that of a facilitator affects their interpersonal and intrapersonal characteristics, as well as the classroom environment. This study also explores the impact of instructors adapting to new roles and how it immediately affects different types of learners. A mathematical model is developed using NCMs to thoroughly investigate the effects of role transformation.

This study [51] explores imaginative play in young children using NCMs. These models are created with input from experts to establish connections between various ideas related to creative play in children aged 1 to 10 years who come from disadvantaged backgrounds in terms of economics, social status, and education. The researchers report that they examined fifteen concepts related to children playing with the same toy and found that multiple perceptions were interconnected, while other relationships between concepts were unclear. This is why NCMs were utilized.

Mary and Merlin [52] are investigating the characteristics that promote comprehensive student–teacher preparation in response to societal changes and advancements. They are assessing course programs, academic outcomes, and the competencies of primary teacher training students to understand the causes of course setbacks and the need to change its content. The researchers are discussing their findings with specialists and deriving elements for analyzing the influencing factors using a comparative analysis of NCM and Triangular FCM.

Research conducted by [53] aims to identify and analyze the challenges faced during the implementation of India's National Education Policy (NEP) 2020. Scholars use the neutrosophic PESTEL analysis technique to pinpoint and prioritize the essential components needed for successful NEP implementation. These connections are accurately depicted through NCMs, followed by the creation of neutrosophic adjacency matrices for quantitative analysis.

2.3.6 Information technology

In the field of information technology (IT), NCMs may be used to describe the dynamic connections between system components such as network infrastructure, software applications, user interactions, and cybersecurity risks. NCMs use neutrosophic logic to represent ambiguous and competing perspectives in IT decision-making processes, allowing IT professionals to analyze risks, foresee weaknesses, and establish robust strategies for system design, implementation, and maintenance.

Current methods and tools for finding, categorizing, and assessing success indicators in IT projects have significant flaws. These limitations can be addressed by employing NCMs to map success and model important success factor perceptions, and their linkages. Bhutani et al. [54] utilized NCMs to discover and evaluate success factors in IT initiatives. Betancourt, Leyva, and Pérez [55] proposed a unique method for modeling project risk interdependencies. Al-Subhi et al. [56] suggested a unique decision-making model based on NCMs for making full judgments utilizing a multi-objective approach (diagnosis, decisions, and prediction), while maintaining an appropriate balance of time, cost, and quality while executing numerous projects concurrently. The suggested technique solves the issue that typical FCMs do not capture uncertain connections between concepts. The same research team conducted a study [57] in which they introduced two new extensions of FCMs and NCMs: Multistage Sequential Triangular NCM and NCM based on Linguistic Data Summarization. These extensions aim to address shortcomings in the existing literature regarding the treatment of indeterminacy and the solution of multistage sequential decision-making problems in project management. The proposed approaches were evaluated using a dataset of 1011 project records, and they were found to outperform classic FCMs and NCMs.

In the field of technology usage, Colorado Pastor et al. [58] present a design plan for Guayaquil's public transit system for disabled individuals based on artificial intelligence, the human aspect, and ergonomic design. While modeling the behavior of persons with impairments is a difficult endeavor, the social and economic consequences are enormous. As a result, NCMs are being researched to better understand the behavior and operation of such complex systems. This soft computing technology enables researchers to make decisions based on travelers' understanding of various modes of transportation properties at various levels of abstraction.

2.3.7 Law

NCMs have the fascinating potential for use in law because they provide a framework for modeling the complexity and uncertainties inherent in legal reasoning and decision-making processes. In law, NCMs may be used to reflect the complex links between numerous legal ideas, principles, and precedents, as well as the varied points of view of stakeholders such as judges, lawyers, and litigants.

NCMs can provide an interesting solution to the challenge of identifying the factors that impact the consolidation of neo-constitutionalism (NC) [59]. The authors utilize the PESTEL approach to conduct an environmental analysis and pinpoint key elements that significantly affect North Carolina. These variables are then quantitatively analyzed using static analysis of NCMs. Consequently, technological, political, and economic factors must be taken into consideration.

Calle Santander et al. [60] constructed a cause-effect tree and used legal causal reasoning to examine the increase in traffic accidents. They integrated the Ishikawa Diagram, Pareto analysis, and NCMs in a hybrid method. Based on their findings, they conclude that the law should consider the requirement of experience for obtaining a driver's license.

The purpose of preventative detention is to ensure the accused's presence, as claimed. Therefore, in their study, colleagues [61] aim to examine the legal framework regarding the proper application of the presumption of innocence. They utilize the neutrosophic Iadov method to evaluate compliance with regulatory guidelines and develop a strategy for their effective implementation using NCMs.

2.4 AN EXTENDED VERSION OF NCMS USING KRIPKE STRUCTURES

2.4.1 The problem of reasoning about time evolution

The representation of time, changes, actions, and events has been extensively studied in philosophy, theoretical computer science, and artificial intelligence. Understanding how systems change over time, predicting future states based on present conditions, and analyzing the impact of events on a temporal scale are fundamental yet intricate problems. One of the disadvantages of classical logic is its inability to represent time in a sufficient way. In classical logic, a fact is simply classified as true or false, without any indication of its temporal context. There is no information about when the fact was true, whether it is still true, or if it will be true in the future. Additionally, there is no knowledge about the absolute duration of the fact or its validity in comparison to other facts.

Considering that a significant portion of human communication is time-dependent, and numerous problem-solving methods rely on temporal relationships between events, it becomes evident that certain knowledge-based systems and Artificial Intelligence systems necessitate the capability to infer based on time. Hence, time-based inference capabilities are required in knowledge-based and Artificial Intelligence systems.

We have already mentioned in Section 2.2 that two basic methodologies are utilized when considering modeling a complex system with an NCM, namely dynamic and static analysis of NCMs. Although the aforesaid methodologies have been used with success in many applications, it is obvious that they cannot model or represent explicitly the dynamics of time. In this context, we will describe, in the current section, a conceptual framework that integrates Kripke structures [62] with NCMs so as to provide a blueprint that will enable us to take into consideration triplets of truth assignments. Kripke structures provide a formal framework for understanding modal logics and reasoning about necessity, possibility, and other modalities. They have been discussed previously in neutrosophic logic, in the work of [63], where scholars examine the ability of neutrosophic logic to process symbolic and numerical statements on belief and knowledge using the possible worlds semantics but without reasoning about time representation and the length of time in which an event is true and the factors that influence it over time. The latter is known as the *frame problem* which is, in plain words, the answer we try to give to the question "what an action does not cause?." Various studies have been proposed to address the frame problem [64–67], which involve the exploration of several approaches such as non-monotonic logics, temporal logics, dynamic logic, and causal models. The objective is to create efficient ways to represent, reason about, and update information that is influenced by actions or changes without having to completely re-specify the environment every time.

In response to this, our approach utilizes neutrosophic modal logic [68], which is an extension of classical modal logic, where some neutrosophic modalities have been included that allow us to draw conclusions concerning the evolution of the world of a problem. These logics are designed to capture and express temporal relationships, events, and propositions that change over time, hence providing a formal and expressive means to capture and reason about the dynamic nature of systems and the temporal progression of propositions and events.

For the scope of this chapter we are only interested in neutrosophic *temporal modalities* (related to *time*) which enables us to generate expressions like the following ($\mathscr{P}$ is a neutrosophic proposition) [68]:

It was the neutrosophic condition that $\mathscr{P}$.
It will neutrosophically be that $\mathscr{P}$.
And analogously:
It has always been so from a neutrosophical perspective.
It will always be $\mathscr{P}$ from a neutrosophical perspective.

2.4.2 Proposed framework

A Kripke model, also referred to as a Kripke structure, is a mathematical structure that resembles a directed labeled graph. The nodes of the graph represent possible worlds from a set S, which are labeled by truth assignments π.

Utilizing the notation from [63], we can define a Kripke structure in a neutrosophic environment as follows:

> **DEFINITION** [63]. A Kripke model for neutrosophic propositions is a triple structure S_K^{NL} of the form $\langle S, R, \vec{\pi} \rangle$ where

- S is a non-empty set (the set of possible worlds);
- $R \subset S \times S$ is the accessibility relation;
- $\vec{\pi} = (\pi_T, \pi_I, \pi_F)$ is a neutrosophic assignment to the propositions per possible world, i.e. π: $(S \rightarrow P) \rightarrow ||\ ^-0,\ 1^+||$ with π being either π_T, or π_I, or π_F.

where $P = \{p_1, \ldots, p_n\}$ is a set of propositional variables.

In a standard Kripke model, the "truth" assignment π is represented as $\vec{\pi} = (\pi_T, \pi_I, \pi_F)$, a three dimensional assignment, with π_T expressing truth, π_F denoting falsity and π_I expressing indeterminacy.

Let us assume that p is a propositional variable. It is known that in classical modal logic we have the following operators:

> $\Diamond\, p$ means that p is possibly true in the future, that is, it will occur sometimes in the future.
> $\Box\, p$ means that p is necessarily true in the future, that is, it will occur always in the future.

Now we extend the above modal operators in a neutrosophic environment:

> **DEFINITION** [68]. $\Diamond_N\, p$ denotes that «It is (t, i, f)-possible that p » where «(t, i, f)-possible» indicates t % possible (chance that p occurs), i % indeterminate (indeterminate-chance that p occurs), and f % impossible (chance that p does not occur).

> **DEFINITION** [68]. $\Box_N$ p means that «It is (t, i, f)-necessary that p » where «(t, i, f)-necessity» implies t % necessary (certainty that p occurs), i % indeterminate (indeterminate-surety that p occurs), and f % unnecessary (unsurely that p occurs).

When modalities are combined with neutrosophic truth-values, it is possible to reason about the necessity and possibility of truth, indeterminacy, or falsity in a given context. This logic extends modal logic's capacity for reasoning into uncertain and partial information domains.

Now let us consider a simple example in order to better understand how we can integrate neutrosophic propositions and modal operators within a Kripke structure.

Example We will examine an example related to weather conditions and specifically how we interpret the proposition "Rainy Weather."

Firstly we represent the possible worlds with neutrosophic elements:

World 1: $(T_{rain}, I_{rain}, F_{rain})$
World 2: $(T_{rain}, I_{rain}, F_{rain})$

Next we define the accessibility relations with neutrosophic transitions:

World 1 can transition to World 2 with a degree of truth = 0.6, indeterminacy = 0.3, and falsehood = 0.1.
World 2 can transition back to World 1 with a degree of truth = 0.5, indeterminacy = 0.2, and falsehood = 0.3.

Neutrosophic Propositions (Rainy Weather):

Proposition $\mathcal{P}$: "It will possibly be a rainy day next Monday." According to the meteorological center and its statistical data, the neutrosophic truth-value of $\mathscr{P}$ is:

In World 1: $\mathscr{P}$ ([0.6, 0.7], (0.3 0.5), {0.4, 0.6}), i.e. [0.6, 0.7] true, (0.3, 0.5) indeterminate, and {0.4, 0.6} false.
In World 2: $\mathscr{P}$ ([0.4, 0.6], (0.2 0.3), {0.3, 0.4}).

Then in World 1, the neutrosophic possibility operator for proposition "it will be rainy next Monday" is:

$\Diamond_N \mathcal{P}$= (sup[0.6, 0.7], inf(0.3, 0.5), inf{0.4, 0.6}) =(0.7, 0.3, 0.4), i.e. 0.7 possible, 0.3 indeterminate-possibility, and 0.4 impossible.

Similarly, we can calculate the neutrosophic necessity operator for the proposition "it will be rainy next Monday" in World 2:

$\Box_N \mathcal{P}$= (inf[0.4, 0.6], sup(0.2, 0.3), sup{0.3, 0.4}) = (0.4, 0.3, 0.4), i.e. 0.4 necessary, 0.3 indeterminate-necessity, and 0.4 unnecessary.

Next we will demonstrate our proposed method in order to model a virtual world presented in [18]. It is noted that the example given in [18] is demonstrated with a simple FCM and dynamic analysis is used so as to "narrate" the life cycle of a dolphin.

Example Suppose now that we want to model a virtual underwater world for a dolphin using an NCM. It represents instances of the life cycle of a dolphin. Figure 2.6 shows a simple NCM with five nodes that depict the above scenario. For example when a threat appears in the form of a shark then the dolphins flee the shark in a tightly packed herd. Dolphins avoid the shark and then rest etc.

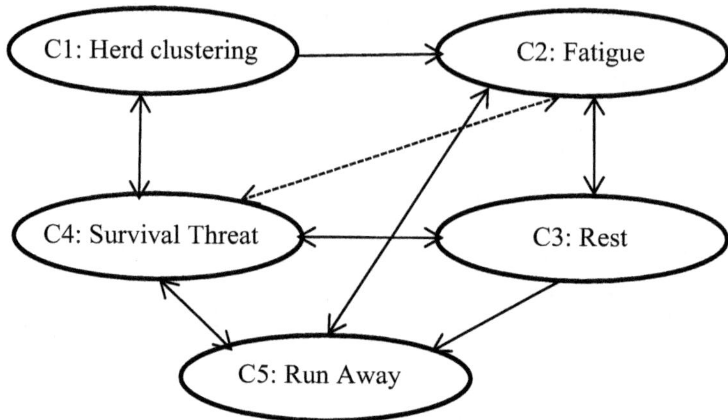

Figure 2.6 Simple NCM modeling of the underwater virtual world.

The utility of temporal logic is to draw conclusions about the world evolution of a problem. In this case we can view the NCM as a Kripke structure and try to model the virtual world. In a neutrosophic environment, which deals with indeterminacy, uncertainty, and incomplete information, the application of Kripke structures can be adapted to accommodate these nuances. Kripke structures provide a formalism for modal logics, which allow for reasoning about necessity, possibility, and belief. By integrating them into neutrosophic logic, we can represent indeterminate or partially true modal statements, thereby enhancing reasoning capabilities in scenarios where uncertainty is inherent.

The figure above illustrates the transitions between states over time. These transitions occur when the dolphin receives the appropriate stimuli from its environment, such as "seeing" a shark or "feeling" tired and "considering" the need to rest after being chased by a shark, among others.

According to our model, neutrosophic modal logic can be used to express various properties of the virtual world. Here, for reasons of brevity, we will illustrate only a few of these situations. For instance, for a dolphin named X, the following expression indicates that it will be true, in a neutrosophic sense, at some point in the future that the dolphin will avoid a shark attack and subsequently rest.

$$\Diamond_N \left\lfloor \neg\, \text{SurvivalThreat}(X) \wedge \text{Rest}(X) \right\rfloor \tag{2.7}$$

In the above equation adding the neutrosophic modal operator $\Diamond_N \mathcal{P}$ we are able to incorporate to the neutrosophic proposition $\mathscr{P}$ a three-dimensional truth-assignment (π_T, π_I, π_F), thus reasoning in scenarios where uncertainty is inherent. This could be very handy when we want to estimate for example the possibility that a dolphin might not be able to survive a shark atack.

This could be done by assigning numerical data in equation (2.7) as shown in example 4.2.1.

In another scenario we could express that it will always be neutrosophically true in the future that a dolphin will play in a herd clustering and then rest or been attacked by a shark and then run away to avoid the threat. This could be expressed as it follows:

$$\Box_N \left[\left(\text{Clustering}(X) \wedge \text{Fatigue}(X) \wedge \text{Rest}(X)\right) \vee \begin{pmatrix} \text{SurvivalThreat}(X) \\ \wedge \text{RunAway}(X) \end{pmatrix} \right] \quad (2.8)$$

By adding the neutrosophic modal operator, we bring our modeling of the virtual world closer to human perception and judgment. This incorporation of indeterminacy into the expression shown in (8) allows us to calculate the possibility of a dolphin avoiding a shark and then clustering with other dolphins to search for food and rest. However, it is important to consider that unexpected underwater environmental conditions, such as pollution, can make the search for food more difficult. Again, this could be achieved by assigning numerical values to expression (2.8).

Lastly, we should note that our proposed framework allows us to handle indeterminate relations in NCMs through the integration of Kripke structures in a neutrosophic environment. To maintain the semantics while accommodating indeterminacy, we need to carefully formulate the accessibility relation and modal operators. In our framework, each possible world in the Kripke structure corresponds to a neutrosophic state in the NCM [69]. Neutrosophic states encapsulate indeterminate relations among elements, expressing truth, indeterminacy, and falsity values. The accessibility relation should account for worlds that are "partially" accessible or relate to each other with certain degrees of indeterminacy. To achieve this, we utilize neutrosophic modal operators in Kripke structures over indeterminate relations. This allows us to reflect on the neutrosophic interpretation.

So in our example the expression:

$$\Diamond_N \left[\text{SurvivalThreat}(X) \wedge \text{Fatigue}(X) \right] \quad (2.9)$$

could be employed in our model so as to indicate that there exists a scenario where a survival threat could cause the dolphin to feel tired, just in the thought of a possible attack by a shark, despite any uncertainties.

From the above scenarios, it is clear that the suggested extended version of an NCM allows us to model a virtual world in a flexible manner that closely aligns with human thinking. Our world is inherently filled with indeterminacy, and developing a neutrosophic framework that integrates it provides us with a useful conceptual tool for reasoning about the significant degree of ambiguity present in real-world problems. In addition, the integration of neutrosophic modal logic in the proposed extended version of NCM enables us to reason about dynamic environments where changes occur frequently over time and where certainty might be difficult to establish. Time dynamics as represented by neutrosophic Kripke allows for the depiction of complicated information networks with changing degrees of truth, falsity, or indeterminacy across distinct propositions or modalities.

2.5 CONCLUSIONS

NCMs are a type of cognitive map that incorporate the concept of neutrosophy, a branch of philosophy that studies neutralities and their interactions with different ideational spectra. Neutrosophy acknowledges the presence of indeterminacy, ambiguity, and contradictions in human cognition and perception, recognizing that phenomena often exist in states of partial truth, falsehood, and indeterminacy simultaneously.

The NCM approach has been well-demonstrated in the literature as a useful tool for modeling and analyzing complicated dynamical systems. It combines neutrosophy and cognitive mapping to create a framework that accommodates the nuances of human cognition. Additionally, NCMs offer a promising avenue for modeling complex systems where uncertainty, ambiguity, and imprecision are inherent.

The key features of NCMs are as follows:

1. Handling uncertainty: NCMs excel in capturing the uncertainties and vagueness prevalent in real-world systems. By incorporating neutrosophic logic, they represent not only true or false relationships but also the indeterminate aspects within cognitive models.
2. Complexity representation: NCMs can represent complex relationships between concepts, making them valuable in analyzing intricate systems where traditional crisp logic may fall short.
3. Applicability: NCMs find applications in diverse fields including decision-making, mathematics, computer science, social sciences, and more, where understanding and modeling uncertain relationships are crucial.

In this book chapter, we suggest a new extension to NCMs that integrates neutrosophic modal logics, which offer advantages over traditional approaches:

1. Decision-making in vague environments: Neutrosophic modal logic allows for reasoning about necessity and possibility under unclear

situations often encountered in real-world decision-making systems or AI applications with partial or ambiguous information.
2. Reasoning flexibility: The capacity to reason about necessity and possibility in a neutrosophic setting provides flexibility in dealing with complicated circumstances where typical binary logics may not adequately capture ambiguity or unclear information.

The contributions of this book chapter can be summarized as follows:

1. Development and analysis of fundamental mathematical theories and methodologies of NCMs, providing a solid background for future research efforts.
2. A review of NCMs' applications in several scientific domains during the last decade, focusing on representative works for each application topic. We identified 78 research studies proposed between January 2013 and November 2023 based on our search in the Scopus database engine.
3. Introduction of a new conceptual framework that extends traditional NCMs by considering the dynamics of time. This framework allows for reasoning about propositions that are true, false, and indeterminate simultaneously, providing a flexible way to represent uncertainty and vagueness in knowledge representation systems. The accessibility relations in Kripke structures are adapted to reflect the degrees of indeterminacy or uncertainty between possible worlds, enabling a detailed analysis of information flow and truth values across these worlds.

We believe that time representation should be included in the NCM model since it can influence the system's dynamics under simulation. Without considering time, model findings may overestimate or underestimate the influence seen by concepts, calling their validity into question. While our model needs to be tested and evaluated on real-world case studies, we hope that this research attempt can serve as an initial step in this direction.

In summary, NCMs provide a unique approach to understanding and analyzing complex systems in the face of ambiguity and incomplete information. The integration of NCMs with Kripke structures offers a promising avenue for addressing complexities in decision-making and knowledge representation. This fusion presents a novel framework that transcends traditional binary logic, accommodating indeterminacy, vagueness, and inconsistency inherent in real-world systems. Its capacity to handle uncertainty, vagueness, and conflicting information offers a powerful tool for modeling complex systems and holds promise for various interdisciplinary applications. While challenges exist, their potential to provide deeper insights into sophisticated relationships within diverse domains makes them a challenging area for further exploration and development.

FUNDING

This work is part of a project that has received funding from the University of Macedonia Research Fund under the "Benchmarking Port Performance: A Multi-criteria Evaluation (POPE)" funding programme.

REFERENCES

1. Peirce, C.S. Existential graphs. *Unpublished manuscript reprinted in (Buchler 1955)*, 1909.
2. Rumelhart, D. E., & Ortony, A. (2017). The representation of knowledge in memory 1. In *Schooling and the acquisition of knowledge* (pp. 99–135). Routledge.
3. Minsky, M. *A framework for representing knowledge*, 1974.
4. Tolman, E.C. Cognitive maps in rats and men. *Psychol Rev*, 1948, vol. 55, pp. 189–208.
5. Sowa, J. F. *Conceptual structures: information processing in mind and machine*. Addison-Wesley Longman Publishing Co., Inc., 1984.
6. Axelrod, R. *Structure of decision: The cognitive maps of political elites*. Princeton University Press, 1976.
7. Kosko, B. Fuzzy cognitive maps. *International Journal of Man-machine Studies*, 1986, vol. 24, pp. 65–75.
8. Kosko, B. Hidden patterns in combined and adaptive knowledge networks. *International Journal of Approximate Reasoning*, 1988, vol. 2, pp. 377–393.
9. Papageorgiou, E. I.; Salmeron, J. L. A review of fuzzy cognitive maps research during the last decade. *IEEE Transactions on Fuzzy Systems*, 2012, vol. 21, pp. 66–79.
10. Smarandache, F. A unifying field in Logics: Neutrosophic Logic. In *Philosophy*, 1999, pp. 1–141. American Research Press.
11. Smarandache, F. *Neutrosophic Perspectives: Triplets, Duplets, Multisets, Hybrid Operators, Modal Logic, Hedge Algebras. And Applications*. Infinite Study, 2017.
12. Kandasamy, W. V.; Smarandache, F. *Fuzzy cognitive maps and neutrosophic cognitive maps*. Infinite Study, 2003.
13. Ricardo, J. E.; Flores, D. F. C.; Díaz, J. A. E.; Teruel, K. P. *An exploration of wisdom of crowds using neutrosophic cognitive maps*, 2020, vol. 37, Infinite Study.
14. Al-Subhi, S. H.; Papageorgiou, E. I.; Pérez, P. P., Mahdi, G. S. S., Acuña, L. A. Triangular neutrosophic cognitive map for multistage sequential decision-making problems. *International Journal of Fuzzy Systems*, 2021, vol. 23, pp. 657–679.
15. Spangler, W. E. The role of artificial intelligence in understanding the strategic decision-making process. *IEEE Transactions on Knowledge and Data Engineering*, 1991, vol. 3, pp. 149–159.
16. Wallis, W. D. *A beginner's guide to finite mathematics: for business, management, and the social sciences*. Springer Science & Business Media, 2012.
17. Aldous, J. M.; Wilson, R. J. *Graphs and applications: an introductory approach*. Springer Science & Business Media, 2003.

18. Kosko, B. *Fuzzy engineering*. Prentice-Hall, Inc., 1996.
19. Kosko, B. Bidirectional associative memories. *IEEE Transactions on Systems, Man, and Cybernetics*, 1988, vol. 18, pp. 49–60.
20. Salmeron, J. L.; Smarandache, F. Redesigning Decision Matrix Method with an indeterminacy-based inference process. *International Journal of Applied Mathematics & Statistics*, 2008, vol. 13, 4–11.
21. Murugesan, R.; Parthiban, Y.; Devi, R.; Mohan, K. R.; Kumarave, S. K. A Comparative Study of Fuzzy Cognitive Maps and Neutrosophic Cognitive Maps on Covid Variants. *Neutrosophic Sets and Systems*, 2023, vol. 55, 20.
22. Zafar, A.; Wajid, M. A. *A mathematical model to analyze the role of uncertain and indeterminate factors in the spread of pandemics like COVID-19 using neutrosophy: a case study of India* (Vol. 38). Infinite Study, 2020.
23. Rodríguez, J. O. C.; Veloz, Á. P. M.; Andi, S. T. T. Neutrosophic Cognitive Maps for the Analysis of the Factors in the proper Diagnosis of Conversion Disorder. *Neutrosophic Sets and Systems*, 2022, vol. 52, p. 36.
24. Al-Subhi, S. H.; Rubio, P. A. R.; Pérez, P. P.; Papageorgiou, E. I.; Vacacela, R. G.; Mahdi, G. S. S. *A New Neutrosophic Clinical Decision Support Model For The Treatment of Pregnant Women With Heart Diseases*. Infinite Study, 2020.
25. Ramalingam, S.; Govindan, K.; Broumi, S. Analysis of covid-19 via fuzzy cognitive maps and neutrosophic cognitive maps. *Neutrosophic Sets and Systems*, 2021, vol. 42, pp. 102–116.
26. Kumar, M., Bhutani, K., & Aggarwal, S. Hybrid model for medical diagnosis using Neutrosophic Cognitive Maps with Genetic Algorithms. In *2015 IEEE international conference on fuzzy systems (FUZZ-IEEE)* (pp. 1–7). IEEE.
27. Angeline, L. A.; Merlin, M. M. M. Threats of cancer due to modernized society–A study on neutrosophic approach. *Advances in Mathematics*, 2020, vol. 9, pp. 1533–1539.
28. Shakil, M. T. A.; Ubaid, S.; Sohail, S. S.; & Alam, M. A. (2021). A Neutrosophic cognitive map based approach to explore the health deterioration factors. *Neutrosophic Sets and Systems*, 2021, vol. 41, 198.
29. Reyes Salgado, L. N.; Ramos Argilagos, M. E.; Sánchez Garrido, A.; Valencia Herrera, A. R.; Saleh Al-Subhi, S. H. Model for the Diagnosis of Autism Based on Neutrosophic Cognitive Maps. *Neutrosophic Sets and Systems*, 2021, vol. 44, p. 15.
30. Raúl, D. C.; Fabricio, T. T.; Elizabeth, D. P. F. Analyzing the impact of social factors on homelessness with neutrosophic cognitive maps. *International Journal of Neutrosophic Science (IJNS)*, 2022, vol. 19(1).
31. Palacios, A. J. P.; Ricardo, J. E.; Piza, I. A. C.; Herrería, M. E. E. Phenomenological hermeneutical method and neutrosophic cognitive maps in the causal analysis of transgressions against the homeless. *Neutrosophic sets and systems*, 2021, vol..44, pp. 147–156.
32. Dimitri, V. B.; De Jesús, M. G.; Enrique, P. A. Neutrosophic cognitive maps for analysis of social problems analysis. *International Journal of Neutrosophic Science (IJNS)*, 2022, vol. 19(1).
33. Rojas, H. E.; Nelson, F. S.; Cabrita Marina, M. Neutrosophic Cognitive Maps for Violence Cause Analysis. *International Journal of Neutrosophic Science (IJNS)*, 2022, vol. 19(1).
34. del Pozo Franco, P. E.; Farias, G. K. A.; Escobar, L. M. O.; Manzo, A. D. M. Method for Recommending Custody to Minors based on Parental

Responsibility using a Neutrosophic Cognitive Map. *Neutrosophic Sets and Systems*, 2021, vol. 44, pp. 324–332.
35. Wajid, M. A., & Zafar, A. (2022). Neutrosophic Image Segmentation: An Approach for the Treatment of Uncertainty in Multimodal Information Systems. *International Journal of Neutrosophic Science (IJNS)*, vol. 19(1).
36. Vasantha, W. B.; Kandasamy, I.; Devvrat, V.; Ghildiyal, S. *Study of imaginative play in children using neutrosophic cognitive maps model*. Infinite Study, 2019.
37. Huera Castro, D. E.; Dávila Castillo, M. R.; Escobar Gonzáles, E. J.; Hidalgo Ruiz, M. R. Method for Recommending Custody to Minors based on Parental Responsibility using a Neutrosophic Cognitive Map. *Neutrosophic Sets and Systems*, 2021, vol. 44, p. 36.
38. Rodríguez, I. X. L.; Diaz, J. A. E.; Salazar, L. D. P. *Analysis of crimonogenic factors in femicide crimes* (Vol. 37). Infinite Study, 2020.
39. Smarandache, F.; Broumi, S.; Singh, P. K.; Liu, C. F.; Rao, V. V.; Yang, H. L.; Elhassouny, A. Introduction to neutrosophy and neutrosophic environment. In *Neutrosophic set in medical image analysi*, 2019 (pp. 3–29). Academic Press.
40. Chithra, B.; Nedunchezhian, R. Dynamic neutrosophic cognitive map with improved cuckoo search algorithm (DNCM-ICSA) and ensemble classifier for rheumatoid arthritis (RA) disease. *Journal of King Saud University-Computer and Information Sciences*, 2022, vol. 34(6), pp. 3236–3246.
41. Dhanusha, C.; Senthil Kumar, A. V. (2020). Enriched neutrosophic clustering with knowledge of chaotic crow search algorithm for alzheimer detection in diverse multidomain environment. *Int J Sci Technol Res (IJSTR)*, 2020, vol. 9, pp. 474–481.
42. Martin, N. Exploration and comparative investigation on neutrosophic cognitive maps non–linear Hebbian learning algorithm in predictive model. In *2022 AIP Conference Proceedings* (Vol. 2516, No. 1). AIP Publishing.
43. Lara, A. D. R.; Tello, C. P. R.; Blacio, J. A.; Guevara, R. H. Neutrosophic interrelationship of Key Performance Indicators in an accounting process. *Neutrosophic Sets and Systems*, 2020, vol. 34, pp. 110–116.
44. Calefariu, E.; Boşcoianu, M.; Smarandache, F.; Buda, T. A. Neutrosophic Modeling of Investment Architectures. *Applied Mechanics and Materials*, 2014, vol. 657, pp. 1011–1015.
45. Rajkumar, A.; Lakshmipathy, N.; Prakash, A. P. (2020). Ranking Method using New Triangular Combined Overlap Neutrosophic Cognitive Maps (TrCOBNCMs). *Advances in Mathematical Scientific Journal*, 2020, vol. 9, pp. 11161–11167.
46. Cortés, L. E. Á.; Armijo, J. G. S.; Sánchez, V. M. Q.; Ortega, R. G. *Evaluation of the impact of informal trade using neutrosophic cognitive maps and weighted power mean operator* (Vol. 37). Infinite Study, 2020.
47. Villalba, M. F. G.; Viteri, M. S. S.; Castro, I. R.; Díaz, F. V. Critical success factors modelling in operational management and the recovery of overdue portfolio of the Babahoyo GAD in The Municipal Market May 4. *Neutrosophic Sets and Systems*, 2020, vol. 34, pp. 57–62.
48. Wajid, M. S., Terashima-Marin, H., Najafirad, P., Pablos, S. E. C., & Wajid, M. A. (2024). DTwin-TEC: An AI-based TEC district digital twin and emulating security events by leveraging knowledge graph. *Journal of Open Innovation: Technology, Market, and Complexity*, vol. 10(2), p. 100297.

49. Villamar, C. M.; Suarez, J.; Coloma, L. D. L.; Vera, C.; Leyva, M. *Analysis of technological innovation contribution to gross domestic product based on neutrosophic cognitive maps and neutrosophic numbers*. Infinite Study, 2019.
50. Martin, N.; Broumi, S. Neutrosophic Cognitive Impact Study on Role Transformation of Teachers to Facilitators. In *Handbook of research on the applications of neutrosophic sets theory and their extensions in education* (pp. 215–234). IGI Global, 2023.
51. Vasantha, W. B.; Kandasamy, I.; Devvrat, V.; Ghildiyal, S. *Study of imaginative play in children using neutrosophic cognitive maps model*. Infinite Study, 2019.
52. Mary, M. F. J.; Merlin, M. M. M. *A Comparative Study Using Neutrosophic Cognitive Map and Triangular Fuzzy Cognitive Map for Analyzing The Factors for Quality Training of Elementary Education Student Teachers In Tamilnadu*. Infinite Study, 2021.
53. Yasir, M.; Zafar, A.; Wajid, M. A. NEP-2020's Implementation & Execution: A Study Conducted Using Neutrosophic PESTEL Analysis. *International Journal of Neutrosophic Science (IJNS)*, 2023, vol. 20(2).
54. Bhutani, K.; Kumar, M.; Garg, G.; Aggarwal, S. Assessing it projects success with extended fuzzy cognitive maps & neutrosophic cognitive maps in comparison to fuzzy cognitive maps, *Neutrosophic Sets and Systems*, 2016, vol. 12(1), 9–19.
55. Betancourt-Vázquez, A.; Leyva-Vázquez, M.; Perez-Teruel, K. Neutrosophic cognitive maps for modeling project portfolio interdependencies, *Critical Review*, 2015, vol. 10, 40–44.
56. Al-Subhi, S. H.; Mahdi, G. S.; Pupo, I. P.; Pérez, P. P.; Vázquez, M. L. Neutrosophic Cognitive Map for a Multi-Objective Decision Making in Software Projects. *In CIbSE* (pp. 58–71), 2019.
57. Al-Subhi, S. S. H.; Pérez Pupo, I.; Piñero Pérez, P. Y.; Mahdi, G. S. S.; Villavicencio Bermúdez, N. New Extensions of Fuzzy Cognitive Maps for Sequential Multistage Decision-Making Problems: Application in Project Management. In *2021 Conferencia científica internacional uciencia* (pp. 171–189). Cham: Springer International Publishing.
58. Colorado Pástor, B. A.; Fois Lugo, M. M.; Leyva Vázquez, M.; Hechavarría Hernández, J. R. Proposal of a technological ergonomic model for people with disabilities in the public transport system in Guayaquil. In *Advances in Usability and User Experience: Proceedings of the AHFE 2019 International Conferences on Usability & User Experience, and Human Factors and Assistive Technology*, July 24–28, 2019, Washington DC, USA 10 (pp. 831–843). Springer International Publishing.
59. Ramírez, M. A. C.; Ahez, J. C. D. J. A.; Riera, O. I. R.; Quezada, R. G. H.; Vera, Á. A. R.; Cegarra, J. C. T.; Sotomayor, P. M. O. (2019). Pestel based on neutrosophic cognitive maps to characterize the factors that influence the consolidation of the neo constitutionalism in Ecuador. *Neutrosophic Sets and Systems*, 2019, Vol. 26. *Special Issue: Social Neutrosophy in Latin America*, 61.
60. Wajid, M. S., & Wajid, M. A. (2021). The importance of indeterminate and unknown factors in nourishing crime: a case study of South Africa using neutrosophy. *Neutrosophic Sets and Systems*, vol. 41(2021), p. 15.
61. Almache, F. A.; Lomas, C. R.; Mina, J. R. Presumption of innocence: analysis of its application using neutrosophic methods. *International Journal of Neutrosophic Science (IJNS)*, 2022, vol. 18(4).

62. Kripke, S. A. Semantical analysis of modal logic I normal modal propositional calculi. *Mathematical Logic Quarterly*, 1963, vol. 9, pp. 67–96.
63. Jousselme, A. L.; Maupin, P. Neutrosophy in situation analysis. In *2004 Proceedings of fusion* (pp. 400–406).
64. Reiter, R. The frame problem in the situation calculus: A simple solution (sometimes) and a completeness result for goal regression. *Artificial and Mathematical Theory of Computation*, 1991, vol. 3.
65. Hölldobler, S.; Schneeberger, J. A new deductive approach to planning. *New Generation Computing*, 1990, vol. 8, pp. 225–244.
66. Thielscher, M. *Introduction to the fluent calculus*. Linköping University Electronic Press, 1998.
67. Pratt, V. R. Semantical considerations on Floyd-Hoare logic. In *17th Annual Symposium on Foundations of Computer Science (sfcs 1976)* (pp. 109–121). IEEE.
68. Smarandache, F. *Neutrosophic modal logic*. Infinite Study, 2017.
69. Wajid, M. A., & Zafar, A. (2022). A multimodal approach of information access and retrieval using neutrosophic sets. In *Emerging trends in IoT and computing technologies* (pp. 382–388). Routledge.

Chapter 3

Every aspect of handling class imbalance in the healthcare sector

Shivani Goswami and Anil Kumar Singh

3.1 INTRODUCTION

These days, there is a sharp rise in the number of instances of disparities in class in real-world machine learning datasets because of a dramatic increase in visual input from numerous sources. The state in which there is a class disparity in a dataset is called a class imbalance. This issue is brought to light when a course's distribution of samples is higher than that of other classes, potentially giving that class priority throughout the learning process (Fernández, et al., 2018). The method would often prioritize minimizing mistakes for the most frequent class at the cost of the fewer frequent classes when one class is significantly more common than the others. As a result, the less common classes may perform worse, which lowers the classification performance as a whole. Incorrect forecasts for the minority classes increase the likelihood of misleading findings, even in cases where a model has high classification accuracy. Machine learning is a hot field for researchers so now problems and challenges in machine learning are also challenging and hot topics for researchers. Figure 3.1 shows publication trends of class imbalance problem in artificial intelligence

Inequality of class is a major problem in the health industry, particularly when it comes to jobs like predicting medical events, diagnosing diseases, and identifying uncommon conditions. Unbalanced distributions, in which certain classes or outcomes are much less frequent than others, are frequently seen in healthcare statistics. The following are some salient features of the class divide in the health sector:

1. Uncommon Illnesses: Numerous illnesses are uncommon, which means that people only get them sometimes. For instance, the frequency of some cancers, hereditary diseases, or autoimmune illnesses may be rather low. The imbalance between healthy people and patients with the condition might cause difficulties during the training and assessment stages of machine learning models used to identify various disorders.

DOI: 10.1201/9781003606055-3

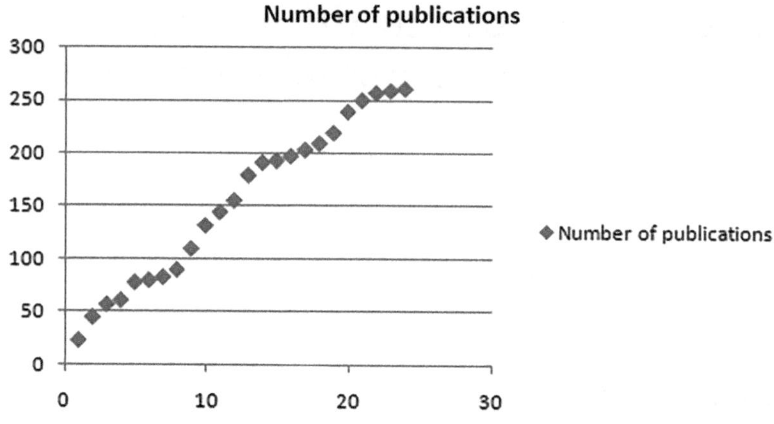

Figure 3.1 Research and publication trend of class imbalance.

2. Imbalanced Outcomes: In medical prediction tasks, such as predicting patient outcomes (like survival after surgery and readmission to the hospital), the likelihood of positive outcomes (like survival and successful treatment) may be noticeably lower than that of negative outcomes (like mortality and complications). This mismatch can affect the model's capacity to accurately detect and prioritize high-risk situations.
3. Deceptive Actions: Two types of fraud that can happen inside the healthcare system are insurance fraud and false claims. While identifying these fraudulent cases is crucial, class imbalance issues in fraud detection algorithms are sometimes challenging to resolve since fraudulent cases often belong to a minority class in the data.
4. Anomaly Detection: A lot of data is produced by healthcare systems, such as sensor readings, medical imaging, and patient records. Early diagnosis and action depend on the ability to identify abnormalities or unexpected patterns in this data, such as irregularities in medical pictures or anomalies in vital signs. However, as anomalies are usually uncommon occurrences, jobs involving anomaly identification are intrinsically unbalanced.

Data practitioners can get a comprehensive grasp of the dataset, unearth insightful information, and make well-informed judgments throughout the data analysis and modeling process by undertaking rigorous exploratory data analysis. In order to prepare data, choose features, validate models, and inform stakeholders of findings, EDA is essential.

It takes specific strategies and thoughtful thought to address class disparity in the health sector:Data Preprocessing: To balance the dataset for training machine learning models, methods including resampling (oversampling minority classes, undersampling majority classes), data augmentation, and synthetic data synthesis can be used.

Selecting Algorithms: Model performance can be enhanced by using algorithms that are resistant to class imbalance, such as ensemble techniques (e.g., Random Forests, Gradient Boosting), anomaly detection techniques (e.g., Isolation Forest, One-Class SVM), and cost-sensitive learning techniques.

Feature Engineering: To improve the model's ability to capture patterns linked to uncommon events or situations, useful features pertaining to the minority class or features created with domain expertise can be crafted.

Evaluation Metrics: A more detailed assessment of the performance of the model is possible, particularly in imbalanced circumstances, when metrics such as accuracy, precision, recall, F1-score, area under the precision-recall curve (AUC-PR), or area under the receiver operating characteristic curve (AUC-ROC) are used in addition to accuracy.

Clinical Approval: To make sure that the models' predictions are in line with medical knowledge and actually improve patient care, it is essential to validate machine learning models in clinical settings with the participation of healthcare professionals.

Effectively addressing the class divide in the healthcare industry can result in machine learning models that are more precise, dependable, and understandable, which will eventually improve patient care and healthcare outcomes.

3.1.1 Some real-world instances of class imbalance

Numerous real-world applications, such as biological diagnostic processes, credit card fraud detection, monitoring, verification of biometric data, etc., are plagued by the class imbalance problem. The minority class is typically the one of more fascination, therefore there is a great deal at stake in figuring out how to effectively learn from such a faulty dataset. Credit card unbalanced datasets have been the focus of Alam et al.'s study. Their study's main conclusion is that, when implemented at the data level, oversampling strategies, such as synthetic minority excessive sampling (SMOTE) (Chawla et al. 2002), perform superior to reducing techniques.

Learning methods are used in several data mining fields. Class imbalance arises in the biomedical realm when some illness states are rare in comparison to healthy persons, such as specific cancer subtypes. In this instance, the model might not correctly identify those who have the illness, which could have detrimental effects on patient treatment. When illicit transactions are significantly less frequent than lawful transactions in the context of identifying credit card fraud, there is a class imbalance.

This type of dataset may cause a model to prioritize minimizing mistakes for the majority class (lawful transactions) over the needs of the minority class (fraudulent transactions), which would ultimately result in the minority class performing poorly. Similarly, class imbalance can develop in monitoring and biometrics systems when certain events or individuals are noticeably less common than others.

A rise in false positives and false negatives might exacerbate the system's class imbalance-related computer vision performance issues. For example, a system of cameras trained to detect illicit activity may underperform on the minority class, or activities that are deemed to be illegal, since it has so few examples from which to learn.

When the system fails to detect instances of illicit activity, this might lead to false negatives. The mismatch between foreground and backdrop in video surveillance systems is a typical illustration of class imbalance.

Zhou et al. present the Attention-loss-driven detection of anomalies method, a novel approach to asymmetry detection that addresses the foreground-background imbalance by giving the foreground and background different weights. This helps to reduce the imbalance. Detecting small objects, like purses and arms, whose training set samples are few contrasted to the background chaos, is another common classification task in surveillance footage. This is because many background objects can resemble the small objects under consideration, increasing the number of errors.

In Pérez-Hernández et al., an improvised method for employing binary classification algorithms to detect tiny objects was put forward.

3.1.2 Factors contributing to class imbalance

In the actual world, unbalanced datasets can arise from several common sources, as the following example of biological picture collections illustrates:

(a) Inappropriate data-gathering techniques can result from a lack of technology or appropriate equipment, which can lead to incorrect data capture. This is typically seen in the medical field when there is a chance of equipment malfunction or a deficiency of contemporary diagnostic techniques.
(b) Obtaining a labeled and annotated dataset requires a significant amount of human labor, which may potentially be a factor in class imbalance. Many biological datasets exhibit an uneven class distribution profile due to the scarcity of medical personnel and expertise for large-scale data gathering and annotation.
(c) A state of imbalance may also arise from a real deficiency of samples in certain classes, which are known as minority classes. This may be seen in pictures of malignant tissues, which are less common than healthy tissues.

(d) Another reason why there could be multiple categories (like different cancer subtypes) is category overlap. Additional noise from class labels and hazy borders between different classes can also have a big influence on the model's performance. Erroneously specified labels may potentially contribute to the imbalance issue. To ensure that class borders do not overlap in situations like these, it is necessary to have well-defined procedures for appropriately labeling the classes. Binary and multiclass datasets are the two general categories into which data sets with imbalances found in the literature fall. In contrast to a binary unbalanced dataset, that is made up of two classes—one dominant and one minority—a multiclass uneven dataset consists of several classes, some of which may be dominant or minority classes.

The literature offers a variety of solutions for binaries and multiclass datasets that are unbalanced, yet it is uncommon to discover research on multiclass data. In both binary versus multiclass scenarios, Figure 3.1 displays examples of balanced and unbalanced class distributions of picture samples from a biological dataset. Addressing the issues surrounding multiclass balance difficulties involves more work and difficulty than with the binary dataset.

Multiclass unbalanced datasets may have one or more of the following mixtures:

fewer minority classes than majority classes
fewer minority classes along with greater population classes
mixtures of both. Furthermore, it is challenging to define the majority and minority classes in a multiclass unbalanced dataset based just on the distribution of samples within every category.

3.1.3 Challenges due to class imbalance

Inequality of class in the health industry can provide several difficulties:

A) A class imbalance may cause a model to perform biasedly, favoring the dominant class while underperforming the minority class. In uncommon diseases, for instance, the dataset may contain extremely few positive cases, which makes it difficult for the machine learning algorithm to identify patterns linked to those occurrences.
B) In the medical field, misdiagnosing or underdiagnosing illnesses can have detrimental effects. This problem can be made worse by class imbalance, since models may not identify uncommon illnesses or ailments that are prevalent in the minority class, which might cause diagnosis to be overlooked or delayed.

C) Models trained on imbalanced datasets may have limited generalizability, particularly when deployed in different demographics or populations. Biases inherent in the dataset may not generalize well to diverse patient populations, leading to inaccurate predictions or recommendations.
D) Limited generalizability may occur from models trained on unbalanced datasets, especially when applied to groups or demographics that differ from one another. Predictions or suggestions may be erroneous due to biases included in the dataset that may not translate well to a variety of patient groups.
E) In healthcare systems, resource allocation may be impacted by class disparities. For instance, if a class imbalance prevents a predictive model from correctly identifying people who are at risk for a given disease, healthcare resources may not be distributed effectively, which might result in worse-than-ideal patient care and results.
F) Unequal class distribution might give rise to ethical questions, especially when it comes to justice and equity in healthcare. Present inequities in healthcare outcomes and access may be unintentionally perpetuated by prediction models if specific demographic groups are neglected in the dataset.

To guarantee that predictive models are reliable, transferable, and egalitarian across a range of patient groups, addressing these issues necessitates careful consideration of data-collecting methodologies, model creation approaches, and assessment criteria. The consequences of class imbalance in healthcare datasets can be lessened by employing strategies including cost-sensitive learning, synthetic data creation, over-sampling, and undersampling. Furthermore, it is crucial to continuously monitor and validate the model's performance in real-world scenarios to spot biases and limits early on and fix them.

3.2 APPROACHES FOR SOLVING CLASS IMBALANCE IN THE HEALTH SECTOR

The mechanism for creating samples can be changed or replaced less often. By adding the Gaussian distribution to the SMOTE sample creation procedure, which promises to address the over-generalization issue brought on by synthetic samples' propensity to fall on one line, Gaussian-SMOTE (Yu & Huang, 2003) minimizes the imbalanced class issue. Similarly, RCSMOTE (Sun et al. 2018) applies an example presented with a range to recognize sample data appropriate for oversampling and considers the margin of precise estimation during the sample creation to avoid over-generalization issues brought on by over-sampling of non-informative and noisy samples. Geometric-SMOTE (Garg et al. 2018), which creates synthetic data around the geometrical region (hyper-sphere) of each chosen minority sample, widens the minority class region to broaden the range of generated samples.

In conclusion, whereas each of the aforementioned oversampling approaches has somewhat remedied SMOTE's flaws but many of them, in particular, neglect interiors and only generate data samples along the line between two minority instances, resulting in dense synthetic samples along the line segments and a generally non-uniform distribution of minority samples. Instead, our plan makes efforts to address this flaw.

The following strategies can be used to address class disparities in the health sector:

a) Resampling Techniques: A group of procedures known as resampling is employed to alter the distribution of classes in a dataset, mainly to correct for imbalances in classes. These are a few popular methods for resampling.
b) Cost-Sensitive Learning: A machine learning technique known as "cost-sensitive learning" accounts for the various expenses linked to distinct mistake kinds (such as false positives and false negatives) in a classification assignment. It's especially helpful in situations when misclassifying some classes has more expensive or negative effects than others. Cost-sensitive learning can be an effective strategy when it comes to healthcare and correcting class disparities.
c) Algorithm Selection: It's critical to find algorithms that can reliably provide forecasts and manage skewed class distributions when addressing class imbalance in the healthcare industry. In this regard, the following algorithms are frequently utilized:
d) Data Augmentation: Data augmentation is the process of applying different changes or alterations to pre-existing data samples to artificially expand the size and variety of a dataset. Data augmentation can be very helpful in the healthcare industry for resolving class imbalance and enhancing the generalization and robustness of machine learning models. These are a few typical data augmentation methods seen in the medical field.
e) Cost-Effective Measures: In the healthcare industry, cost-effective strategies seek to maximize resource use while preserving or enhancing patient outcomes. Cost-effective solutions concentrate on resource allocation techniques that minimize the influence of unbalanced data on model performance while resolving class imbalance.

Here are a few affordable ways to address healthcare class disparities:

Ensemble Methods: Three Solutions for imbalanced data in the health sector. To overcome the SMOTE shortcomings, some modified methods were proposed that try to generate synthetic samples along the line segment of selected minority samples. Most of these methods adopt one of the two policies for selecting minority samples to generate synthetic samples: borderline region sampling or safe region sampling. However, they both suffer from over-generalization problem.

3.2.1 Resampling technique

An essential phase in the data analysis process is exploratory data analysis (EDA), which entails examining and comprehending the traits, patterns, and connections present in a dataset. Before using more complex modeling techniques, EDA assists analysts and data scientists in gaining insights, seeing patterns, spotting anomalies, and developing hypotheses. The following are important components of exploratory data analysis:

Data cleaning:

Address missing values: Determine and address missing data by methods like deletion (e.g., eliminating rows or columns containing missing data) or imputation (e.g., substituting mean, median, or mode for missing values).

Find data outliers that could distort the outcomes of modeling or analysis. Outliers can be found using methods such as box plots, scatter plots, and statistical techniques (z-score, IQR, etc.).

Characteristic Statistics:

A synopsis of the statistics: To characterize the central tendency, dispersion, and distribution of numerical data, compute fundamental statistics such as mean, median, mode, standard deviation, range, and quartiles.

Tables of frequencies: To see the distribution of categorical data and determine the frequency of various categories, make frequency tables or histograms.

Data Visualization:

Univariate Analysis: To comprehend the distribution of a variable and spot any patterns or anomalies, visualize it separately using histograms, box plots, bar charts, or pie charts.

Bivariate analysis: To find correlations, trends, or dependencies between pairs of data, examine their relationships using scatter plots, line graphs, heatmaps, or correlation matrices.

Multivariate analysis: To examine intricate interactions and patterns, and visualize relationships between several variables at once using methods like 3D charts, parallel coordinate plots, or dimensionality reduction techniques (like PCA).

Feature engineering:

Produce new features: To gather more data and enhance model performance, produce new features based on variables already in place or domain expertise.

Variable transformation: To address skewed distributions, heteroscedasticity, or nonlinearity in the data, apply transformations such as scaling, normalization, or log transformations.

Identification of Patterns:

Recognize trends: To find naturally occurring groups or clusters within the data, use clustering techniques (e.g., K-means clustering, hierarchical clustering).

Time series analysis: Use methods such as trend analysis, seasonality decomposition, autocorrelation plots, and forecasting models to examine temporal patterns and trends in time series data.

The mechanism for creating samples can be changed or replaced less often. By adding the Gaussian distribution to the SMOTE sample creation procedure, which promises to address the over-generalization issue brought on by synthetic samples' propensity to fall on one line, Gaussian-SMOTE (Yu & Huang, 2003) minimizes the imbalanced class issue. Similarly, RCSMOTE (Sun et al. 2018) applies an example presented with a range to recognize sample data appropriate for oversampling and considers the margin of precise estimation during the sample creation to avoid over-generalization issues brought on by over-sampling of non-informative and noisy samples. Geometric-SMOTE (Garg et al. 2018), which creates synthetic data around the geometrical region (hyper-sphere) of each chosen minority sample, widens the minority class region to broaden the range of generated samples.

In conclusion, whereas each of the aforementioned oversampling approaches has somewhat remedied SMOTE's flaws but many of them, in particular, neglect interiors and only generate data samples along the line between two minority instances, resulting in dense synthetic samples along the line segments and a generally non-uniform distribution of minority samples. Instead, our plan makes efforts to address this flaw.

A collection of approaches known as resampling techniques is used to alter a dataset's class distribution, mostly to correct for imbalances in classes. Here are a few popular resampling methods: Excessive sampling Until a balanced class distribution is reached, instances from the minority class are randomly duplicated through the process of random oversampling. By interpolating between instances of the minority class that already exist, SMOTE (Synthetic Minority Over-sampling Technique) creates synthetic instances for the minority class. SMOTE generates synthetic samples along the line segments connecting each minority class sample by randomly picking one or more of the k-nearest neighbors. Comparable to SMOTE, ADASYN (Adaptive Synthetic Sampling) adds additional synthetic samples to areas with sparse class distributions (Figure 3.2).

It is possible to alter or update the sample creation method less frequently.

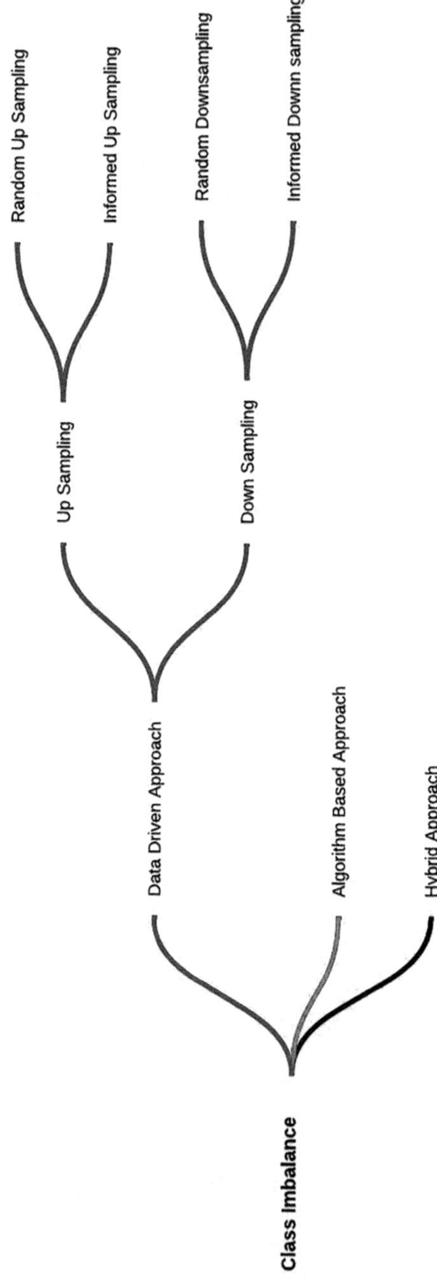

Figure 3.2 Approaches to solve imbalance issue.

The SMOTE sample production procedure will incorporate the Gaussian distribution, which should alleviate the over-generalization problem caused.

The uneven class problem is minimized by Gaussian-SMOTE (Yu & Huang, 2003), which takes into account synthetic samples' tendency to fall on one line. Similar to this, RCSMOTE (Sun et al. 2018) uses a range-presented example to identify sample data suitable for oversampling and takes into account the margin of exact estimation at sample formation to prevent problems with overgeneralization caused by oversampling of noisy and non-informative samples. To increase the variety of generated samples, Geometric-SMOTE (Garg et al. 2018) expands the minority class territory by generating synthetic data around the geometrical region (hyper-sphere) of each selected minority sample.

Undersampling:

Random Undersampling: Randomly delete instances from the dominant class until a balanced class distribution is attained. Based on distance measurements, Near Miss selects instances from the majority class that are close examples in the minority class, reducing the majority class while maintaining important information.

Tomek Connections: It locates pairs of instances that are the nearest neighbors—one from the majority class and one from the minority class—to minimize class overlaps. Next, the instance of the majority class is

Updated The Nearest Neighbors (ENN) algorithm is used to exclude instances of the majority class whose class label differs from that of the majority of its k-nearest neighbors. Using the SMOTE oversampling technique with the ENN under-sampling approach results in SMITTEN, which combines the two techniques.

SMOTE Tomek: Inte grates the Tomek Links under-sampling method with the SMOTE oversampling method. Generating Synthetic Data: Generative Adversarial Networks, or GANs, are trained to produce artificial samples of data that closely mimic occurrences of the minority class.

Hybrid Methods: Cluster-oriented Oversampling: To create synthetic samples, identify clusters in the minority class and use oversampling techniques within these clusters.

Ensemble Methods with Resampling: To increase predictive performance, ensemble learning strategies like boosting or bagging are combined with resampling. The particulars of the dataset, available computing power, and the analytic objectives must all be taken into account when choosing a resampling approach. To guarantee the efficacy of the selected resampling strategy, it is also essential to assess the model's performance using the proper metrics and validation methods.

It suggests that the model for identifying fraudulent transactions is created by machine learning (ML) using the negative class. To overcome this challenge, a number of techniques and algorithms were applied to close the gap between the two categories. By randomly reproducing the examples, oversampling is a technique to increase the instance count in the positive class, whereas undersampling is a technique to reduce the number of examples in the negative class.

There are a number of challenges when using classification techniques to identify fraudulent transactions. These challenges include: (a) class imbalance (fraud transactions are very small) and (b) concept drift, which is the idea that class conditional proportions change over time and require feature preprocessing to be up-dimensional in the trained model.

Author	*Approach*	*Contribution*
Chawla et al. (2002)	SMOTE: Synthetic Minority Over-sampling Technique	Introduced SMOTE algorithm for addressing class imbalance
Weiss & Provost (2003)	Cost-sensitive learning	Examined the impact of class distribution on decision trees
Van Hulse et al. (2007)	Experimental perspectives on learning from imbalanced data	Investigated performance of different algorithms on imbalanced data
Lemaitre et al. (2017)	Imbalanced-learn toolbox	Developed Python toolbox for imbalanced data handling
Drummond & Holte (2003)	Comparison of under-sampling and over-sampling method	Demonstrated the superiority of under-sampling over over-sampling
Sun et al. (2009)	Cost-sensitive boosting	Proposed boosting algorithm for imbalanced data
Saez et al. (2015)	SMOTE-IPF: SMOTE with Instance Priority Filtering	Introduced SMOTE-IPF for handling noisy examples
Batista et al. (2004)	Comparison of different balancing methods	Investigated behavior of various methods on imbalanced data
Haixiang & Garcia	Learning from imbalanced data	Reviewed various approaches to address class imbalance
Fernández et al. (2018)	Learning from Imbalanced Data Sets	Comprehensive overview of techniques for imbalanced data
Shivani et al.	Imbalanced data	Comprehensive overview of techniques for imbalanced data and future challenges too.

Cost-Sensitive Learning: A machine learning technique known as "cost-sensitive learning" accounts for the various expenses linked to distinct mistake kinds (such as false positives and false negatives) in a classification assignment. It's especially helpful in situations when misclassifying some classes has more expensive or negative effects than others. Cost-sensitive learning can be an effective strategy when it comes to healthcare and correcting class disparities. This is the operation of cost-sensitive learning.

Cost Assignment: Rather than treating all misclassifications identically, charges are allocated to various mistake categories according to their significance. In a medical diagnosis task, for instance, misdiagnosing a patient as healthy when they have a dangerous ailment (false negative) might have more dire repercussions than misdiagnosing a healthy patient as having the condition (false positive). Modifying the Learning Algorithm: Conventional machine learning methods seek to reduce the total amount of categorization error (Haibo & Garcia, 2008).

The algorithm is adjusted for cost-sensitive learning to minimize a cost function that takes the allocated costs into account. Instead of only minimizing the misclassification rate, the algorithm is taught to make judgments that minimize the projected overall cost.

Class Weighting: Giving several classes distinct weights throughout training is a popular strategy in cost-sensitive learning. For instance, the weight given to the minority class typically the positive class an issue of binary categorization with class imbalance might be raised to represent its significance in comparison to the majority class.

Threshold Adjustment: The anticipated class label is often determined by a decision threshold in categorization algorithms. This threshold can be changed in cost-sensitive learning according to the relative costs of various mistake kinds. For instance, the threshold can be changed to decrease the possibility of false negatives at the risk of perhaps raising false positives if false negatives are more expensive.

Evaluation Measures: When evaluating the efficacy of cost-sensitive designs, especially in unbalanced datasets, traditional measures like accuracy may not be appropriate. Instead, cost-sensitive versions of accuracy, recall, F1-score, and ROC curves—as well as other metrics that account for the costs of various types of errors—are frequently employed. Machine learning models may be trained to produce judgments that are more in line with the unique requirements and limitations of the healthcare industry by implementing cost-sensitive learning approaches, particularly in situations where disparities in class and variable error costs are common.

3.3 ALGORITHM SELECTION

When developing machine learning models, choosing an algorithm is an essential first step, particularly when tackling the issue of class disparity in the healthcare industry. Here are some things to think about while choosing an algorithm: Ensemble Methods:

Random Forest: Random Forest is an ensemble learning method that can handle imbalanced datasets well. It builds multiple decision trees and combines their predictions, which can help mitigate the effects of class imbalance.

Gradient Boosting: Algorithms like Gradient Boosting (e.g., XGBoost, LightGBM) iteratively train weak learners to minimize errors, making them effective for imbalanced datasets. They can assign higher weights to misclassified instances, focusing on improving performance in minority classes.

Algorithms Weighted by Class: By giving each class a distinct weight during instruction, logistic regression may be modified to address class imbalance. This enables the model to modify decision boundaries to better represent the minority class. Class weights can be used in Support Vector Machines (SVMs) to penalize misclassifications in the minority class more severely. SVMs can manage nonlinear connections in unbalanced data with suitable kernel functions. Convolutional neural networks, more commonly known as CNNs, are a type of neural network that is frequently used in the interpretation of medical images. CNNs may be trained on unbalanced datasets by applying the proper changes, such as weights for classes or specialized loss functions.

Recurrent Neural Networks (RNNs): RNNs, especially Long Short-Term Memory (LSTM) systems, operate well with sequential data in the medical field, including electronic medical records and time series data. They may be modified to account for class imbalance by adding class weights or changing the loss algorithms.

Cost-Sensitive Learning: Decision Trees: By modifying the decision criteria according to the costs connected with various mistake kinds, decision trees may be modified for cost-sensitive learning. Decision trees for unbalanced datasets can be optimized with the use of strategies such as cost-complexity pruning. K-Nearest Neighbors (KNN): When identifying the category of a query instance, KNN algorithms may be adjusted to take the costs of misinterpretation into account. To counteract the impacts of class imbalance, the majority class can, for instance, be given a larger weight in the distance computation.

Algorithms for Anomaly Detection: Isolation Forest: These algorithms are effective in spotting unusual occurrences or anomalies, which makes them appropriate for spotting uncommon illnesses or ailments in medical datasets.

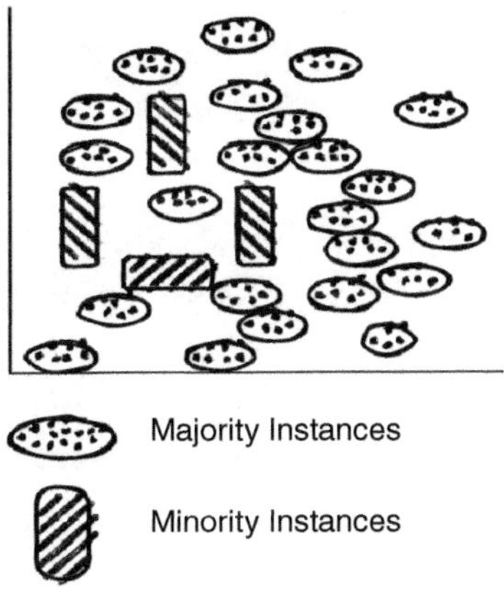

Figure 3.3 Types of cybercrime which have class imbalance.

Hybrid Methods: Group of Various Algorithms: Overall performance can be enhanced by combining predictions from several algorithms trained on various subsets of data or using various methodologies, particularly when there is a class imbalance. It is crucial to take into account several aspects when choosing an algorithm, including the kind of data (unstructured or structured), the complexity of the task, the available computing resources, the need for interpretability, and the unique difficulties caused by a class imbalance in the healthcare industry. To find the best algorithm for a particular task, extensive testing and verification with suitable assessment measures are essential (Figure 3.3).

3.4 DATA AUGMENTATION

Data Augmentation is the process of applying different changes or alterations to pre-existing data samples to artificially expand the size and variety of a dataset. Data augmentation can be very helpful in the healthcare industry for resolving class imbalance and enhancing the generalization and robustness of machine learning models (Figure 3.4).

The following are some typical methods for data augmentation in the medical field: Transformations of Geometry:

Rotation: To replicate various perspectives or orientations, medical pictures (such as MRIs and X-rays) can be rotated by a certain angle. Image resizing to mimic changes in the size of lesions or anatomical structures is known as scaling. Translation is the process of moving an image either vertically or horizontally to represent various locations inside the visual frame.

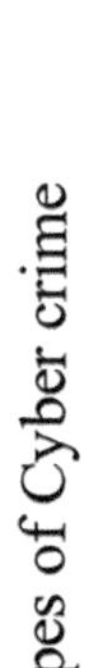

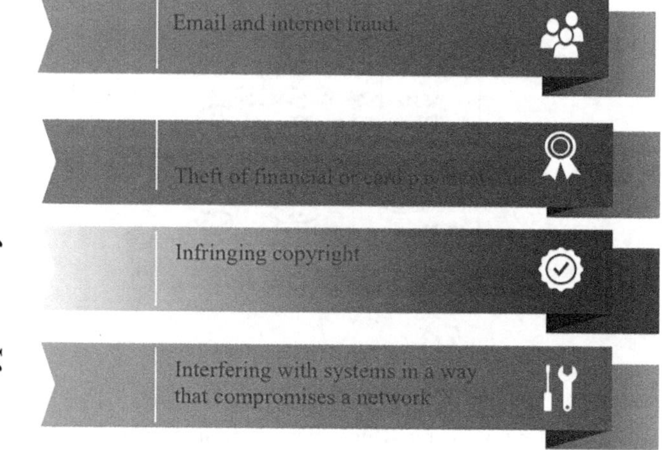

Figure 3.4 Sampling need due to minority instances.

3.4.1 Flip and reflection

Horizontal Flip: This technique simulates changes in patient location or anatomical symmetry by flipping medical pictures horizontally to produce mirrored images.

Vertical Flip: This technique, which can be very helpful for some imaging modalities, involves flipping pictures vertically. Gaussian noise is added randomly to medical pictures by noise injection to mimic imaging artifacts or fluctuations in image quality (Jovic et al., 2015).

Using sporadic black-and-white pixels to imitate the random noise sometimes seen in medical imaging is known as "salt-and-pepper noise."

3.4.2 Modification of intensity

Contrast Adjustment: Changing the contrast of medical pictures to mimic changes in the imaging environment or equipment configuration.

Brightness Adjustment: Changing an image's brightness to replicate variations in exposure or lighting.

Elastic Deformations: Using elastic deformations on medical pictures, one may replicate distortions or deformations brought on by imaging artifacts or anatomic variation. Patch extraction is the process of randomly selecting patches or areas of interest from medical pictures to produce more training examples (López et al., 2013). This technique is very helpful for segmentation tasks.

Temporal Augmentation: The introduction of temporal changes, such as time warping or time shifting, to enhance time-series data, such as physiological signals or electronic health records.

3.4.3 Domain-specialized augmentation

Creating augmentation methods that are specialized to a given healthcare application, including modeling changes in patient demographics, the course of a disease, or the results of a therapy by applying data augmentation techniques, machine learning models can be trained on more diverse and representative datasets, leading to improved performance, robustness, and generalization.

However, it's essential to carefully design and validate augmentation strategies to ensure that augmented data retains its relevance and fidelity to real-world scenarios. Additionally, augmentation should be applied judiciously to avoid introducing biases or artifacts into the dataset.

3.5 COST-EFFECTIVE MEASURES

In the healthcare industry, cost-effective strategies seek to maximize resource use while preserving or enhancing outcomes for patients. Cost-effective solutions concentrate on resource allocation techniques that minimize the influence of unbalanced data on the model's effectiveness while resolving class imbalance. Here are a few affordable ways to address healthcare class disparities: Targeted sampling is the process of obtaining data from certain subgroups or populations if there is a significant class imbalance or underrepresentation of particular classes, as opposed to indiscriminately collecting data. By giving priority to samples that provide the greatest information on fixing class imbalance, this strategy maximizes the usefulness of data-gathering efforts.

Active Learning: To minimize the amount of annotation required while optimizing the caliber and variety of labeled data, active learning entails repeatedly choosing the most instructive examples for annotation or labeling. Active learning can enhance model performance in the case of class imbalance by giving priority to labeling examples from minority classes, hence avoiding the need for comprehensive labeling of the complete dataset.

Transfer Learning: This technique uses the information obtained from training on one dataset to enhance performance on a different dataset that is similar. Transfer learning may be applied to healthcare to enhance the performance of models trained on unbalanced datasets by transferring information from related tasks or well-balanced datasets (Liu et al., 2018). This method lessens the requirement for laborious data labeling and gathering for unbalanced datasets.

Data Fusion: To enhance model performance, data fusion brings together information from several different sources of data. To develop more complete and well-rounded datasets in the healthcare industry, data fusion techniques may merge data from a variety of sources, including wearable technology, medical imaging, electronic health records, and genetic data. Through the use of complementary information from many sources, data fusion can assist in reducing class disparities and improving.

Semi-Supervised Learning: By training prediction models with both labeled and unlabeled data, this method reduces the need for fully labeled datasets (Rani et al., 2017). In the situation of class imbalance, semi-supervised learning can leverage the abundance of unlabeled data to complement the limited quantity of labeled data for minority classes. By effectively using unlabeled data, semi-supervised learning can improve model robustness and generalization without the need for costly labeling.

Cost-Sensitive Sampling: By selecting examples according to their relative costs or significance, cost-sensitive sampling strategies seek to achieve a balance in the dataset's class distribution. Cost-sensitive sampling gives preference to cases from minority classes or areas of the feature space where misclassifications result in higher costs, as opposed to randomly picking instances (Yu et al., 2006).

By ensuring that scarce resources are distributed to cases that significantly affect model performance, this method maximizes the economic efficiency of data-collecting activities. Healthcare organizations and academics may optimize resource utilization and improve the efficacy of predictive modeling methodologies while mitigating the issues associated with a class imbalance within medical datasets by putting these affordable strategies into practice.

3.6 ENSEMBLE METHODS

The predictions of several different models are combined using ensemble methods, which are strong machine-learning approaches that enhance overall predictive performance. Ensemble approaches have the potential to be very useful in reducing the impacts of unbalanced data and improving the generalization and robustness of prediction models when it comes to managing class imbalance in healthcare.

The following ensemble techniques are frequently applied in the medical field:

Bagging (Bootstrap Aggregating): Bagging involves training several base models (such as decision trees). They are using bootstrapping samples of the original data (sampling with replacements). Each initial model is trained independently, and then the final prediction is obtained by aggregating the predictions made by each model (e.g., voting for classification, averaging for regression). When bagging is used on unstable base models, it can help reduce over-fitting and improve the stability of the model.

Random Forest: Using decision trees as its underlying models, Random Forest is a bagging-based ensemble learning approach. Alongside sampling with replacement, Random Forest selects a subset of characteristics at random at each decision tree split. Random Forest is robust against imbalanced and noisy databases and is capable of handling high-dimensional data effectively.

Boosting: Boosting techniques train a series of weak learners (such as decision trees) repeatedly, with each new model concentrating on fixing the mistakes caused by the preceding ones. Usually, models are trained one after the other, with every new one giving more weight to cases that previous models misclassified (Wajid et al., 2024).

Boosting Gradients Machines (GBM), AdaBoost, XGBoost, and LightGBM are a few examples of boosting algorithms. Assigning larger weights to incorrectly classified cases can boost prediction performance, especially when there is a class imbalance and more emphasis is placed on minority classes.

Stacking: Utilizing a meta-learner (such as logistic regression or neural network), stacking (also known as a stacked generalization) integrates predictions from several diverse base models. Stacking learns how to mix base model forecasts in a way that maximizes overall success on a validation set, as opposed to merely average or voting. Stacking is frequently used to outperform individual models because it may capture intricate correlations between the predictions made by base models (Batista & Monard, 2003).

Ensemble of Diverse Models: Integrating predictions from several base models trained using various methods, feature depictions, or hyper parameters can help reduce model bias and improve generalization. Ensemble techniques may combine nonlinear as well as linear models, models trained on multiple data subsets, and shallow and deep machine learning architectures. Variety in ensemble models is crucial for obtaining complementary data and improving overall prediction accuracy, particularly in challenging tasks like class imbalance (Huang et al., 2016).

The flexibility of ensemble techniques allows them to be applied to a wide range of healthcare applications, such as medical imaging analysis, risk assessment, prognosis, and illness diagnosis. Ensemble techniques provide reliable solutions for managing class imbalance and enhancing the accuracy and reliability of forecasting algorithms in the healthcare sector by using the capabilities of many models.

3.7 PERFORMANCE METRICS

Estimating the outcome once therapy has started is a crucial part of treating diseases. Two key factors determine an outcome the patient's response and the healthcare providers' effective treatment methods. One of the biggest challenges facing healthcare professionals is continuing to develop efficient and effective methods (Chawla et al., 2002) for handling critically sick patients. Growing morbidity and death rates are an unwanted outcome of people's uncontrolled blood pressure caused by inadequate treatment methods. This provides a compelling argument in favor of using predictive learning techniques that can find significant correlated variables linked to the prevalence of hypertension. The use of predictive learning algorithms helps to address poor detection rates in several societal categories in real time. Large amounts of information are now available due in part to automated systems like the Internet of Things (IoT), a new model (Weiss & Provost, 2003; Gao et al., 2019) incorporating human interaction and device connectivity. Large quantities of information are typically generated by linked medical devices, such as AI devices for pulse, blood pressure, heart rate, and diabetes monitors, and remote monitoring of patients.

This is a characteristic of healthcare systems. Additional linked gadgets consist of wearables, glucose monitors, virtual healthcare staff members, virtual dispensing assistants, connected contact lenses, and fitness trackers, among others. Numerous studies have examined the data produced by these apps to find patterns for change using various predictive machine learning (ML) techniques, including non-clinical (Fernández & García, 2017) methods, to improve illness diagnosis and treatment outcomes. Predictive modeling performance has been the subject of numerous study initiatives, including monitoring a review of techniques for selecting features and prediction model application to cancer-causing radio microphones (Gu & Zhu, 2019). The use of random forests along with support vector algorithms was shown to be useful in task classification in the review study that it examined. Furthermore, the use of outside variables to improve deep learning model achievement concludes that the inclusion of the environment. variables resulted in enhanced model achievement (Huang et al., 2016).

This was achieved by using a variety of recurrent neural network models (LSTM) to predict COVID-19 daily cases in nine cities throughout three countries in different climate zones. When diabetes was predicted using data mining techniques such as logistic regression, randomly generated forests, naive Bayes, and support vector neural networks, it was shown that logistic regression was achieved.

3.8 CONCLUSION

To sum up, resolving the class divide in the healthcare industry is essential to creating precise and trustworthy prediction models for a range of medical uses. This research makes it clear that class disparity presents serious difficulties when analyzing healthcare data, especially when some medical illnesses are uncommon or underreported.

To lessen the effects of class imbalance, a wide range of approaches and techniques have been put forth and investigated. These include algorithmic approaches like cost-sensitive learning, ensemble methods, and anomaly detection, as well as data-level approaches like oversampling, undersampling, and synthetic data generation. Every strategy has advantages and disadvantages, and the best strategy is typically determined by the particulars of the dataset as well as the nature of the healthcare issue being addressed.

Large-scale healthcare datasets are readily available, and the advent of artificial intelligence and machine learning technology presents exciting prospects for more study and growth in this field. Subsequent endeavors need to concentrate on investigating innovative techniques that may proficiently manage class imbalance, all the while guaranteeing the dependability, comprehensibility, and applicability of predictive models in actual healthcare environments.

All things considered, correcting the class disparity in the healthcare industry is essential not only to boost the accuracy of predictive models but also to enable improved decision-making, improve patient care, and ultimately save lives. Researchers and practitioners may help progress healthcare analytics and the provision of more individualized and efficient medical solutions by carrying out more research and developing new ideas in this area.

BIBILOGRAPHY

Batista, G. E., & Monard, M. C. (2003). An analysis of four missing data treatment methods for supervised learning. *Applied Artificial Intelligence*, 17(5–6), 519–533.

Batista, G. E., Prati, R. C., & Monard, M. C. (2004). A study of the behavior of several methods for balancing machine learning training data. *SIGKDD Explorations Newsletter*, 6(1), 20–29.

Becker, T., & Lewis, P. (2000). HeatMapGenerator: High throughput primer design, sequence analysis, and genotyping. *In Abstracts of Papers Presented at the 2000 Meeting on Genome Mapping and Sequencing.*

Bhattacharya, U., Chakraborty, S., & Nath, B. (2019). Comparative Study of Techniques for Handling Class Imbalance in Healthcare Data. In *2019 IEEE 20th International Conference on Information Reuse and Integration for Data Science (IRI)* (pp. 227–233). IEEE.

Chawla, N. V., Bowyer, K. W., Hall, L. O., & Kegelmeyer, W. P. (2002). SMOTE: Synthetic Minority Over-sampling Technique. *Journal of Artificial Intelligence Research*, 16, 321–357.

Daskalaki, S., Kopanas, I., & Avouris, N. (2006). Evaluation of classifiers for an uneven class distribution problem. *Applied Artificial Intelligence*, 20(4), 381–417.

Drummond, C., & Holte, R. C. (2003). C4.5, class imbalance, and cost sensitivity: Why under-sampling beats over-sampling. In *Workshop on Learning from Imbalanced Data Sets II*, ICML, 1–8).

Fernández, A., & García, S. (2015). Gene selection in cancer classification using PSO/SVM and GA/SVM hybrid algorithms. In *International Conference on Intelligent Data Engineering and Automated Learning* (pp. 60–

Fernández, A., & García, S. (2017). A study of the behavior of several methods for balancing machine learning training data. *ACM Sigkdd Explorations Newsletter*, 19(1), 20–29.

Fernández, A., García, S., Galar, M., Prati, R. C., Krawczyk, B., & Herrera, F. (2018). *Learning from Imbalanced Data Sets*. Cham: Springer International Publishing.

Gao, X., Shen, F., Wang, X., Zhu, X., & Xu, L. (2019). A Comprehensive Survey on Class Imbalance Learning. *ACM Transactions on Intelligent Systems and Technology (TIST)*, 10(3), 1–42.

García, S., Fernández, A., Luengo, J., & Herrera, F. (2009). A study of statistical techniques and performance measures for genetics-based machine learning: Accuracy and interpretability. *Soft Computing*, 13(10), 959–977.

García, S., Luengo, J., & Herrera, F. (2010a). Advanced nonparametric tests for multiple comparisons in the design of experiments in computational intelligence and data mining: Experimental analysis of power. *Information Sciences*, 180(10), 2044–2064.

García, S., Luengo, J., Sáez, J. A., López, V., & Herrera, F. (2010b). A survey of discretization techniques: Taxonomy and empirical analysis in supervised learning. *IEEE Transactions on Knowledge and Data Engineering*, 25(4), 734–750.

Garg, R., Kumar, D., & Khanna, A. (2018). A survey on handling class imbalance problem. In *2018 Second International Conference on Computing Methodologies and Communication (ICCMC)* (pp. 815–820). IEEE.

Gu, Y., & Zhu, H. (2019). Deep Learning with Class Imbalance Problem. *IEEE Access*, 7, 76787–76805.

Haibo, H., & Garcia, E. A. (2008). Learning from imbalanced data. *IEEE Transactions on Knowledge and Data Engineering*, 21(9), 1263–1284.

Haixiang, G., Yijing, L., Shang, J., Mingyun, G., Yuanyue, H., & Bing, G. (2017). Learning from class-imbalanced data: Review of methods and applications. *Expert Systems with Applications*, 73, 220–239.

He, H., & Garcia, E. A. (2009). Learning from imbalanced data. *IEEE Transactions on Knowledge and Data Engineering*, 21(9), 1263–1284.

Hripcsak, G., & Rothschild, A. S. (2005). Agreement, the f-measure, and reliability in information retrieval. *Journal of the American Medical Informatics Association*, 12(3), 296–298.

Huang, C., Li, Y., Yu, L., & Wang, X. (2016). A review of ensemble learning algorithms for classification. *Journal of Information Processing Systems*, 12(5), 757–777.

Joachims, T. (1998). Text categorization with support vector machines: Learning with many relevant features. In *European conference on machine learning* (pp. 137–142). Springer, Berlin, Heidelberg.

Jovic, A., Brkic, K., & Bogunovic, N. (2015). A review of feature selection methods with applications. In *2015 38th International Convention on Information and Communication Technology, Electronics and Microelectronics (MIPRO)* (pp. 1200–1205). IEEE.

Khalilia, M., Chakraborty, S., & Popescu, M. (2011). Predicting disease risks from highly imbalanced data using random forest. *BMC Medical Informatics and Decision Making*, 11(1), 51.

Kubat, M., Holte, R. C., & Matwin, S. (1998). Machine learning for the detection of oil spills in satellite radar images. *Machine Learning*, 30(2–3), 195–215.

Kubat, M., & Matwin, S. (1997). *Addressing the curse of imbalanced training sets: One-sided selection*. In *Proceedings of the Fourteenth International Conference on Machine Learning ICML* (Vol. 97 (pp. 179–186).

Lee, Y. K., & Cha, S. H. (2007). A neural network model for prognosis of breast cancer based on feature extraction with fine needle aspiration. *Expert Systems with Applications*, 32(1), 214–222.

Lemaitre, G., Nogueira, F., & Aridas, C. K. (2017). Imbalanced-learn: A Python toolbox to tackle the curse of imbalanced datasets in machine learning. *Journal of Machine Learning Research*, 18(17), 1–5.

Liu, J., Zhang, Y., & Luo, M. (2018). Learning from imbalanced data: A review. *Artificial Intelligence Review*, 50(1), 91–103.

Liu, Y., Wu, J., & Sui, Y. (2009). Anomaly detection in crowd scenes: A review. In *International Conference on Image and Graphics* (pp. 169–173). Springer, Berlin, Heidelberg.

Liu, Y., Wu, J., & Sui, Y. (2010). Crowded scenes analysis: A review. In *International Conference on Pattern Recognition* (pp. 1–4). IEEE.

Liu, Y., & Zhou, S. (2010). Ensemble learning algorithms: A survey. In *2010 International Conference on Artificial Intelligence and Computational Intelligence* (pp. 94–98). IEEE.

López, V., Fernández, A., García, S., Palade, V., & Herrera, F. (2013). An insight into classification with imbalanced data: Empirical results and current trends on using data intrinsic characteristics. *Information Sciences*, 250, 113–141.

Lu, J., & Zhao, H. (2015). A survey of evolutionary algorithms for decision tree induction. *IEEE Transactions on Systems, Man, and Cybernetics: Systems*, 45(2), 141–156.

Martínez, V., & Kak, A. C. (2001). Pca versus lda. *Pattern Analysis and Machine Intelligence, IEEE Transactions on*, 23(2), 228–233.

Massart, D. L., Vandeginste, B. G., Buydens, L. M., De Jong, S., Lewi, P. J., & Smeyers-Verbeke, J. (1997). *Handbook of chemometrics and qualimetrics*. Elsevier.

Matthews, B. W. (1975). Comparison of the predicted and observed secondary structure of T4 phage lysozyme. *Biochimica et Biophysica Acta (BBA)-Protein Structure*, 405(2), 442–451.

McClish, D. K. (1989). *Analyzing a portion of the ROC curve*.

Moradi, M., Shaban, M., & Kermani, F. (2018). Deep learning framework for imbalanced biomedical classification: A case study in Alzheimer's disease. *Healthcare Technology Letters*, 5(2), 47–53.

Namazi, B., Karimi, F., & Kermani, F. (2020). Attention-based convolutional neural network for imbalanced bi-classification of Alzheimer's disease. *Cognitive Neurodynamics*, 14(6), 847–855.

Perneczky, R., Drzezga, A., Diehl-Schmid, J., Schmid, G., Wohlschläger, A., Kars, S., … & Kurz, A. (2007). Schooling mediates brain reserve in Alzheimer's disease:

findings of fluoro-deoxy-glucose-positron emission tomography. *Journal of Neurology*, 254(4), 532–539.

Ramos, J., Sousa, R., Almeida, J. S., & Neves, J. (2017). Learning from imbalanced data in healthcare: Strategies and methods. *Progress in Artificial Intelligence*, 6(4), 281–297.

Rani, M. S., Gadamsetty, S., & Kaul, P. (2017). Addressing data imbalance problem for Alzheimer's disease diagnosis with generative adversarial networks. In *International Conference on Intelligent Data Engineering and Automated Learning* (pp. 364–371). Springer, Cham.

Ren, X., & Sun, J. (2016). A Review on Ensemble Learning Algorithms for Big Data. *IEEE Access*, 4, 6185–6200.

Saez, J. A., Luengo, J., & Herrera, F. (2015). SMOTE-IPF: Addressing the noisy and borderline examples problem in imbalanced classification by a re-sampling method with filtering. *Information Sciences*, 291, 184–203.

Sun, L., Shen, J., & Luo, Z. (2018). A Review of Class Imbalance Learning in Medical Diagnosis: Current Status and Future Directions. *IEEE Access*, 6, 11336–11347.

Sun, Y., Kamel, M. S., Wong, A. K., & Wang, Y. (2009). Cost-sensitive boosting for classification of imbalanced data. *Pattern Recognition*, 40(12), 3358–3378.

Van Hulse, J., Khoshgoftaar, T. M., & Napolitano, A. (2007). Experimental perspectives on learning from imbalanced data. In *Proceedings of the 24th International Conference on Machine Learning (ICML)*, 935–942.

Wajid, M. A., & Zafar, A. (2022a). A multimodal approach of information access and retrieval using neutrosophic sets. In *Emerging Trends in IoT and Computing Technologies* (pp. 382–388). Routledge.

Wajid, M. A., & Zafar, A. (2022b). Neutrosophic Image Segmentation: An Approach for the Treatment of Uncertainty in Multimodal Information Systems.*International Journal of Neutrosophic Science (IJNS)*, 19(1).

Wajid, M. S., Terashima-Marin, H., Najafirad, P., Pablos, S. E. C., & Wajid, M. A. (2024). DTwin-TEC: An AI-based TEC district digital twin and emulating security events by leveraging knowledge graph. *Journal of Open Innovation: Technology, Market, and Complexity*, 10(2), 100297.

Wajid, M. S., & Wajid, M. A. (2021). The importance of indeterminate and unknown factors in nourishing crime: a case study of South Africa using neutrosophy. *Neutrosophic Sets and Systems*, 41(2021), 15.

Weiss, G. M., & Provost, F. (2003). Learning when training data are costly: The effect of class distribution on tree induction. *Journal of Artificial Intelligence Research*, 19, 315–354.

Yao, X., Sun, X., Liu, X., & Hu, X. (2018). A survey of ensemble learning algorithms. *Journal of Computer and System Sciences*, 84, 48–75.

Yu, H. F., & Huang, F. L. (2003). Weighted majority voting methods for protein secondary structure prediction. *In Pacific Symposium on Biocomputing* (pp. 564–575).

Yu, L., Liu, H., & Wang, S. (2006). A review of accuracy evaluation metrics of recommendation tasks. *Journal of Convergence Information Technology*, 1(1), 1–10.

Chapter 4

Introduction to Neutrosophic Logic in the Narrow Sense

A fuzzy and neutrosophic approach to soft computing applications

Mohammad Ishrat, Wasim Khan, Syed Mohd Faisal, and Monia Mohammed Al Farsi

4.1 INTRODUCTION

Artificial intelligence (AI) is a transformative force, revolutionizing industries by offering unprecedented capabilities in data processing, decision-making, automation, and optimization. Across various sectors such as healthcare, agriculture, and business, AI's ability to analyze vast amounts of data and provide actionable insights has led to groundbreaking advancements (Figure 4.1). However, as powerful as AI is, traditional methods often struggle with real-world complexity, where data is ambiguous, uncertain, or incomplete. This is where soft computing techniques, particularly fuzzy logic and neutrosophic logic, come into play, offering flexible, robust solutions for addressing the inherent uncertainty in practical applications. AI's transformative impact spans diverse sectors. In healthcare, AI is being used for predictive diagnostics, personalized treatment plans, and patient monitoring [1]. In agriculture, AI systems optimize resource use and enhance crop yields through data-driven decisions, while in business, AI powers everything from customer service chatbots to advanced financial modeling. Despite AI's impressive capabilities, real-world data is often noisy, incomplete, or imprecise. Traditional AI techniques typically rely on crisp, exact data inputs to generate precise outputs, but many real-world problems, especially those involving human behaviors, environmental factors, or large-scale operations, are filled with uncertainties. These uncertainties make it difficult for conventional AI models to provide reliable or flexible solutions. For example, in healthcare, the symptoms of a disease may not always be clear-cut, leading to diagnostic ambiguity[2]. In agriculture, weather patterns and soil conditions are highly variable, complicating decision-making scenarios [3].

In business, customer preferences and behaviors are often unpredictable, creating challenges for targeted marketing and personalized services. Soft computing addresses these challenges by allowing for more nuanced decision-making. Unlike traditional AI, which operates on rigid, binary logic, soft computing can handle uncertainty, vagueness, or incomplete data, providing adaptive and resilient solutions in real-world scenarios [4–6].

DOI: 10.1201/9781003606055-4

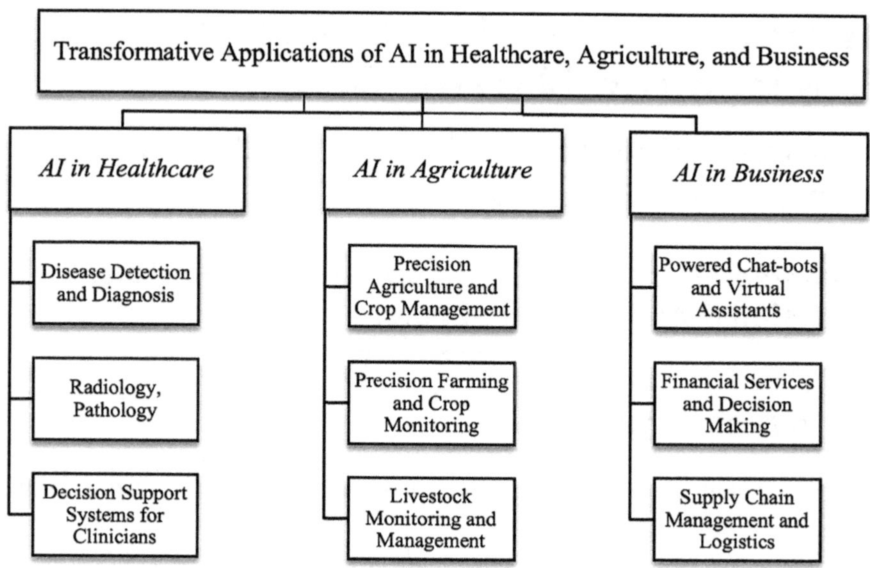

Figure 4.1 AI in diverse sectors (healthcare, agriculture, and business).

In healthcare, AI enhances disease detection and diagnosis by analyzing large datasets from medical records, imaging, and test results. Machine learning models are used to detect early signs of diseases such as cancer, heart disease, and diabetes, improving diagnostic accuracy and speed while reducing human error. In radiology and pathology, AI-driven tools assist in interpreting medical images and tissue samples, detecting anomalies like tumors and fractures with high precision, enabling faster and more accurate diagnosis. Additionally, AI-powered decision support systems offer clinicians data-driven recommendations for diagnosis and treatment planning, providing personalized treatment options, predicting patient outcomes, and optimizing resource allocation.

In agriculture, AI is revolutionizing precision agriculture by analyzing environmental and crop data to optimize farming practices. AI models can predict weather patterns, assess soil conditions, and monitor crop health, helping farmers make informed decisions about planting, watering, and fertilization. AI-driven precision farming enables real-time crop monitoring through sensors and satellite imagery, allowing for early detection of diseases or pests and recommending corrective actions, leading to higher crop yields and improved sustainability. AI also supports livestock monitoring by tracking animal health, movement, and behavior using wearable devices and cameras [5]. These AI systems detect health issues and optimize feeding and breeding practices, resulting in better herd management and increased productivity.

In business, AI enhances customer service through powered chatbots and virtual assistants, which automate responses to common queries, provide personalized recommendations, and streamline interactions between customers and businesses, offering 24/7 customer support [7, 8]. In the financial industry, AI is used for decision-making in areas such as risk management, fraud detection, and investment strategies, analyzing large volumes of financial data to identify patterns and predict trends for more informed decisions. AI also plays a critical role in supply chain management, optimizing demand forecasting, inventory management, and logistics planning, resulting in streamlined operations, reduced costs, and improved delivery times.

4.1.1 Overview of soft computing

Soft computing is an umbrella term that refers to a collection of computational techniques that are tolerant of imprecision, uncertainty, and approximation. These methods mimic the way humans process information, enabling systems to make decisions in environments where data may be incomplete, contradictory, or ambiguous. The main components of soft computing include [9–11]:

- *Fuzzy Logic*: Developed by Lotfi Zadeh in 1965, fuzzy logic allows systems to work with partial truths. In contrast to traditional binary logic, where variables must be either true or false, fuzzy logic permits values to range between 0 and 1. This makes it ideal for handling the imprecision often found in real-world data [9, 12, 13]. For example, in healthcare, a patient may exhibit symptoms that are "mild," "moderate," or "severe" rather than simply "present" or "absent."
- *Neutrosophic Logic*: An extension of fuzzy logic, neutrosophic logic introduces the concept of indeterminacy in addition to truth and falsehood. Proposed by Florentin Smarandache in the late 1990s, neutrosophic logic is particularly suited for handling situations where information is incomplete, contradictory, or unclear. This makes it highly relevant for applications in fields like agriculture, where environmental conditions can be unpredictable, or in business, where customer behavior may defy straightforward categorization [14].

  Neural Networks and Evolutionary Algorithms: These are other key elements of soft computing. Neural networks, inspired by the human brain, excel at pattern recognition and learning from data. Evolutionary algorithms simulate natural selection, optimizing solutions over time. While these techniques are crucial to AI's ability to learn and adapt, they are often used in tandem with fuzzy and neutrosophic systems to enhance decision-making under uncertainty.

The ability of soft computing to approximate complex, nonlinear functions and to deal with imprecision makes it especially useful in environments where traditional AI techniques might fall short. By incorporating soft computing into AI systems, we can better handle the complexities of real-world data, improving the adaptability, accuracy, and reliability of AI-driven applications.

4.1.2 Importance of addressing uncertainty

Uncertainty is a fundamental challenge in many real-world applications. Whether it's diagnosing a medical condition, predicting crop yields, or understanding customer preferences, decision-makers must often work with incomplete, imprecise, or conflicting information. Traditional AI systems, which rely on clear-cut data and rigid rules, struggle in these situations. This is where the flexibility of soft computing, particularly fuzzy and neutrosophic logic, becomes essential [15, 16].

- *In Healthcare*: Medical diagnosis is rarely straightforward. Patients often present symptoms that vary in severity, overlap with multiple conditions, or are influenced by factors like age, lifestyle, and genetics. For example, consider the difficulty of diagnosing a condition like diabetes, which may manifest through a combination of high blood sugar, fatigue, and blurred vision symptoms that are also common in other conditions. Fuzzy logic can assess these symptoms in degrees, allowing for a more accurate and flexible diagnostic process. Similarly, neutrosophic logic can manage the indeterminacy that arises when medical data is conflicting or incomplete, providing healthcare professionals with more nuanced insights [17].
- *In Agriculture*: Farmers must make decisions based on highly variable and uncertain data, such as soil moisture levels, weather forecasts, and crop health. These variables are rarely binary or deterministic. For instance, soil moisture might be described as "dry," "optimal," or "wet," but these categories are not always clear-cut. Fuzzy logic systems can assess soil conditions with more granularity, helping farmers make informed decisions about irrigation or fertilization. Additionally, neutrosophic logic helps in handling indeterminacies in crop yield predictions due to changing environmental conditions, pests, or diseases.
- *In Business*: In business, uncertainty manifests in many forms, from fluctuating market trends to unpredictable customer behaviors. Traditional AI systems may fall short in personalizing customer experiences because they require crisp data inputs like specific customer preferences that are not always available. Fuzzy logic helps businesses navigate this uncertainty by analyzing customer behavior in shades of gray. For example, a fuzzy system could assess how likely a customer is to purchase a product based on a combination of factors such as browsing history, past purchases, and social media activity. Neutrosophic logic further enhances decision-making by accounting for the indeterminate aspects of market trends or consumer sentiments, enabling businesses to make more robust and adaptive decisions.

4.2 BRIDGING AI WITH SOFT COMPUTING FOR ENHANCED DECISION-MAKING

By integrating soft computing techniques into AI, we can bridge the gap between the rigid, deterministic nature of traditional AI and the complex, uncertain realities of many industrial sectors. Soft computing enables AI systems to function more effectively in environments where data is not only incomplete but also inherently ambiguous [18, 19].

For example, a traditional AI system designed to optimize resource allocation in a hospital might rely on specific, well-defined inputs like the number of available beds or staff. However, real-world hospital management involves many uncertain variables, such as patient inflows or the severity of cases. Fuzzy and neutrosophic systems can factor in this uncertainty, providing recommendations that better reflect the hospital's dynamic needs. Similarly, in agriculture, a fuzzy system can adjust irrigation schedules in response to uncertain weather patterns, while neutrosophic systems help farmers plan for unpredictable events like droughts or pest infestations.

In business, integrating soft computing into customer relationship management (CRM) systems allows for more personalized and adaptive interactions with customers. For instance, instead of recommending products based solely on past purchases, a fuzzy system can factor in less definitive but still valuable data, like browsing behavior or social media engagement, to offer more accurate product recommendations.

4.2.1 AI in healthcare: a fuzzy and neutrosophic approach

4.2.1.1 Fuzzy systems in diagnostics

Healthcare diagnostics are often characterized by uncertainty and imprecision. Patients rarely present with clear-cut symptoms, and the complexity of medical conditions demands sophisticated systems that can deal with partial information. Fuzzy logic has become an invaluable tool in medical diagnostics, enabling more flexible and nuanced decision-making processes [15, 18, 19].

4.2.1.2 Application in medical imaging

In medical imaging, fuzzy logic helps interpret ambiguous data. For instance, when a radiologist examines an MRI scan to detect a tumor, the boundaries of the tumor may not be sharply defined. This can result from noise in the image, the contrast between tissues, or other factors that make it hard to differentiate between healthy and malignant tissue[15].

Fuzzy algorithms excel in this context because they allow for intermediate states. A pixel in an MRI image is not just "tumor" or "no tumor" but can belong to the tumor with a certain probability (e.g., 60% tumor, 40% healthy). This gradation in the classification of image regions allows doctors to identify abnormalities with greater precision.

4.2.1.3 Case study: Brain tumor detection with fuzzy logic

In brain tumor detection, fuzzy logic plays a crucial role in identifying and segmenting the tumor from surrounding healthy tissue. Traditional imaging systems often struggle with ambiguous regions, such as where a tumor's boundary is not clearly distinguishable due to noise or overlapping structures.

- *Fuzzy Logic Implementation*: In this case, a fuzzy logic-based system assigns probabilities to each pixel in the MRI image, indicating the likelihood that the pixel belongs to a tumor. This system doesn't classify a pixel as definitively healthy or tumorous but instead places it on a continuum of health and abnormality (e.g., "0.8" indicates an 80% likelihood that the pixel represents a tumor).
- *Outcome*: This fuzzy classification allows for more accurate segmentation of the tumor, leading to better planning for surgical interventions or radiation therapy. The system's flexibility in handling ambiguous data makes it an invaluable tool for radiologists.

4.2.1.4 Personalized treatment plans

Every patient is different, and personalized treatment is increasingly recognized as a key to better healthcare outcomes. In cases where standardized treatment protocols fail to account for individual patient variability, *fuzzy logic* can be used to tailor treatments to each patient's unique characteristics, such as age, health history, genetic predispositions, and lifestyle.

- *Fuzzy Decision Support Systems*: A fuzzy logic-based *decision support system* takes into account multiple inputs—such as symptom severity, patient health indicators, and potential side effects—to recommend personalized treatment strategies. For instance, a system might evaluate the severity of a patient's symptoms on a continuum from mild to severe, helping doctors adjust medication doses accordingly.

4.2.1.5 Case study: Fuzzy logic for tailoring chemotherapy doses in oncology

In oncology, chemotherapy doses need to be carefully calibrated to balance *maximum efficacy* against *minimal side effects*. Since patients respond differently to chemotherapy, rigid dose recommendations can lead to under-treatment or dangerous levels of toxicity. *Fuzzy logic* provides a more adaptable approach [20, 21].

- *Fuzzy Logic Implementation*: A *fuzzy decision system* assesses several factors, such as *tumor size*, *rate of shrinkage*, *blood toxicity levels*, and *patient weight*. The system then recommends a chemotherapy dose that accounts for these factors on a sliding scale, rather than using binary thresholds.
- *Outcome*: By using *fuzzy-based recommendations*, oncologists can tailor chemotherapy treatments more precisely, optimizing the dose for each patient's unique situation, improving outcomes, and minimizing adverse side effects.

4.2.1.6 Transition: Moving from healthcare to agriculture managing uncertainty in biological systems

Just as healthcare must deal with uncertainties in patient data, agriculture must manage variability in biological systems, such as unpredictable weather patterns, soil conditions, and crop health. The fuzzy logic systems used to interpret medical data and personalize treatments can similarly be applied to precision farming, where decisions about irrigation, fertilization, and harvesting must be made based on incomplete or uncertain data.

4.2.2 Neutrosophic logic in predictive healthcare

While fuzzy logic handles degrees of truth, neutrosophic logic goes a step further by incorporating three distinct values: truth, falsehood, and indeterminacy. This approach is particularly effective in predictive healthcare, where medical data is often incomplete, contradictory, or ambiguous [22,23].

4.2.2.1 Introduction to neutrosophic logic in healthcare

Neutrosophic logic introduces an additional dimension—indeterminacy—that allows healthcare systems to quantify uncertainty more effectively. While fuzzy logic can tell us that a patient might be "moderately sick," neutrosophic logic goes further by also stating the level of confidence in this diagnosis. This is particularly useful in scenarios where medical tests provide conflicting or inconclusive results.

For example, a patient might show symptoms that partially match two different conditions, such as heart disease and a respiratory illness. Neutrosophic logic allows the system to express indeterminacy, indicating that while the patient is likely to have heart disease, there is also some probability that the diagnosis is incomplete or unclear.

4.2.2.2 Handling conflicting data with neutrosophic logic

Medical records are often filled with conflicting data, for example, lab results that contradict a patient's symptoms or diagnostic images that suggest multiple possible conditions. Traditional AI systems, which rely on clean, binary data, struggle in these situations. Neutrosophic systems, on the other hand, can process conflicting data and provide a more nuanced conclusion.

4.2.2.3 Predictive analytics in remote monitoring systems

Neutrosophic logic is also highly effective in *Remote Patient Monitoring*, where IoT devices continuously collect health data. Often, these devices produce noisy or incomplete data due to technical issues or patient non-compliance. Neutrosophic systems handle these uncertainties by assigning confidence levels to the data and helping healthcare providers make informed predictions about a patient's future health.

4.2.2.4 Case study: Predicting heart disease in remote monitoring systems

Remote heart monitoring systems use IoT devices to continuously track a patient's heart rate, blood pressure, and physical activity. In many cases, these systems face issues with data reliability due to sensor malfunctions or patient non-compliance. Neutrosophic logic helps predict heart disease risk by handling indeterminate data effectively.

- *Neutrosophic Logic Implementation*: In this case, a neutrosophic system assesses conflicting data, such as irregular heart rates and missing blood pressure readings, providing healthcare professionals with a confidence interval regarding the likelihood of a cardiac event.
- *Outcome*: This approach allows doctors to predict heart disease more accurately, even when the data is incomplete or contradictory. The system provides a graded risk assessment, helping doctors prioritize interventions for high-risk patients.

4.2.2.5 Transition: From healthcare to agriculture leveraging neutrosophic logic for environmental uncertainty

Much like healthcare, agriculture deals with uncertain and incomplete data, such as changing weather conditions, soil quality, and crop health. Neutrosophic logic, which quantifies the indeterminacy in medical data, can be adapted to precision agriculture, where farmers must make decisions based on variable environmental factors. By managing these uncertainties, neutrosophic systems enable data-driven farming decisions, helping farmers optimize their practices under unpredictable conditions [24].

4.2.3 AI-driven remote monitoring in healthcare: IoT integration with fuzzy and neutrosophic logic in healthcare monitoring

IoT devices play a critical role in remote patient monitoring, continuously collecting data such as heart rate, blood pressure, glucose levels, and physical activity. However, these devices often produce noisy or incomplete data, making it difficult for traditional AI models to interpret effectively. Fuzzy and neutrosophic logic are essential for managing this variability [25, 26].

- *Fuzzy Logic*: Fuzzy systems interpret the data in degrees, allowing for partial truths rather than binary conclusions. For instance, a patient's blood pressure might be classified as "moderately elevated" rather than simply "normal" or "high." This allows for more nuanced and actionable insights.
- *Neutrosophic Logic*: Neutrosophic logic handles indeterminacy, which is particularly useful when IoT devices fail to record data or when readings are inconsistent. It assigns a degree of uncertainty to the data, enabling healthcare professionals to make informed decisions, even when the data is incomplete.

4.2.3.1 Case study: Continuous monitoring for chronic diseases

Chronic diseases like diabetes and heart disease require constant monitoring to prevent complications. AI-driven remote monitoring systems, enhanced by fuzzy logic and neutrosophic logic, offer a sophisticated solution for managing these conditions.

- *Example: Diabetes Management*: A continuous glucose monitor tracks a patient's blood sugar levels throughout the day. Fuzzy logic processes the fluctuations in glucose levels, categorizing them into risk levels (e.g., "low risk," "moderate risk," or "high risk") instead of binary categories, allowing for more precise insulin adjustments.
- *Example: Heart Disease Monitoring*: Patients with heart disease benefit from AI-driven systems that continuously monitor heart rate and rhythm. Neutrosophic logic manages inconsistent data from heart rate monitors, enabling early detection of abnormalities despite incomplete data.

4.2.3.2 Dynamic adjustments in patient care through fuzzy systems

The real power of fuzzy systems lies in their ability to make dynamic adjustments to treatment plans in real time. As new data flows in, the system can adjust medication dosages or trigger alerts if certain thresholds are met [27, 28].

- *Example: Adaptive Insulin Dosing for Diabetic Patients*: Fuzzy logic can dynamically adjust insulin doses based on real-time blood sugar levels, ensuring optimal glucose control while minimizing the risk of hypoglycemia.
- *Example: Heart Rate Monitoring and Adjustment*: Fuzzy systems monitor heart rate variability and adjust treatment plans in real time, providing dynamic alerts when certain thresholds are crossed.

4.2.3.3 *Transition: How AI-driven monitoring systems in healthcare inspire smart agriculture solutions*

The dynamic, context-aware decision-making that powers AI-driven healthcare monitoring systems can also be applied to smart agriculture. Just as fuzzy and neutrosophic logic systems adapt to patient data in real time, they can also be used in precision farming to make dynamic adjustments to irrigation, fertilization, and pest control, ensuring that crops receive the right number of resources based on real-time environmental data. The cross-industry applications of these soft computing techniques highlight the versatility and power of AI in managing uncertainty across various domains.

4.3 AI IN AGRICULTURE: PRECISION FARMING WITH SOFT COMPUTING

Agriculture is one of the most important industries that benefit from the integration of AI and soft computing techniques, particularly fuzzy logic and neutrosophic logic. These methods provide robust solutions for handling uncertainty, variability, and incomplete data in farming, which are common due to unpredictable environmental conditions, changing weather patterns, and soil diversity. Precision farming, an AI-driven approach, focuses on optimizing farming practices by analyzing large amounts of data related to soil, crops, climate, and market trends, enabling farmers to make informed decisions that maximize yields while reducing resource usage [8, 29].

4.3.1 Fuzzy logic in precision agriculture: Introduction to precision agriculture [8]

Precision agriculture refers to the use of data-driven technologies to optimize farming practices. It involves the application of advanced AI tools, remote sensing, IoT devices, and fuzzy logic to make real-time decisions about irrigation, fertilization, pest control, and harvesting. By analyzing data from various sources, such as soil sensors, weather stations, and drones, precision agriculture systems provide recommendations that are tailored to specific areas of the field, ensuring that crops receive the right resources at the right time.

- *Fuzzy logic* is particularly valuable in precision agriculture because it can process uncertain or imprecise data and provide solutions that are flexible and adaptable. In farming, factors such as soil moisture, temperature, and nutrient levels rarely fit into binary categories of "good" or "bad." Instead, they vary along a continuum, which fuzzy logic handles effectively by assigning degrees of membership to different states (e.g., "slightly dry," "moderately moist").

4.3.1.1 Irrigation and fertilization optimization using fuzzy logic

One of the primary applications of fuzzy logic in precision agriculture is the optimization of irrigation and fertilization. Water and fertilizer are critical resources for crop growth, but they need to be applied in the right amounts and at the right time. Over-irrigation leads to water waste and nutrient leaching, while under-irrigation can stress crops. Similarly, improper fertilization can lead to poor yields or environmental damage.

- *Fuzzy systems* are used to evaluate a range of factors such as soil moisture, temperature, humidity, and crop type—and recommend precise irrigation schedules. Rather than giving binary recommendations like "irrigate" or "don't irrigate," fuzzy systems provide nuanced outputs, such as "apply a moderate amount of water."

4.3.1.2 Crop health monitoring and disease detection with fuzzy systems

Fuzzy logic is also applied in crop health monitoring and disease detection. Crops are vulnerable to a variety of stressors, such as nutrient deficiencies, pest infestations, and diseases, which often manifest in subtle changes in plant appearance or growth patterns. These changes are not always immediately obvious, and the boundaries between healthy and unhealthy crops are often fuzzy.

- *Fuzzy systems* can analyze data from sensors and aerial imagery to detect early signs of crop stress. For instance, changes in leaf color might indicate nutrient deficiencies or the onset of a disease, but these changes are often gradual and difficult to categorize. Fuzzy logic can assess the degree of abnormality and provide early warnings to farmers, allowing them to take preventive actions before the problem becomes severe.

4.3.1.3 Fuzzy decision models for harvesting and fertilization

Timing is critical in agriculture, particularly when it comes to harvesting and fertilization. Harvesting too early or too late can significantly affect the quality and quantity of the yield. Fuzzy decision models help farmers make more informed decisions by considering multiple variables, such as crop maturity, weather forecasts, and market conditions.

- *Fuzzy logic-based models* evaluate crop readiness based on a combination of factors, such as growth stage, weather predictions, and market prices. These models can provide recommendations on the best time to harvest crops or apply fertilizers, considering the uncertainties in environmental conditions.
- *Case Study*: Optimizing Water Usage with Fuzzy Systems in Agriculture

In regions where water is scarce, efficient water management is crucial for ensuring crop productivity. Fuzzy systems have been implemented to help farmers optimize water usage by continuously monitoring soil moisture and weather conditions.

4.3.1.4 Transition: From agriculture to business: The role of soft computing in decision-making systems

The use of fuzzy logic in agriculture for handling variability and optimizing resources demonstrates how soft computing can be applied across different industries. In business, similar fuzzy decision-making systems can be used to optimize pricing strategies, risk management, and CRM, where data is often uncertain and requires flexible decision models.

4.3.2 Neutrosophic logic in agricultural decision-making

4.3.2.1 Handling uncertainty in crop yield predictions with neutrosophic logic

While fuzzy logic handles degrees of certainty, neutrosophic logic extends this by accounting for indeterminacy in data. This is particularly useful in agriculture, where predictions about crop yields are often based on incomplete or conflicting information. Factors such as weather, soil conditions, and market demand can vary widely, making accurate yield predictions difficult [17, 30, 31].

> *Neutrosophic systems* incorporate truth, falsehood, and indeterminacy, enabling farmers to make more informed decisions even when some data points are missing or contradictory. For instance, a neutrosophic model might predict a crop yield of 70% certainty, with 20% indeterminacy due to uncertain weather patterns.

4.3.2.2 *Managing environmental variability and pest control with neutrosophic models*

- *Neutrosophic logic* is also applied in managing environmental variability and pest control. Farmers must constantly adapt to changing conditions, such as sudden temperature drops, unexpected rainfall, or pest outbreaks, which can drastically affect crop health and yields.
- *Neutrosophic models* can process incomplete data about pest activity, such as sensor data that indicates pest presence in some areas but lacks data for others. The model will assign a degree of indeterminacy to the predictions, allowing farmers to make better decisions about when and where to apply pesticides.

4.3.2.3 *Case study: Precision farming and the use of neutrosophic logic in crop management*

Precision farming in developing regions faces the challenge of uncertain environmental data due to lack of consistent weather monitoring infrastructure. Neutrosophic logic helps overcome this issue by compensating for missing weather data and offering reliable predictions.

4.3.2.4 *Transition: Similarities in managing business uncertainty using neutrosophic logic*

In the business world, neutrosophic logic can be used to handle market volatility and financial risk in much the same way it is used in agriculture to manage environmental variability. Business leaders can make more informed decisions by factoring in the indeterminacy of market data and consumer behavior, using neutrosophic models to provide graded predictions about market trends or investment outcomes.

4.4 CHALLENGES AND FUTURE TRENDS IN AI ADOPTION IN AGRICULTURE: BARRIERS TO IMPLEMENTING AI IN AGRICULTURE: COST, INFRASTRUCTURE, AND TECHNICAL EXPERTISE

Despite the clear benefits of *AI* and *soft computing* in agriculture, several barriers remain that prevent widespread adoption (Figure 4.2) [32]:

- *Cost*: The initial investment in *AI technologies*, such as IoT sensors, drones, and data analytics platforms, can be prohibitively expensive, particularly for small-scale farmers.
- *Infrastructure*: In many rural areas, the lack of reliable *internet connectivity* and *data infrastructure* makes it difficult to implement AI systems effectively.

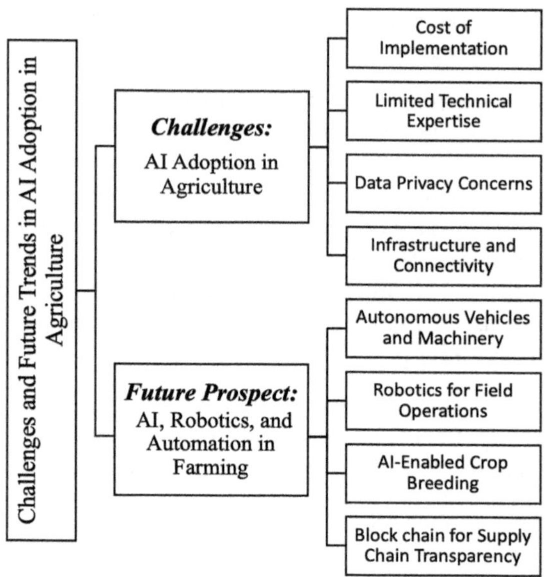

Figure 4.2 Challenges and future trends in AI adoption in agriculture.

- *Technical Expertise*: Implementing and maintaining AI systems requires specialized knowledge, and many farmers lack the necessary training to fully leverage these technologies.

4.4.1 Future trends: Robotics, automation, and AI-enabled crop breeding

The future of AI in agriculture points toward an even greater integration of robotics and automation. AI-driven robots are already being used for tasks such as weeding, harvesting, and crop spraying, reducing the need for manual labor and improving efficiency [32, 33].

- AI-enabled crop breeding is another emerging trend. Using AI algorithms, researchers can analyze genetic data to develop new crop varieties that are more resistant to pests, diseases, and environmental stresses. This will help farmers grow crops that are more productive and resilient to climate change.

4.4.2 Case study: Blockchain and AI for supply chain transparency

In the global food supply chain, ensuring transparency and traceability is becoming increasingly important. AI and blockchain technology are being used together to track the entire lifecycle of agricultural products, from the farm to the consumer's table.

- *Example*: In Europe, AI-driven systems and blockchain technology have been integrated to provide consumers with detailed information about where their food comes from, how it was produced, and the environmental impact of its production. By scanning a QR code, consumers can access a digital record of the product's entire journey through the supply chain.

4.4.3 Transition: From agricultural transparency to business efficiency the impact of AI

The same principles of transparency and traceability being used in agriculture can also be applied in business to improve supply chain efficiency and reduce costs. AI-driven decision systems, coupled with blockchain, can help businesses track their products, optimize logistics, and ensure that supply chains are sustainable and transparent from production to delivery.

4.5 BRIDGING AI AND SOFT COMPUTING ACROSS INDUSTRIES

AI's potential goes beyond individual industries like healthcare, agriculture, and business, offering cross-industry solutions through soft computing techniques such as fuzzy logic and neutrosophic logic. These approaches help manage uncertainty, imprecision, and incomplete data across various domains. In healthcare, fuzzy logic improves diagnostics and treatments, while neutrosophic logic handles conflicting or unclear patient data. In agriculture, AI optimizes resources, monitors crop health, and predicts yields using these techniques to manage variable conditions like weather. In business, fuzzy logic enhances CRM, and neutrosophic logic helps handle market volatility and financial uncertainty, providing more accurate risk assessments [34–36].

The synergies between these industries arise from the shared need to manage uncertainty, making these soft computing techniques transferable. Fuzzy logic introduces degrees of truth, while neutrosophic logic adds indeterminacy to decision-making, allowing AI to handle incomplete or conflicting data. These approaches support AI systems in making flexible, reliable decisions across sectors.

In cross-industry applications, AI systems use fuzzy logic for healthcare monitoring, optimizing precision farming, and enabling business automation through personalized marketing and dynamic pricing. As AI evolves, the ethical and societal implications of these intelligent systems, such as data privacy and fairness, must be considered. AI's future will depend on addressing these challenges while leveraging its ability to bridge gaps across industries.

4.6 FUTURE DIRECTIONS AND CHALLENGES FOR AI AND SOFT COMPUTING

The future of AI and soft computing holds great potential, but several technological and ethical challenges must be addressed. Fuzzy logic and neutrosophic logic will continue evolving to handle more complex, uncertain environments. Advancements like hierarchical fuzzy systems and multivalued decision-making with neutrosophic logic are emerging, particularly in healthcare, agriculture, and business. However, barriers to adoption such as infrastructure limitations, high costs, and the need for technical expertise pose significant challenges, especially in developing regions.

As AI expands into new areas, ethical concerns around data privacy, bias, and transparency become critical. Systems must ensure fairness and accountability in decision-making, particularly in sensitive sectors like healthcare and finance [37, 38].

Moving forward, cross-disciplinary collaboration will be essential for advancing AI technologies. Global policies must be developed to ensure ethical AI deployment, focusing on data protection, fairness, and equitable access. International cooperation in sectors like healthcare has already shown promise, where collaborations have led to AI guidelines that ensure ethical and transparent patient care.

In conclusion, the successful future of AI and soft computing depends on innovation, collaboration, and the establishment of global policies that promote ethical development and responsible implementation across all industries.

4.7 CONCLUSION

AI, enhanced by soft computing techniques like fuzzy logic and neutrosophic logic, has proven to be transformative across sectors such as healthcare, agriculture, and business. These methods have allowed AI systems to manage real-world uncertainty, imprecision, and incomplete data, offering more robust decision-making capabilities in complex environments.

In healthcare, AI systems equipped with soft computing techniques improve diagnostics, tailor treatment plans, and enable continuous patient monitoring by handling uncertain or conflicting data. In agriculture, precision farming powered by AI optimizes resource management, improves crop yield predictions, and adapts to changing environmental conditions. In business, AI systems enable personalized customer experiences, dynamic pricing, and more accurate financial risk management by addressing market volatility and customer behavior with nuanced decision models.

The impact of fuzzy and neutrosophic logic in these fields is evident in their ability to handle uncertainty and make decisions based on degrees of truth and indeterminacy. This adaptability ensures more reliable and context-aware outcomes, whether in healthcare diagnostics, agricultural practices, or business operations.

As AI continues to evolve, soft computing will play an increasingly important role in future systems, allowing them to navigate complex, uncertain scenarios more effectively. However, with this growth comes the need for careful consideration of ethical concerns such as data privacy, algorithmic bias, and accountability. Ensuring that AI operates transparently and fairly will be critical for maintaining public trust and ensuring equitable benefits across all sectors.

Looking forward, the future of AI will require continued innovation and collaboration across industries and sectors. By responsibly integrating AI into critical systems, we can unlock its full potential to solve real-world challenges while safeguarding ethical principles. As AI advances, its role in shaping industries and society will depend on innovation, collaboration, and responsible implementation.

REFERENCES

[1] Deveci M. Effective use of artificial intelligence in healthcare supply chain resilience using fuzzy decision-making model. *Soft Comput* 2023:1–14.

[2] Qahtan S, Zaidan AA, Ibrahim HA, Deveci M, Ding W, Pamucar D. A decision modeling approach for smart training environment with motor Imagery-based brain computer interface under neutrosophic cubic fuzzy set. *Expert Syst Appl* 2023;224:119991.

[3] Jafar MN, Saqlain M, Shafiq AR, Khalid M, Akbar H, Naveed A. New technology in agriculture using neutrosophic soft matrices with the help of score function. *International Journal of Neutrosophic Science* 2020;3:78–88.

[4] Lee D, Yoon SN. Application of artificial intelligence-based technologies in the healthcare industry: Opportunities and challenges. *Int J Environ Res Public Health* 2021;18:271.

[5] Javaid M, Haleem A, Khan IH, Suman R. Understanding the potential applications of Artificial Intelligence in Agriculture Sector. *Advanced Agrochem* 2023;2:15–30.

[6] Singh S, Jain P. Applications of artificial intelligence for the development of sustainable agriculture. *Agro-biodiversity and Agri-ecosystem Management*, Springer; 2022, p. 303–22.

[7] Zaidi T, Syed Faisal. An Overview: Various Attacks in VANET. *2018 4th International Conference on Computing Communication and Automation (ICCCA)*, 2018, p. 1–6. https://doi.org/10.1109/CCAA.2018.8777538.

[8] Sharma A, Sharma A, Tselykh A, Bozhenyuk A, Choudhury T, Alomar MA, et al. Artificial intelligence and internet of things oriented sustainable precision farming: Towards modern agriculture. *Open Life Sci* 2023;18:20220713.

[9] Ramik J. Soft computing: overview and recent developments in fuzzy optimization. Ostravská Univerzita, Listopad 2001:33–42.

[10] Rao KK, Svp Raju G. An overview on soft computing techniques. *International Conference on High Performance Architecture and Grid Computing*, Springer; 2011, p. 9–23.

[11] Ibrahim D. An overview of soft computing. *Procedia Comput Sci* 2016;102:34–8.
[12] Deb AK. Introduction to soft computing techniques: artificial neural networks, fuzzy logic and genetic algorithms. *Soft computing in textile engineering*, Elsevier; 2011, p. 3–24.
[13] Ibrahim D. An overview of soft computing. *Procedia Comput Sci* 2016;102:34–8.
[14] Muzaffar A, Nafis MT, Sohail SS. Neutrosophy logic and its classification: an overview. *Neutrosophic Sets and Systems* 2020;35:239–51.
[15] Peters G, Weber R, Crespo F. Uncertainty modeling in dynamic clustering—a soft computing perspective. *International Conference on Fuzzy Systems*, IEEE; 2010, p. 1–6.
[16] Saridakis KM, Dentsoras AJ. Soft computing in engineering design–A review. *Advanced Engineering Informatics* 2008;22:202–21.
[17] Abidin S, Raghunath MP, Rajasekar P, Kumar A, Ghosal D, Ishrat M. Identification of Disease based on Symptoms by Employing ML. *2022 International Conference on Inventive Computation Technologies (ICICT)*, IEEE; 2022, p. 1357–62.
[18] Konar A. *Artificial intelligence and soft computing: behavioral and cognitive modeling of the human brain*. CRC press; 2018.
[19] Jafari-Marandi R, Davarzani S, Gharibdousti MS, Smith BK. An optimum ANN-based breast cancer diagnosis: Bridging gaps between ANN learning and decision-making goals. *Appl Soft Comput* 2018;72:108–20.
[20] Nobile MS, Votta G, Palorini R, Spolaor S, De Vitto H, Cazzaniga P, et al. Fuzzy modeling and global optimization to predict novel therapeutic targets in cancer cells. *Bioinformatics* 2020;36:2181–8.
[21] Garibaldi JM, Zhou S-M, Wang X-Y, John RI, Ellis IO. Incorporation of expert variability into breast cancer treatment recommendation in designing clinical protocol guided fuzzy rule system models. *J Biomed Inform* 2012;45:447–59.
[22] Ahmed R, Nasiri F, Zayed T. A novel Neutrosophic-based machine learning approach for maintenance prioritization in healthcare facilities. *Journal of Building Engineering* 2021;42:102480.
[23] Okpako AE, Ako RE. A novel framework for predicting and managing comorbid diseases using neutrosophic logic and machine learning. *Int J Comput Appl* n.d.;975:8887.
[24] AbdelMouty AM, Abdel-Monem A, Aal SIA, Ismail MM. Analysis the Role of the Internet of Things and Industry 4.0 in Healthcare Supply Chain Using Neutrosophic Sets. *Neutrosophic Systems with Applications* 2023;4:33–42.
[25] Deveci M. Effective use of artificial intelligence in healthcare supply chain resilience using fuzzy decision-making model. *Soft Comput* 2023:1–14.
[26] Alenizi JA, Alrashdi I. Neutrosophic Intelligence Approach Safeguarding Patient Data in Blockchain-based Smart Healthcare. *Neutrosophic Sets and Systems* 2023;58:409–24.
[27] Jatobá A, de Castro Nunes P, de Carvalho PVR. A framework to assess potential health system resilience using fuzzy logic. *Revista Panamericana de Salud Pública* 2023;47:e73.
[28] Nazarian-Jashnabadi J, Bonab SR, Haseli G, Tomaskova H, Hajiaghaei-Keshteli M. A dynamic expert system to increase patient satisfaction with an integrated approach of system dynamics, ISM, and ANP methods. *Expert Syst Appl* 2023;234:121010.

[29] Ghazal S, Munir A, Qureshi WS. Computer vision in smart agriculture and precision farming: Techniques and applications. *Artificial Intelligence in Agriculture* 2024, Elsevier.
[30] Meng Q, Pang N, Zhao S, Gao J. Two-stage optimal site selection for waste-to-energy plant using single-valued neutrosophic sets and geographic information system based multi-criteria decision-making approach: A case study of Beijing, *China. Waste Management* 2023;156:283–96.
[31] Mohagheghi V, Mousavi SM. A new model for resilient-sustainable energy project portfolio with bi-level budgeting and project manager skill utilization under neutrosophic fuzzy uncertainty: A case study. *Eng Appl Artif Intell* 2024;131:107821.
[32] Mishra D, Muduli K, Raut R, Narkhede BE, Shee H, Jana SK. Challenges facing artificial intelligence adoption during COVID-19 pandemic: an investigation into the agriculture and agri-food supply chain in India. *Sustainability* 2023;15:6377.
[33] Da Silveira F, Da Silva SLC, Machado FM, Barbedo JGA, Amaral FG. Farmers' perception of the barriers that hinder the implementation of agriculture 4.0. *Agric Syst* 2023;208:103656.
[34] Perera YS, Ratnaweera D, Dasanayaka CH, Abeykoon C. The role of artificial intelligence-driven soft sensors in advanced sustainable process industries: A critical review. *Eng Appl Artif Intell* 2023;121:105988.
[35] Jatothu R, Lal JD, Bhavani NPG, Sharada KA, Balraj E, Vaigandla KK, & Sahile, K End-to-end latency analysis for data transmission via optimum path allocation in industrial sensor networks. *Wireless Communications and Mobile Computing*;2022, 1, 1697746.
[36] Khan MA, Khan GA, Khan J, Anwar T, Ashraf Z, Atoum I, et al. Adaptive Weighted Low-rank Sparse Representation for Multi-view Clustering. *IEEE Access* 2023.
[37] Khan AN, Jabeen F, Mehmood K, Soomro MA, Bresciani S. Paving the way for technological innovation through adoption of artificial intelligence in conservative industries. *J Bus Res* 2023;165:114019.
[38] Perera YS, Ratnaweera D, Dasanayaka CH, Abeykoon C. The role of artificial intelligence-driven soft sensors in advanced sustainable process industries: A critical review. *Eng Appl Artif Intell* 2023;121:105988.

Chapter 5

Machine learning vs. neutrosophic machine learning

Shabbir Hassan

5.1 INTRODUCTION

The theory of machine learning (ML) is an interdisciplinary domain that converges statistical, probabilistic, computer science, and algorithmic elements. It involves iterative learning from data, unveiling concealed insights to construct intelligent applications. In the dynamic domain of artificial intelligence (AI), ML has been the cornerstone of numerous technological advancements, revolutionizing industries and reshaping how we perceive data analysis and predictive modeling. However, amidst this robust framework, a newer paradigm known as Neutrosophic Machine Learning (NML) has emerged, offering a novel perspective on handling uncertainty, imprecision, and indeterminacy within datasets.

ML, with its formidable algorithms and data-driven methodologies, has thrived on the principles of learning patterns from vast datasets to make predictions, classifications, and decisions. Its power lies in recognizing patterns, yet it often operates within the constraints of well-defined, crisp data, struggling when faced with ambiguity or incomplete information [1].

Contrarily, NML transcends these limitations by embracing and manipulating uncertainty. It acknowledges not only the presence of true or false values but also considers a third parameter indeterminacy—where the truth value is neither true nor false but lies in between, accommodating partial truths or ambiguities within the dataset. This innovative framework introduces a paradigm shift, allowing AI systems to navigate and process information that would traditionally confound conventional ML models [2].

The distinction between ML and NML lies not just in their methodologies but also in their adaptability to handle complex, uncertain, and incomplete datasets. While traditional ML models excel in well-defined scenarios, NML stands poised to revolutionize industries where ambiguity is prevalent, such as healthcare diagnostics, financial forecasting, or natural language processing.

Understanding the divergence between these two approaches holds immense significance in unlocking the true potential of AI; this chapter

 DOI: 10.1201/9781003606055-5

presents an elaborative study of ML and NML and compares them in respective parameters [3, 4].

The major contributions of this chapter are to explain the concepts of ML and NML, their classification, possible applications, underlying mathematics, and pros and cons from scratch. At the end of the chapter, a comparison of both ML and NML is carried out on various effective parameters.

Section 5.1 explains the introduction to ML and NML, Section 5.2 explains the classification of ML, application of ML, limitations, Mathematical Model of a System, ML Process, and underlying Mathematics behind ML. Section 5.3 comprises an introduction to NML, Classification of NML, Applications of NML, limitations, NML Process, and underlying Mathematics behind NML. Section 5.4 provides a fair comparison of ML and NML in various parameters such as origin, Mathematical modeling and role-play, Handling Uncertainty, Representation of Knowledge, Algorithm Design, and many more.

5.2 MACHINE LEARNING

Machine learning is a field of AI that involves creating algorithms and models that enable computers to learn and make predictions or decisions without being explicitly programmed to do so. It's a process where machines learn from data, identify patterns, and improve their performance over time through experience as shown in Figure 5.1.

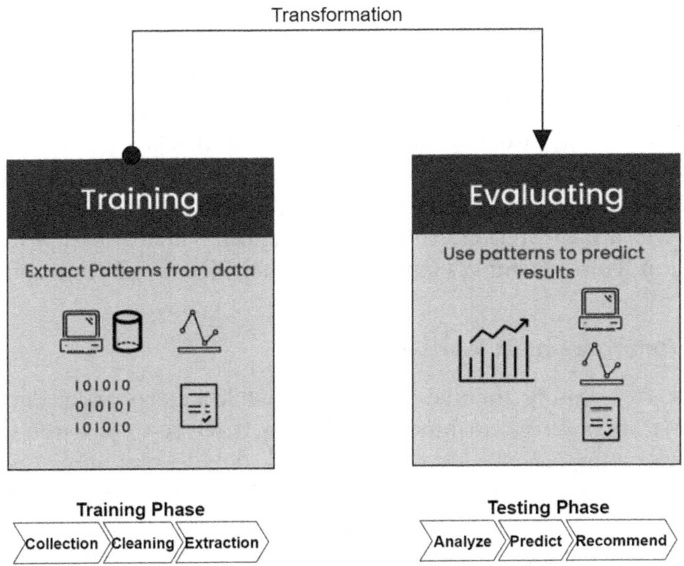

Figure 5.1 A simple machine learning (ML) model.

5.2.1 Classification of ML

Based on the idea of teaching *(such as Features, Labels/Classes, Training Data, Training Phase, algorithms, Prediction, and Model Optimization)* a computer to categorize data into predefined classes or categories based on their characteristics or features. The ML algorithms are classified into four parts as mentioned below [5].

- *Supervised Learning*
- *Unsupervised Learning*
- *Reinforcement Learning*
- *Deep learning*

5.2.1.1 Supervised learning

In this approach, the algorithm learns from labeled data, which means it's provided with input–output pairs. The model makes predictions or learns patterns based on the provided examples as shown in the Figure 5.2. For instance, in a supervised learning algorithm for recognizing handwritten digits, the model is trained on a dataset of labeled images (inputs) and their corresponding digits (outputs). Examples of Supervised ML are classifications, Regression, Natural Language Processing (NLP), Recommendation Systems, Speech Recognition, and fraud detection.

5.2.1.2 Unsupervised learning

Unsupervised learning involves training algorithms on unlabeled data. The system tries to find patterns or relationships within the data by itself. As shown in the Figure 5.3, how the model separates the shapes (rectangles, circles, polygons, etc.) into a number of clusters of the class (shape). Clustering algorithms, for example, group similar data points together without any prior information about which group they should belong to. Examples of unsupervised ML are Clustering, Anomaly Detection, Dimensionality Reduction, Association Rule Learning, Generative Models, Density Estimation, etc.

5.2.1.3 Reinforcement learning

This type of learning involves an agent that learns to make decisions by interacting with an environment. It learns from the consequences of its actions, receiving rewards or penalties, and aims to maximize the cumulative reward over time as shown in Figure 5.4. Reinforcement learning is used in various applications where an agent learns to make sequential decisions to achieve specific goals by interacting with an environment. Some examples of reinforcement learning in different domains are Game Playing, Robotics, Autonomous Vehicles, Recommendation Systems, Finance, Resource Management, Healthcare, etc.

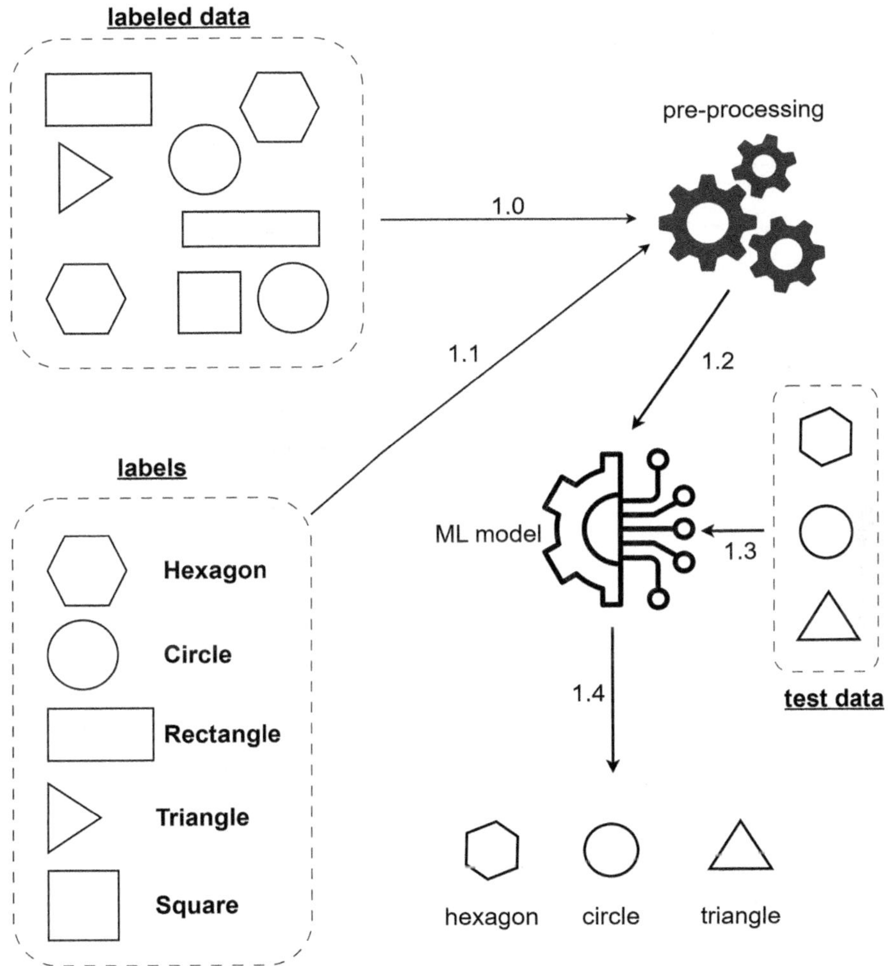

Figure 5.2 Supervised machine learning (ML) model.

It is noted that an unsupervised learning model tries to find hidden patterns in order to draw conclusions from the unlabeled dataset. It does this by reasoning and does not require supervision. Whereas a Reinforcement Learning model performs experiments and utilizes various actions to optimize long-term benefits, it gains knowledge through experience [6].

5.2.1.4 Deep learning

It is a sub-field of ML that uses artificial neural networks to mimic the learning process of the human brain. The basic building block of a deep learning model is neural networks which are made up of neurons. Besides neurons,

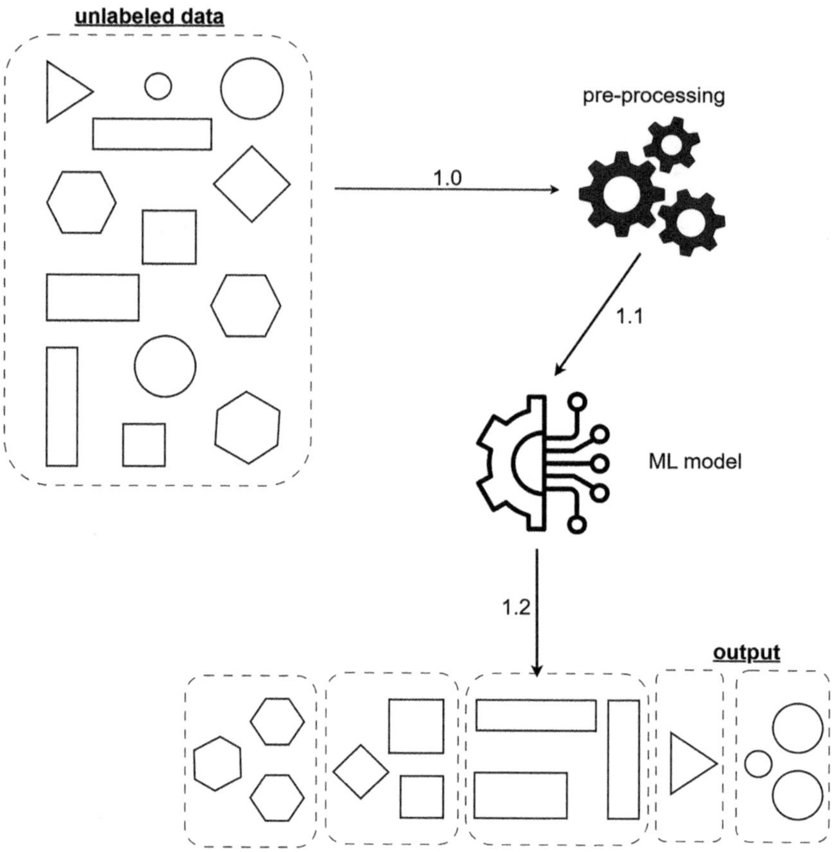

Figure 5.3 Unsupervised machine learning (ML) model.

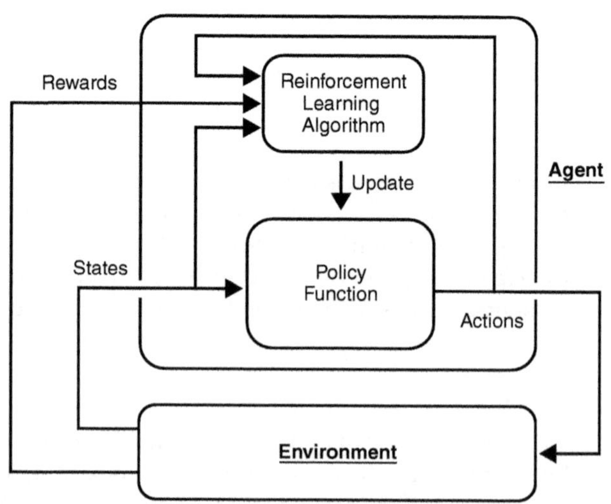

Figure 5.4 Schematic diagram of a reinforcement learning model.

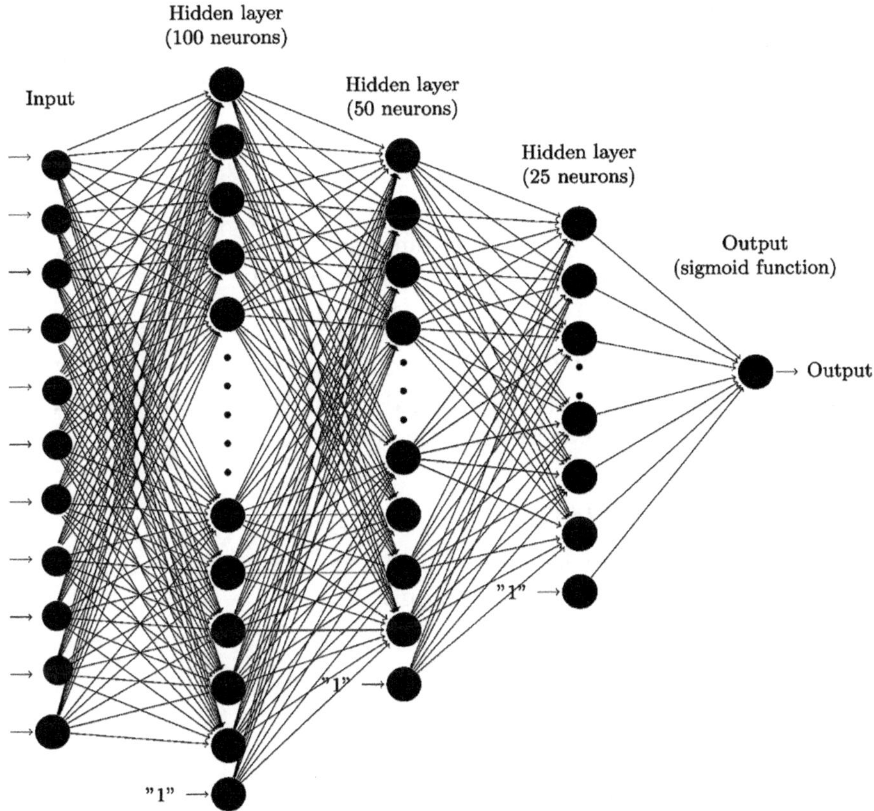

Figure 5.5 Schematic diagram of a deep learning model.

layers, weights, biases, and training are the crucial part of a neural network (see Figure 5.5). ML algorithms can be applied to various tasks such as image and speech recognition, natural language processing, recommendation systems, medical diagnosis, and many other domains where learning from data and making predictions are valuable [7, 8].

The goal of ML is to develop models and algorithms that can generalize patterns from data and make accurate predictions or decisions when presented with new, unseen data. Convolutional Neural Networks (CNNs), Recurrent Neural Networks (RNNs), Generative Adversarial Networks (GANs), Transformer Models, Autoencoders, and Deep Reinforcement Learning are examples of deep learning models [9, 10].

Deep learning models find applications across various domains due to their ability to learn intricate patterns and representations directly from data. Some key applications include: Computer Vision, Natural Language Processing (NLP), Speech Recognition, Healthcare, Autonomous Vehicles, Finance, Gaming & Entertainment, Recommendation Systems, Robotics, Cybersecurity, etc. [11].

5.2.2 Applications of ML

ML has been applied to various fields and industries, offering numerous benefits and improvements to our lives. Some examples of ML applications include [12, 13].

5.2.2.1 Social media

Platforms like Facebook use ML for automatic friend tagging suggestions, analyzing user interactions, and personalizing content.

5.2.2.2 Product recommendations

Retailers like Amazon and Netflix use ML to analyze user preferences and provide personalized product recommendations.

5.2.2.3 Image recognition

This application is used in various domains, such as facial recognition, image classification, and object detection.

5.2.2.4 Speech recognition

Voice assistants like Google Assistant, Alexa, and Siri use ML algorithms to understand and respond to user queries.

5.2.2.5 Traffic prediction

ML can be used to analyze traffic patterns and predict future traffic conditions, helping authorities plan road improvements and reduce congestion.

5.2.2.6 Self-driving cars

Autonomous vehicles use ML algorithms to navigate roads, avoid obstacles, and maintain safety.

5.2.2.7 Email automation and spam filtering

ML helps sort and filter emails, reducing spam and improving email management.

5.2.2.8 Healthcare

ML can be used to analyze medical data, predict disease outcomes, and assist in drug discovery.

5.2.2.9 Financial services

ML can help detect fraud transactions, analyze customer spending patterns, and improve investment decision-making.

5.2.2.10 Marine wildlife preservation

ML algorithms are used to develop behavior models for endangered species, helping scientists regulate and monitor their populations.

These examples demonstrate the versatility and potential of ML in various industries and applications, enhancing our lives and making the world more efficient and secure.

5.2.3 Limitation of ML

ML has various limitations that need to be considered when developing and using ML algorithms. Some of the limitations of ML include [14].

5.2.3.1 Data dependency

ML models heavily rely on the quality, quantity, and representativeness of data. Biased or incomplete datasets can lead to biased or inaccurate predictions. Obtaining and labeling large amounts of high-quality data can be expensive and time-consuming.

5.2.3.2 Overfitting and underfitting

Overfitting occurs when a model learns too much from the training data, capturing noise and irrelevant patterns, which makes it perform poorly on new data. Underfitting, on the other hand, happens when a model is too simple to capture the underlying patterns in the data, resulting in low performance on both training and new data.

5.2.3.3 Interpretability

Some ML models, like deep neural networks, can be complex and function as "black boxes," making it challenging to understand and interpret their decisions or predictions. This lack of interpretability can be a concern, especially in critical applications like healthcare or finance.

5.2.3.4 Computational resources

Training complex models, especially deep learning models, requires significant computational power and resources. This can limit the accessibility of advanced ML techniques to individuals or organizations with limited computational capabilities.

5.2.3.5 *Ethical and privacy concerns*

ML models can perpetuate biases present in the training data, leading to biased decisions or outcomes, which can exacerbate societal inequalities. Moreover, the use of personal data raises concerns about privacy and data security.

5.2.3.6 *Lack of transparency*

ML models might struggle to generalize well to unseen data if the distribution of the test data significantly differs from the training data. Adapting to new environments or scenarios that are different from the training data can be challenging.

5.2.3.7 *Continuous learning and adaptability*

Some ML models might not easily adapt to changes in the data distribution over time. Continuous learning and adaptation to new information without forgetting previously learned knowledge is an ongoing challenge.

Addressing these limitations involves ongoing research and advancements in the field of ML. Techniques like regularization, data augmentation, transfer learning, and the development of more interpretable models are among the efforts aimed at mitigating these challenges. These limitations demonstrate the need for careful consideration and evaluation when developing and using ML algorithms. While ML has numerous benefits and applications, it is important to understand its limitations to ensure its effective and ethical use.

5.2.4 Mathematical model of a system

In computer science, a "Mathematical Model of a System" refers to a representation of a system using mathematical structures and expressions. The purpose of creating such a model is to gain a better understanding of the system's behavior, analyze its properties, and make predictions or decisions based on this understanding. Mathematical modeling is widely used in various fields, including computer science, to simulate and study the behavior of systems [5, 8].

Following are some key aspects and concepts related to the mathematical modeling of a system in computer science:

5.2.4.1 *System representation*

A system can be anything from a software application to a complex network. The first step in creating a mathematical model is to identify the components of the system and their interactions.

5.2.4.2 Variables and parameters

Variables represent the changing quantities within the system, while parameters are constants that affect the system's behavior. These are often expressed using mathematical symbols and equations.

5.2.4.3 Equations and formulas

Mathematical equations and formulas are used to describe the relationships between variables and parameters. These equations may be based on principles from various branches of mathematics, such as algebra, calculus, or probability theory.

5.2.4.4 Differential equations

In dynamic systems, especially those involving time-dependent changes, differential equations are commonly used. These equations describe how the variables change with respect to time.

5.2.4.5 Simulation and analysis

Once the mathematical model is formulated, it can be used for simulation and analysis. Simulation involves running the model to observe how the system behaves under different conditions. Analysis may include studying stability, convergence, and other properties.

5.2.4.6 Optimization

Mathematical models can be employed to optimize certain aspects of a system. For example, in computer networks, models might be used to optimize routing algorithms for better performance.

5.2.4.7 Decision-making

Mathematical models assist in decision-making processes by providing insights into the consequences of different choices. This is particularly important in fields like AI and ML.

5.2.4.8 Validation and verification

It's crucial to validate and verify mathematical models to ensure that they accurately represent the real-world system. This involves comparing model predictions with observed or measured data.

5.2.4.9 Abstraction

Mathematical models often involve a level of abstraction, where certain details of the system are simplified to make the model more manageable while still capturing essential characteristics.

In general, a mathematical model of a system in computer science is a formal representation that allows researchers and practitioners to analyze and understand the behavior of the system, make predictions, and facilitate decision-making processes. The level of complexity and detail in the model depends on the specific goals and requirements of the analysis.

5.2.5 Machine learning process

The ML process involves several steps that a data science team executes to create and deliver a ML model. The high-level steps include collecting and preparing data, choosing a relevant mathematical model, training the model, and evaluating its performance. There are mainly six steps, they are [5, 6, 8]:

- *Collection of Data from various data source*
- *Data cleaning and Feature Engineering*
- *Splitting the Dataset*
- *Model building and Selection of ML Algorithm*
- *Model Evaluation*
- *Model Deployment*

Figure 5.6 is designed to assist in comprehending the workflow of the ML process.

5.2.5.1 Collection of data from various data sources

Typically, data collection serves as a fundamental step in the field of ML. Depending on the business problem at hand, it becomes necessary to acquire data from diverse sources, whether internal or external, aligning with customer trends in the markets. Certain data may originate from sources like social media and public opinions. Ensure that the data is representative of the real-world scenarios the model will encounter.

5.2.5.2 Data cleaning and feature engineering

Feature engineering stands as a crucial step in the realm of ML. Before delving into the selection of an appropriate algorithm for a given dataset, it is essential to thoroughly comprehend the dataset and initiate a cleansing process to refine it, thereby optimizing results. The data-cleaning procedure provides a deeper insight into features and their interrelationships.

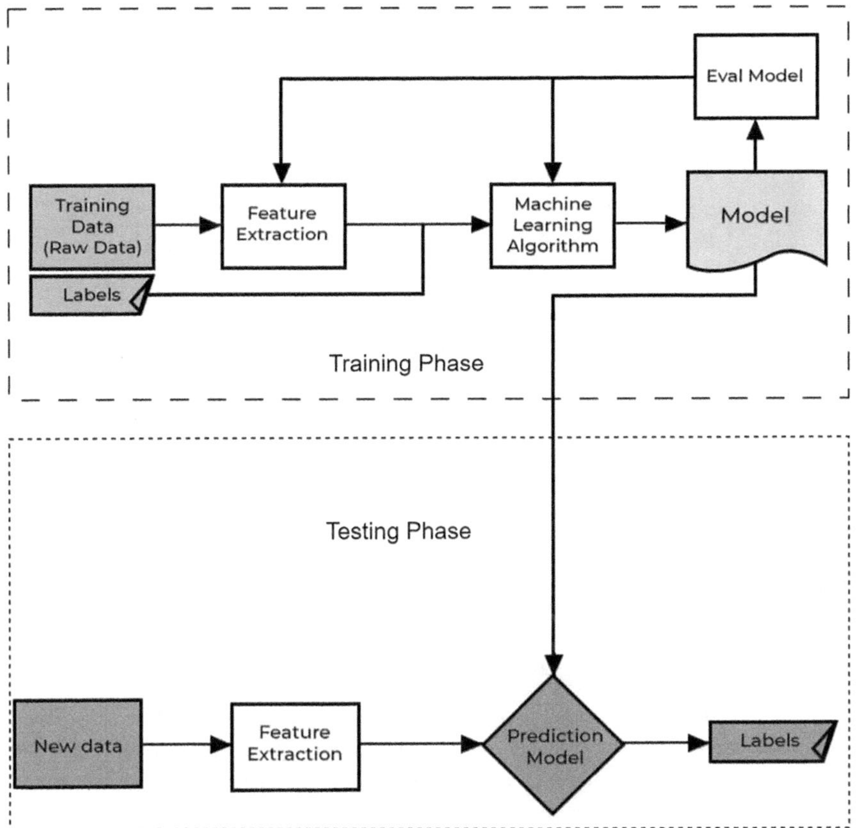

Figure 5.6 Workflow of a machine learning model.

Effectively addressing missing values or human errors in the dataset contributes significantly to enhancing the overall modeling performance, resulting in a considerable improvement in data quality as shown in Figure 5.7. In certain instances, identifying and managing outliers is imperative to prevent skewed or unrealistic outcomes. This can involve either truncating data beyond a specified threshold or employing transformations such as log transformation shown in Figure 5.8. Various methodologies within the ML domain exist to handle such situations.

A critical approach involves converting categorical columns into numerical ones, as many ML algorithms necessitate numerical features for optimal functioning. While feature engineering demands time and resources, it remains an indispensable and worthwhile endeavor to ensure "data integrity."

#	Height	Weight	Country	Place	Days	Remarks
0	10.0	56.4	India	Aligarh	8.0	NaN
1	NaN	35.0	Dubai	Abu Dhabi	12.3	NaN
2	12.0	52.6	France	Paris	NaN	10.5
3	15.3	55	India	Hazaribag	1.0	NaN
4	13.7	55.05	Germany	Berlin	5.6	12.2
5	NaN	NaN	NaN	NaN	2.6	NaN
6	NaN	72.5	USA	New York	NaN	0.1

Figure 5.7 Missing value representation in a dataset.

5.2.5.3 Splitting the dataset

Divide the dataset into training, validation, and testing sets. The training set is used to train the model, the validation set helps tune hyperparameters, and the testing set evaluates the model's performance on unseen data. A frequently employed ratio is 80:20, indicating that 80% of the data is allocated for training purposes, while the remaining 20% is designated for testing. Alternatives such as 70:30, 60:40, and even 50:50 (not recommended) are also practical choices in real-world applications as shown in Figure 5.9 [15, 16].

5.2.5.4 Model building and selection of ML algorithm

Select a ML algorithm or model building that is appropriate for your problem. The choice depends on the nature of the data and the task (e.g., classification, regression, clustering).

5.2.5.5 Model evaluation

Assess the model's performance on the testing set to understand how well it generalizes to new, unseen data. Use metrics relevant to the specific task, such as accuracy, precision, recall, or mean squared error.

#	Name	D.O.B	SSN
0	Amit Kumar Ranjan	03.09.1971	897
1	Ramiz	22.02.1987	211
2	Alice	**11.02.2018**	**036**
3	Aasim Z.	15.01.1968	179
4	Iqbal A.	**09.11.1877**	427
5	Shabbir H.	10.02.1988	**924**
6	Bokhari M. U.	10.03.1964	**052**
7	Shamsuddin A.	**26.06.1945**	**988**
8	Smith	01.01.1978	155
9	Arshad I.	20.08.1975	340
A	Roaaiya P.	06.06.1997	226
B	Bob	15.12.1976	**816**
C	Faridi A. R.	11.05.1971	166
D	Sumaiya H.	**01.12.2021**	232
E	Eve	03.11.1991	**065**
F	Prity Sinha	17.09.1984	485

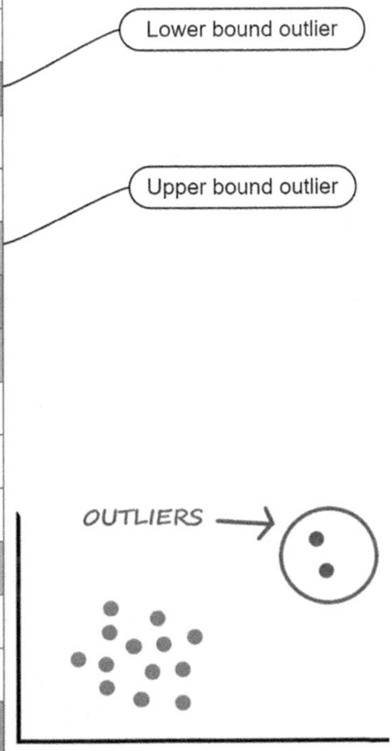

Lower bound outlier

Upper bound outlier

OUTLIERS

Figure 5.8 Outliers data points and their impact.

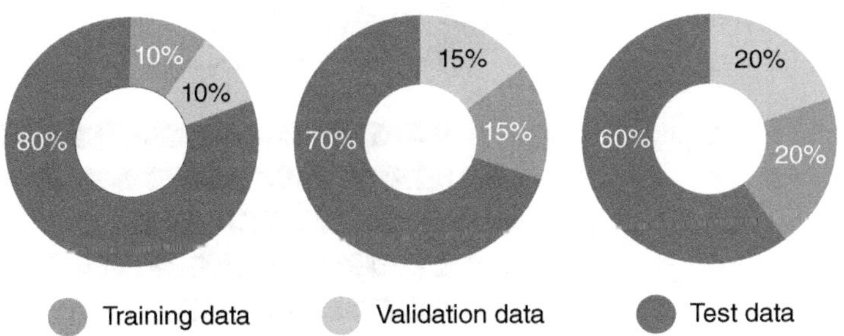

Figure 5.9 Dataset splitting ratios in an ML model.

5.2.5.6 Model deployment

Once satisfied with the model's performance, deploy it to a production environment where it can make predictions on new, real-world data.

5.2.6 Mathematics behind ML

ML draws its strength from four fundamental principles: Statistics, Linear Algebra, Probability, and Calculus. Statistical concepts form the backbone of each model, while calculus aids in the learning and optimization processes. Linear algebra proves invaluable when handling extensive datasets, and probability plays a crucial role in forecasting the likelihood of future events. These mathematical principles are recurrent in your journey through data science and ML careers. The use case of such branches of mathematics is shown in Figure 5.10 [17].

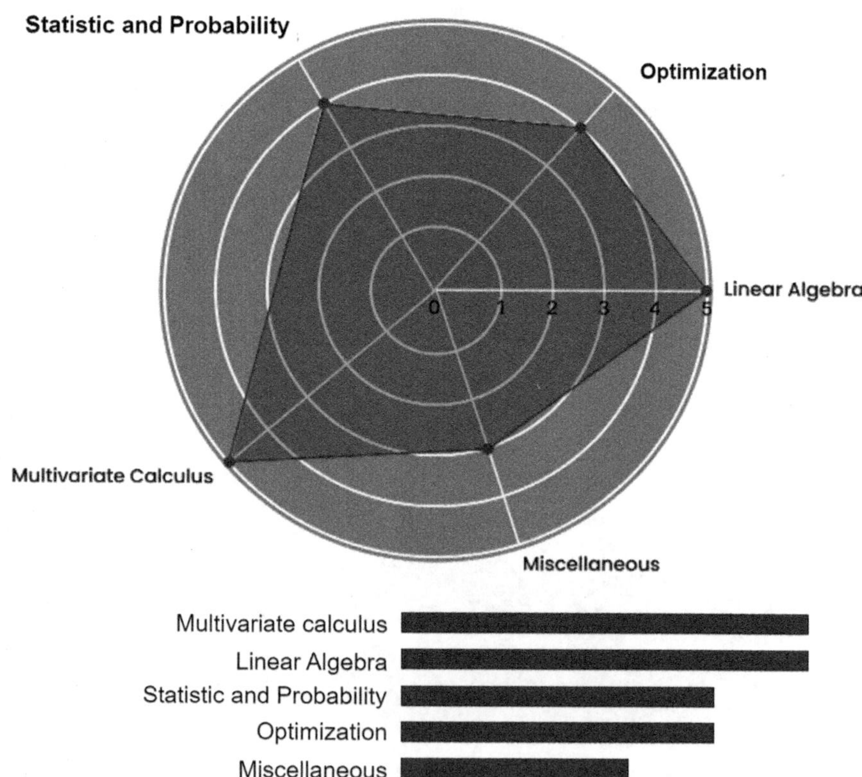

Figure 5.10 Role of mathematics in ML.

5.2.6.1 Mathematical concepts important for machine learning

- *Linear Algebra*
- *Calculus*
- *Probability Theory*
- *Discrete Mathematics*
- *Statistics*

5.2.6.1.1 Linear algebra

Linear algebra finds application in various aspects of ML algorithms, including loss functions, regularization, covariance matrices, Singular Value Decomposition, Matrix Operations, and support vector machine classification. It is integral to algorithms like linear regression and plays a crucial role in understanding the optimization methods employed in ML. Principal Component Analysis (PCA), utilized for dimensionality reduction in data, relies on linear algebra. Moreover, linear algebra is a foundational element in neural networks, contributing to the processing and representation of network structures. Therefore, a keen interest in linear algebra is essential, given its extensive use in the field of data science [18].

5.2.6.1.2 Calculus

Many learners who may have been uninterested in studying calculus during their advanced studies may find it surprising that it plays a crucial role in ML. Fortunately, one doesn't necessarily have to become a calculus expert; it is more important to grasp the fundamental principles and understand the practical applications of calculus in the context of building ML models.

For instance, if you comprehend how derivatives in calculus signify the rate of change of a function, you can grasp the concept of gradient descent in a better way. In gradient descent, the goal is to locate local minima for a function. However, dealing with saddle points or multiple minima poses challenges, and gradient descent might converge to a local minimum instead of the global minimum unless initiated from multiple points. Essential topics for mastering calculus in data science include Differential and Integral Calculus, Partial Derivatives, Vector-Valued Functions, and Directional Gradients [19].

In the realm of algorithm training and gradient descent, multivariate calculus comes into play. Understanding derivatives, divergence, curvature, and quadratic approximations is crucial for effective learning and implementation.

5.2.6.1.3 Probability theory

To effectively navigate a ML predictive modeling project, it becomes evident that a solid understanding of probability is indispensable. ML involves constructing predictive models from ambiguous data, where dealing with imperfect or incomplete information introduces an element of uncertainty. While uncertainty is a fundamental aspect of ML, it poses significant challenges, especially for those entering the field from a programming background.

Within the realm of ML, three primary sources of uncertainty are encountered: noisy data, limited coverage of the problem domain, and inherent imperfections in models. However, leveraging appropriate probability tools enables us to approximate solutions to these challenges. Probability plays a vital role in tasks such as hypothesis testing and working with distributions like the Gaussian distribution and probability density functions [20, 21].

5.2.6.1.4 Discrete mathematics

Discrete mathematics deals with non-continuous numbers, primarily integers, which are often essential for various applications. In scenarios like organizing a taxi fleet schedule, one cannot dispatch 0.34 taxis; a complete unit is required. Similarly, in tasks such as postal deliveries, a postman cannot be divided into halves or visit fractional places to distribute letters.

AI structures frequently rely on discrete elements. Take a neural network, for instance, where the number of nodes and interconnections is an integer value. It cannot have fractional nodes or a fraction of a link. Consequently, the mathematical framework employed in constructing a neural network incorporates a discrete component, where integers represent the count of nodes and interconnections.

While a basic understanding of discrete math may suffice for many aspects of ML, delving into more advanced concepts becomes necessary for tasks involving relational domains, graphical models, combinatorial problems, structured prediction, and similar domains [22].

5.2.6.1.5 Statistics

A foundational understanding of descriptive statistics is crucial for any aspiring data scientist aiming to comprehend ML, especially when dealing with classifications such as logistic regression, distributions, discrimination analysis, and hypothesis testing.

Statistics holds a central position in the mathematical framework of ML. Key statistical concepts essential for ML include Combinatorics, Axioms, Bayes' Theorem, Variance and Expectation, Random Variables, Conditional, and Joint Distributions [23, 24].

5.3 NEUTROSOPHIC MACHINE LEARNING

NML is an extension of traditional ML methodologies designed to handle uncertainty, imprecision, vagueness, and indeterminacy in data. It incorporates the principles of neutrosophy, a theory introduced by mathematician Florentin Smarandache in the 1990s to handle indeterminate and contradictory information.

In NML, data and knowledge are represented using neutrosophic sets, which include three components: truth-membership (T), indeterminacy-membership (I), and falsity-membership (F). Each component represents the degree of truth, uncertainty, and falsity associated with a particular element in the dataset. These elements could be features, classes, rules, or any other entities relevant to the problem being solved [2, 3, 23].

NML algorithms are designed to work with these neutrosophic sets, allowing the handling of incomplete, imprecise, or contradictory information. The goal is to develop ML models that are more robust and adaptable in scenarios where traditional models might struggle due to uncertainty or ambiguity in the data. NML has applications in various domains where uncertainty plays a significant role, such as medical diagnosis, decision-making systems, financial forecasting, image recognition in unclear conditions, and more. It aims to provide a framework that can effectively process and utilize uncertain information to make informed decisions or predictions [9, 25].

5.3.1 Classification of NML

NML is an extension of traditional ML that incorporates neutrosophic logic, which deals with indeterminate, imprecise, and incomplete information. Neutrosophic logic allows for the representation of truth, falsehood, and indeterminacy simultaneously. In the context of ML, NML is used to handle uncertainty and vagueness in data. Figure 5.11 represents a concise classification of NML [7, 26].

5.3.1.1 Neutrosophic set theory

Neutrosophic Sets (NS): Basic building blocks in NML. Neutrosophic sets generalize classical sets by allowing degrees of membership, non-membership, and indeterminacy.

Neutrosophic Soft Sets (NSS): Extension of neutrosophic sets that incorporates the concept of soft sets. Soft sets are used to handle uncertainty in a more flexible way.

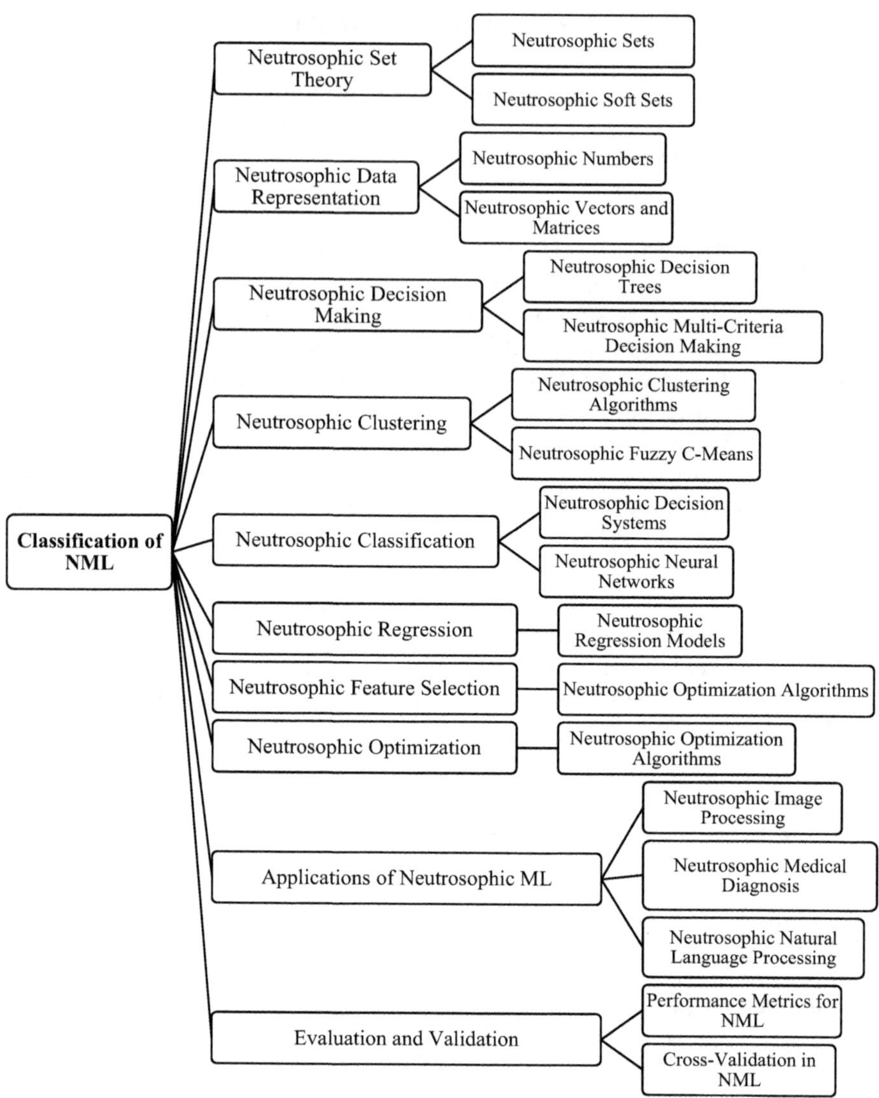

Figure 5.11 Classification of neutrosophic machine learning [5, 7, 9, 13, 23, 25, 26, 27].

5.3.1.2 Neutrosophic data representation

Neutrosophic Numbers (NN): Represent numbers with indeterminacy, allowing for a more expressive way to model uncertainty.

Neutrosophic Vectors and Matrices: Extension of traditional vectors and matrices to handle neutrosophic elements.

5.3.1.3 Neutrosophic decision-making

Neutrosophic Decision Trees: Decision trees incorporate neutrosophic logic to make decisions in the presence of uncertainty.

Neutrosophic Multi-Criteria Decision Making (NMCDM): Extends traditional decision-making methods to handle multiple criteria under neutrosophic uncertainty.

5.3.1.4 Neutrosophic clustering

Neutrosophic Clustering Algorithms: Clustering methods that consider neutrosophic information in the clustering process.

Neutrosophic Fuzzy C-Means (NFCM): An extension of the traditional fuzzy C-means algorithm to handle neutrosophic information.

5.3.1.5 Neutrosophic classification

Neutrosophic Decision Systems: Systems that use neutrosophic logic for classification tasks.

Neutrosophic Neural Networks: Neural network architectures that incorporate neutrosophic elements.

5.3.1.6 Neutrosophic regression

Neutrosophic Regression Models: Regression models that account for indeterminacy in the input and output variables.

5.3.1.7 Neutrosophic feature selection

Neutrosophic Feature Selection Methods: Techniques for selecting relevant features in the presence of uncertainty.

5.3.1.8 Neutrosophic optimization

Neutrosophic Optimization Algorithms: Optimization algorithms that take into account neutrosophic information.

5.3.1.9 Applications of neutrosophic machine learning

Neutrosophic Image Processing: Using NML for image analysis and understanding.

Neutrosophic Medical Diagnosis: Applying NML in healthcare for diagnosis and decision support.

Neutrosophic Natural Language Processing: Handling uncertainty in language understanding and processing.

5.3.1.10 Evaluation and validation

Performance Metrics for NML: Developing metrics that account for indeterminacy in the evaluation of NML models.

Cross-Validation in NML: Adapting traditional cross-validation techniques to NML scenarios.

The classification above outlines different aspects and components of NML, covering its theoretical foundations, data representation, decision-making, clustering, classification, regression, feature selection, optimization, applications, and evaluation [4, 9].

5.3.2 Applications of NML

NML is a relatively new field that combines neutrosophic sets and ML techniques. Neutrosophic sets are sets that allow for the representation of indeterminate or uncertain information. The combination of these two fields has been applied in various areas such as medical diagnosis, image processing, and sentiment analysis. Some examples of practical applications of NML include a neutrosophic recommender system for medical diagnosis, automatic segmentation of choroid layer in EDI OCT images, and a novel approach to image categorization using a diverse density support vector machine. However, the practicality of neutrosophic sets in real-life applications is still being explored. Some potential applications of NML include Medical Diagnosis, Financial Forecasting, Image and Signal Processing, Natural Language Processing, Robotics and Autonomous Systems, Environmental Monitoring, Social Media Analysis, Internet of Things, and many more [13, 27].

5.3.3 Limitation of NML

The limitations of NML are still being actively researched. Some current limitations and challenges include the need for further generalization and robust testing of the combined models, such as Q learning with a neutrosophic set, and the rich research area of combining ML with neutrosophic sets. Additionally, the practical examples or real-life use cases to prove the effectiveness and reliability of neutrosophic sets in various application areas are still not widely available. Furthermore, while neutrosophic sets have shown effectiveness in handling inconsistency and indeterminacy in medical image analysis, there is ongoing research to address the challenges and future directions in this area. Overall, the practicality and generalizability of NML in real-life applications are still being explored, and further research is needed to overcome these limitations such as Computational Complexity, Interpretability, Lack of Standardization, Limited Adoption and Awareness, Algorithmic Development, Parameter Sensitivity, Integration with Existing Systems, Risk of Overfitting, Limited Tool Support and many more [16, 28].

5.3.4 NML Process

The NML process involves the combination of neutrosophic sets and ML techniques to handle indeterminate or uncertain information in various applications. The process typically involves the following steps [13, 17, 29, 30].

5.3.4.1 Define the problem

Identify the problem or task that needs to be solved, such as medical diagnosis, image processing, or sentiment analysis.

5.3.4.2 Collect data

Gather relevant data for the problem, which may include labeled examples, images, or text.

5.3.4.3 Preprocess the data

Clean and preprocess the data to ensure it is in a suitable format for the ML model.

5.3.4.4 Apply neutrosophic sets

Convert the data into neutrosophic sets, which allow for the representation of indeterminate or uncertain information.

5.3.4.5 Choose a machine learning model

Select an appropriate ML model, such as a neural network, support vector machine, or decision tree, to solve the problem.

5.3.4.6 Train the model

Train the ML model using the neutrosophic data.

5.3.4.7 *Evaluate the model*

Assess the performance of the model using appropriate metrics, such as accuracy, precision, recall, or F1 score.

5.3.4.8 *Refine the model*

If necessary, refine the model by adjusting its parameters or architecture to improve its performance.

5.3.4.9 *Apply the model to new data*

Use the trained model to make predictions or classifications on new, unseen data.

5.3.4.10 *Analyze the results*

Analyze the results to understand the performance of the model and identify any potential improvements or limitations.

The NML process can be applied to various domains, such as medical diagnosis, image processing, and sentiment analysis, to handle indeterminate or uncertain information and improve the performance of ML models.

5.3.5 Mathematics behind NML

NML is based on neutrosophic sets and logic, which extend classical set theory and classical logic to handle indeterminacy, uncertainty, and imprecision. The mathematics behind NML involves the integration of neutrosophic sets, logic, numbers, and probability into various ML algorithms and methodologies. Here are the key mathematical concepts behind NML [17, 25, 31]:

5.3.5.1 *Neutrosophic sets*

A neutrosophic set is denoted by $A = (T, I, F)$ and is characterized by three membership functions:

- $T(x)$ represents the degree of truth or membership of an element x in the set.
- $I(x)$ represents the degree of indeterminacy or neutrality of x.
- $F(x)$ represents the degree of falsity or non-membership of x in the set.

The value of $T(x)$, $I(x)$ and $F(x)$ are in the range [0, 1], and they satisfy the condition $T(x) + I(x) + F(x) \leq 1$

5.3.5.2 Neutrosophic logic

Neutrosophic logic extends classical logic by introducing the concept of indeterminacy. It involves three truth values: true, indeterminate, and false. Neutrosophic propositions are expressed using neutrosophic truth values. Logical operations such as conjunction (∧), disjunction (∨), and negation (~) are extended to neutrosophic logic. These operations take into account the degrees of truth, indeterminacy, and falsity associated with the involved propositions.

5.3.5.3 Neutrosophic number

A neutrosophic number is represented as $N = a + bi + cj$, where a, b, and c are real numbers, and i and j are imaginary units. Neutrosophic numbers are used to represent indeterminate and uncertain quantities.

5.3.5.4 Neutrosophic relations

Neutrosophic relations extend classical relations to handle uncertainty. A neutrosophic relation R between two neutrosophic sets A and B is defined as a triplet (T_R, I_R, F_R), where T_R, I_R, and F_R are the truth, indeterminacy, and falsity degrees, respectively.

5.3.5.5 Neutrosophic probability

Neutrosophic probability extends classical probability theory by incorporating indeterminacy. It is used to represent the likelihood of an event occurring within a neutrosophic set. Neutrosophic probability distributions can be defined over neutrosophic sets.

5.3.5.6 Neutrosophic statistics

Neutrosophic statistics involves the application of statistical methods to neutrosophic data. Descriptive statistics, inferential statistics, and hypothesis testing can be adapted to handle neutrosophic information.

5.3.5.7 Neutrosophic clustering and classification

Neutrosophic clustering and classification algorithms are developed to work with neutrosophic data. These algorithms consider the indeterminacy and uncertainty in the data points to group them or make predictions.

5.3.5.8 *Neutrosophic logic programming*

Neutrosophic logic programming extends traditional logic programming to handle neutrosophic information. It is used in knowledge representation and reasoning tasks in NML.

5.4 COMPARISON OF ML VS. NML

ML and NML are both techniques used in data analysis and decision-making, but they have different approaches and principles. On the basis of *origin, Mathematical modeling and role-play, Handling Uncertainty, Representation of Knowledge, Algorithm Design, Data Types and Applicability, Interpretability and Explainability, Complexity and Computational Cost, Applications, Efficiency and Limitations* ML and NML can be compared by the following parameters [1, 3, 4, 13, 14, 29, 30, 31, 32, 33, 34, 35]:

5.4.1 Origin

- *Origin*: ML has its roots in the broader field of AI, which emerged in the mid-20th century. The primary goal of early AI researchers was to create systems capable of learning from data and improving their performance over time.
- *Philosophical Foundation*: The philosophical foundation of classical ML is rooted in the idea of learning from experience and data. It draws inspiration from statistics, probability theory, optimization, and cognitive science.
- *Historical Development*: The historical development of ML can be traced through several phases, including rule-based systems in the 1950s, symbolic AI in the 1960s, and the resurgence of interest in statistical approaches, neural networks, and deep learning in more recent decades.
- *Key Contributors*: Key contributors to the development of classical ML include Alan Turing, Arthur Samuel, John McCarthy, Tom Mitchell, Geoffrey Hinton, and others.

- *Origin*: NML is based on the principles of neutrosophy, a branch of philosophy introduced by Florentin Smarandache in the late 20th century. Neutrosophy extends classical logic to handle indeterminacy, vagueness, and inconsistency.
- *Philosophical Foundation*: The philosophical foundation of NML is grounded in neutrosophic logic and sets, providing a mathematical framework to deal with incomplete, imprecise, and indeterminate information. Neutrosophy introduces the concept of neutralities, acknowledging the existence of neutral elements between truth and falsity.
- *Historical Development*: The introduction of neutrosophic logic and sets dates back to the 1990s, and the application of these concepts to machine learning is a more recent development. The motivation behind NML is to address the challenges posed by uncertain and incomplete information in data.
- *Key Contributors*: The key contributor to neutrosophy and NML is Florentin Smarandache, who proposed and developed the neutrosophic framework.

5.4.2 Mathematical modeling and role-play

- *Foundation*: ML is grounded in statistical and probabilistic models. Common mathematical models include linear regression, decision trees, support vector machines, and neural networks.
- *Role-Play*: In ML, algorithms learn patterns from labeled data and make predictions or classifications based on these learned patterns. The role-play involves the algorithm acting as a function approximator, mapping input data to output predictions.
- *Uncertainty Handling*: Traditional ML models assume crisp, deterministic values for input features and labels. While there are methods to handle uncertainty (e.g., using probability distributions), the primary focus is on precise modeling.
- *Objective*: The primary objective of ML is to develop models that generalize well to unseen data, make accurate predictions, and perform specific tasks without being explicitly programmed.

- *Foundation*: NML extends classical ML by incorporating neutrosophic sets and logic. Mathematical models in NML involve representing truth-membership, indeterminacy-membership, and falsity-membership functions within the neutrosophic set.
- *Role-Play*: NML introduces the concept of neutralities, allowing for the representation of indeterminacy and uncertainty explicitly. In role-play, NML models act as systems that can handle imprecise and incomplete information in a more nuanced way.
- *Uncertainty Handling*: The key strength of NML lies in its ability to explicitly handle uncertainty, vagueness, and inconsistencies in data. The neutrosophic set allows for the representation of elements that may simultaneously belong, not belong, and be indeterminate.
- *Objective*: NML aims to develop models that not only make predictions but also explicitly account for the indeterminacy and neutral elements within the data. It provides a more flexible framework for modeling real-world uncertainties.

5.4.3 Handling uncertainty

- *Representation*: Traditional ML models often represent uncertainty through probability distributions. However, these distributions may not explicitly capture all forms of uncertainty, such as vagueness or indeterminacy.
- *Methods*: ML models may use techniques like bootstrapping, ensemble methods, or Bayesian approaches to incorporate uncertainty. However, the focus is often on providing probabilistic confidence intervals rather than explicitly addressing multiple truth values.
- *Crisp vs. Probabilistic*: ML assumes a crisp distinction between classes or values, and uncertainty is typically modeled using probabilities. The uncertainty is often represented by a single probability value associated with each class or outcome.

- *Representation*: NML explicitly incorporates the concept of neutrosophic sets, allowing for the representation of truth-membership, indeterminacy-membership, and falsity-membership functions. This enables a more comprehensive representation of uncertainty.
- *Methods*: NML introduces methodologies specifically designed to handle neutrosophic information, providing a structured way to represent and reason about uncertainty, vagueness, and inconsistency in the data.
- *Multiple Truth Values*: NML acknowledges the existence of neutral elements, allowing for the simultaneous representation of truth, falsity, and indeterminacy. This provides a richer and more flexible framework for handling a wide range of uncertainties.

- *Common Approaches*: Common ML approaches for uncertainty include dropout in neural networks, Monte Carlo simulations, and ensemble methods like bagging and boosting.

- *Applications*: Neutrosophic sets are particularly well-suited for applications where the uncertainty is not only probabilistic but also involves vagueness and imprecision. NML has applications in areas such as medical diagnosis, robotics, and decision-making where handling multiple truth values is crucial.

5.4.4 Representation of knowledge

- *Data-Driven*: ML relies on data-driven approaches, where knowledge is extracted from large datasets through learning algorithms. The representation of knowledge is implicitly encoded in the learned model parameters.
- *Feature-Based*: Knowledge in ML models is often represented through features extracted from the data. These features can be engineered or automatically learned, and the model's understanding is based on patterns within these features.
- *Implicit Knowledge*: The knowledge acquired by ML models is implicit and embedded in the model's parameters. Interpretability of these parameters might be challenging, especially in complex models like deep neural networks.
- *Task-Specific*: The representation of knowledge is task-specific, meaning that the learned knowledge is optimized for the particular task the model was trained on.

- *Explicit Neutrosophic Sets*: NML explicitly represents knowledge using neutrosophic sets, which consist of truth, indeterminacy, and falsity membership functions. This allows for the explicit modeling of uncertainty and neutrality.
- *Handling Indeterminacy*: NML provides a structured way to represent indeterminacy in knowledge. The indeterminacy component acknowledges situations where the truth and falsity of a statement cannot be precisely determined.
- *Explicit Neutral Elements*: The concept of neutrality in neutrosophic sets allows for the explicit representation of elements that may simultaneously belong or not belong, and be in between, offering a more nuanced representation of knowledge.
- *Interpretability*: While interpretability is a challenge in traditional ML, NML's explicit representation of knowledge allows for more transparent and interpretable models. The components of neutrosophic sets provide insights into the model's understanding of uncertainty.

5.4.5 Algorithm design

- *Model Architecture*: ML algorithm design involves defining the architecture of a model. Common ML models include linear regression, decision trees, support vector machines, and neural networks. The choice of the model depends on the problem at hand and the characteristics of the data.

- *Neutrosophic Sets Integration*: NML integrates the concept of neutrosophic sets into algorithm design. This involves adapting traditional ML algorithms to accommodate neutrosophic representations of data and knowledge.

- *Training Process*: ML algorithms are trained using labeled data, where the model learns patterns and relationships within the data to make predictions. The training process typically involves optimizing parameters through techniques like gradient descent.
- *Supervised, Unsupervised, Reinforcement Learning*: ML algorithms are categorized into supervised, unsupervised, and reinforcement learning based on the nature of the learning task. Supervised learning requires labeled data, unsupervised learning deals with unlabeled data, and reinforcement learning involves learning from feedback in an environment.
- *Objective Functions*: The design of ML algorithms includes specifying an objective function that the model aims to optimize during the training process. This function reflects the goal of the learning task, such as minimizing prediction errors or maximizing likelihood.

- *Explicit Handling of Neutrality*: NML algorithms are designed to explicitly handle neutrosophic sets, incorporating truth-membership, indeterminacy-membership, and falsity-membership functions. The design ensures that neutrality and indeterminacy are considered during the learning process.
- *Extended Learning Objectives*: NML algorithm design may involve extending learning objectives to explicitly account for the three components of neutrosophic sets. This includes modifying objective functions to optimize for truth, indeterminacy, and falsity.
- *Enhanced Flexibility*: Algorithm design in NML is focused on enhancing the flexibility of models to handle uncertainties in a more nuanced way. This includes adjustments to traditional learning algorithms to capture and represent indeterminacy and neutrality.

5.4.6 Data types and applicability

- *Data Types*: ML primarily works with crisp and deterministic data types. Features and labels are often represented as precise values, and the algorithms assume well-defined relationships within the data.
- *Applicability*: ML is widely applicable across various domains, including image and speech recognition, natural language processing, recommendation systems, and predictive modeling. It is effective when dealing with structured and well-defined data.
- *Supervised and Unsupervised Learning*: ML algorithms are applied in both supervised and unsupervised learning scenarios. Supervised learning requires labeled data, while unsupervised learning explores patterns in unlabeled data.

- *Data Types*: NML is designed to handle more complex and uncertain data types. It explicitly deals with neutrosophic sets, where elements can have truth, indeterminacy, and falsity membership degrees simultaneously.
- *Applicability*: NML is particularly applicable in domains where uncertainty, vagueness, and inconsistency are prevalent. It finds use in applications where data is inherently imprecise, such as medical diagnosis, decision-making under uncertainty, and situations involving subjective judgments.
- *Explicit Handling of Uncertainty*: NML is specifically tailored to address uncertainties through the explicit handling of neutrosophic sets. It provides a better-nuanced representation of information, allowing for the modeling of indeterminacy and neutrality.

- *Common Algorithms*: ML algorithms such as linear regression, support vector machines, decision trees, and neural networks are commonly used for tasks like classification, regression, clustering, and dimensionality reduction.

- *Applications*: NML can be applied in areas such as medical diagnosis, robotics, decision support systems, and social sciences, where uncertainties may arise from imprecise measurements, subjective assessments, or incomplete information.

5.4.7 Interpretability and explainability

- *Model Complexity*: ML models, especially complex ones like deep neural networks, can lack interpretability due to the large number of parameters and intricate relationships between features.
- *Black-Box Nature*: Deep learning models, in particular, are often considered black boxes, which means understanding how they arrive at a specific decision can be challenging.
- *Feature Importance*: While some ML models provide feature importance metrics, the interpretability is often limited, and understanding the decision-making process may require expert knowledge.
- *Explainability Challenges*: The high dimensionality and non-linearity of some ML models make it difficult to explain their decisions in a straightforward manner, especially to non-experts.

- *Structured Representation*: NML, by explicitly representing truth, indeterminacy, and falsity in neutrosophic sets, provides a more structured and interpretable framework.
- *Transparent Components*: The components of neutrosophic sets allow for transparency in understanding how uncertainty, indeterminacy, and neutrality contribute to the decision-making process.
- *Explicit Handling of Uncertainty*: NML, through its explicit handling of uncertainties, makes it easier to explain how the model considers different degrees of truth, indeterminacy, and falsity in arriving at a decision.
- *Subjective Interpretations*: NML can handle subjective interpretations more explicitly, providing a clearer understanding of how subjective judgments contribute to the overall decision process.

5.4.8 Complexity and computational cost

- *Model Complexity*: ML models can vary widely in complexity, ranging from simple linear models to highly complex deep neural networks. The complexity often depends on the task and the nature of the data.
- *Training Complexity*: Training complex ML models, especially deep neural networks, can be computationally expensive and time-consuming. The optimization of numerous parameters requires substantial computational resources.

- *Model Complexity*: NML models introduce additional complexity due to the explicit representation of neutrosophic sets and the handling of multiple truth values.
- *Training Complexity*: Training NML models involves dealing with the complexities of neutrosophic sets, which may require modifications to traditional training algorithms. The computational cost can be influenced by the need to optimize truth, indeterminacy, and falsity components.

- *Inference Complexity*: Inference, or making predictions on new data, can be computationally efficient for some ML models, especially simpler ones. However, complex models may require significant computational power during inference.
- *Computational Resources*: ML models, particularly deep learning models, often benefit from specialized hardware such as GPUs and TPUs to accelerate training and inference processes.

- *Inference Complexity*: Inference in NML may involve additional computations related to the explicit representation of indeterminacy and neutrality. The complexity can depend on the specific NML algorithm and its adaptation for handling neutrosophic sets.
- *Computational Resources*: Similar to ML, NML models may benefit from specialized hardware, especially when dealing with large-scale and computationally intensive applications. The handling of neutrosophic sets may have additional computational requirements.

5.4.9 Applications

- *Classification and Prediction*: ML is widely used for classification-related tasks, such as image and speech recognition, natural language processing, and sentiment analysis. Predictive modeling is another common application, including financial predictions and weather forecasting.
- *Recommendation Systems*: ML powers recommendation systems, providing personalized suggestions in areas like online shopping, streaming services, and social media.
- *Clustering and Unsupervised Learning*: Unsupervised learning in ML is applied to clustering tasks, helping identify patterns and groups within data. This is useful in customer segmentation, anomaly detection, and market basket analysis.
- *Healthcare*: ML is employed in healthcare for disease diagnosis, medical imaging analysis, patient risk prediction, and personalized medicine.

- *Medical Diagnosis under Uncertainty*: NML is applied in medical diagnosis where uncertainties in patient data and test results need to be explicitly considered. It provides a structured framework for handling indeterminacy and vagueness.
- *Robotics*: In robotics, NML can be used for decision-making under uncertainty, allowing robots to make decisions when faced with incomplete or imprecise information.
- *Decision Support Systems*: NML is employed in decision support systems where decisions are made in environment with varying degrees of uncertainty. It can assist in areas like risk assessment, financial planning, and strategic decision-making.
- *Subjective Decision-Making*: NML is particularly useful in applications involving subjective assessments and human judgments, where explicit representation of indeterminacy and neutrality enhances the decision-making process.

5.4.10 Efficiency

- *Training Efficiency*: The efficiency of ML models during training depends on factors such as the model architecture, optimization algorithms, and the size and quality of the training dataset. Training complex models can be computationally intensive.

- *Training Efficiency*: NML introduces additional complexity in the form of neutrosophic sets, which may impact the efficiency of training algorithms. The handling of indeterminacy and neutrality components can contribute to computational costs.

- *Inference Efficiency*: Inference efficiency varies across ML models. Simple models, such as linear regression, are generally computationally efficient during inference. However, more complex models, like deep neural networks, may require more computational resources.
- *Scalability*: Scalability of ML models depends on the algorithm and the ability to handle larger datasets and computing resources. Deep learning models often benefit from parallel processing using GPUs or TPUs.
- *Real-time Applications*: ML models may face challenges in real-time applications where low latency is crucial. The computational demands of certain models may limit their suitability for real-time deployment.

- *Inference Efficiency*: Inference in NML may involve additional computations related to the explicit representation of neutrosophic sets. The efficiency can depend on the specific NML algorithm and its adaptation for handling neutrosophic representations.
- *Scalability*: Scalability in NML may be influenced by the computational cost associated with neutrosophic set representations. The ability to scale efficiently depends on the algorithmic optimizations and parallelization techniques.
- *Real-time Applications*: NML may face challenges in real-time applications where low latency is crucial. The additional complexity introduced by neutrosophic sets may impact the suitability for real-time deployment.

5.4.11 Limitations

- *Sensitivity to Data Quality*: ML models can be sensitive to the quality of training data. Biases, inaccuracies, or outliers in the data may lead to biased or inaccurate model predictions.
- *Lack of Explainability*: Complex ML models, especially deep neural networks, often lack interpretability and explainability. Understanding the reasoning behind specific predictions can be challenging.
- *Need for Large Datasets*: Deep learning models, in particular, often require large amounts of labeled data for training. Obtaining and annotating such datasets can be resource-intensive.
- *Overfitting and Generalization*: ML models may suffer from overfitting, especially when the model is too complex and learns noise in the training data. Ensuring good generalization to unseen data is a challenge.

- *Complexity in Representation*: The explicit representation of neutrosophic sets introduces additional complexity, which can make the learning algorithms and models in NML more intricate.
- *Computational Cost*: Handling neutrosophic sets may lead to increased computational costs in terms of both training and inference, particularly when compared to simpler ML models.
- *Limited Adoption*: NML is a specialized field, and its adoption is not as widespread as traditional ML. This limitation is partly due to the relatively recent development of NML and the need for specific expertise.
- *Interpretability Challenges*: While NML models provide a more structured representation of uncertainties, interpreting and explaining decisions involving neutrosophic sets may still pose challenges, especially to non-experts.

REFERENCES

[1] Elhassouny, A., Idbrahim, S., & Smarandache, F. (2019). *Machine learning in Neutrosophic Environment: A Survey*. Infinite Study.

[2] Zafar, A., & Wajid, M. A. (2019). *Neutrosophic cognitive maps for situation analysis*. Infinite Study.

[3] Wajid, M. A., & Zafar, A. (2021). Pestel analysis to identify key barriers to smart cities development in india. *Neutrosophic Sets and Systems*, *42*, 39–48.

[4] Ahmed, R., Nasiri, F., & Zayed, T. (2021). A novel neutrosophic-based machine learning approach for maintenance prioritization in healthcare facilities. *Journal of Building Engineering*, 42, 102480.

[5] Osisanwo, F. Y., Akinsola, J. E. T., Awodele, O., Hinmikaiye, J. O., Olakanmi, O., & Akinjobi, J. (2017). Supervised machine learning algorithms: classification and comparison. *International Journal of Computer Trends and Technology (IJCTT)*, 48(3), 128–138.

[6] Wiering, M. A., & Van Otterlo, M. (2012). Reinforcement learning. *Adaptation, learning, and optimization*, 12(3), 729.

[7] Wajid, M. A., & Zafar, A. (2021). Multimodal Fusion: A Review, Taxonomy, Open Challenges, Research Roadmap and Future Directions. *Neutrosophic Sets and Systems*, 45(1), 8.

[8] Kelleher, J. D. (2019). *Deep learning*. MIT press.

[9] Zafar, A., & Wajid, M. A. (2020). *A Mathematical Model to Analyze the Role of Uncertain and Indeterminate Factors in the Spread of Pandemics like COVID-19 Using Neutrosophy: A Case Study of India* (Vol. 38). Infinite Study.

[10] Kamilaris, A., & Prenafeta-Boldú, F. X. (2018). Deep learning in agriculture: A survey. *Computers and electronics in agriculture*, 147, 70–90.

[11] Wajid, M. A., & Zafar, A. (2019, July). Multimodal Information Access and Retrieval Notable Work and Milestones. In *2019 10th International Conference on Computing, Communication and Networking Technologies (ICCCNT)* (pp. 1–6). IEEE.

[12] Buduma, N., Buduma, N., & Papa, J. (2022). *Fundamentals of deep learning*. O'Reilly Media, Inc..

[13] Hegde, J., & Rokseth, B. (2020). Applications of machine learning methods for engineering risk assessment–A review. *Safety science*, 122, 104492.

[14] Malik, M. M. (2020). A hierarchy of limitations in machine learning. arXiv preprint arXiv:2002.05193.

[15] Kahloot, K. M., & Ekler, P. (2021). Algorithmic splitting: A method for dataset preparation. *IEEE Access*, 9, 125229–125237.

[16] Wajid, Mohd A. , Zafar, Aasim, Neutrosophic Image Segmentation: An Approach for the Treatment of Uncertainty in Multimodal Information Systems, *International Journal of Neutrosophic Science*, Vol. 19 , No. 1 , (2022): 217–230 (DOI: https://doi.org/10.54216/IJNS.190117)

[17] Hammoud, R. (2023). *Mathematics Behind Machine Learning*, Electronic Theses, Projects, and Dissertations. 1778. https://scholarworks.lib.csusb.edu/etd/1778

[18] Dhanalakshmi, P. (2021). Linear algebra for machine learning. In *Artificial intelligence theory, models, and applications* (pp. 405–428). Auerbach Publications.

[19] Brownlee, J., Cristina, S., & Saeed, M. (2022). Calculus for machine learning. *Machine Learning Mastery*.
[20] Wajid, M. A., Zafar, A., Terashima-Marín, H., & Wajid, M. S. (2023). Neutrosophic-CNN-based image and text fusion for multimodal classification. *Journal of Intelligent & Fuzzy Systems*, *45*(1), 1039–1055.
[21] Marvasti, A. E., Marvasti, E. E., Bagci, U., & Foroosh, H. (2021). Maximum probability theorem: A framework for probabilistic machine learning. *IEEE Transactions on Artificial Intelligence*, 2(3), 214–227.
[22] Deisenroth, M. P., Faisal, A. A., & Ong, C. S. (2020). *Mathematics for machine learning*. Cambridge University Press.
[23] Wajid, M. A., Zafar, A., & Wajid, M. S. (2024). A deep learning approach for image and text classification using neutrosophy. *International Journal of Information Technology*, 16(2), 853–859.
[24] Ij, H. (2018). Statistics versus machine learning. *Nat Methods*, 15(4), 233.
[25] Alshikho, M., Jdid, M., & Broumi, S. (2023). Artificial Intelligence and Neutrosophic Machine learning in the Diagnosis and Detection of COVID 19. *Journal Prospects for Applied Mathematics and Data Analysis*, 1(2).
[26] Wajid, M. A., & Zafar, A. (2022). A multimodal approach of information access and retrieval using neutrosophic sets. In *Emerging Trends in IoT and Computing Technologies* (pp. 382–388). Routledge.
[27] Smarandache, F., & Jdid, M. (2023). A*n Overview of Neutrosophic and Plithogenic Theories and Applications*, American Science Publishing Group.
[28] El-Hefenawy, N., Metwally, M. A., Ahmed, Z. M., & El-Henawy, I. M. (2016). A review on the applications of neutrosophic sets. *Journal of Computational and Theoretical Nanoscience*, 13(1), 936–944.
[29] Sharma, M., Kandasamy, I., & Vasantha, W. B. (2021). Comparison of neutrosophic approach to various deep learning models for sentiment analysis. *Knowledge-Based Systems*, 223, 107058.
[30] Computing, N., & Learning, M. (2018). Neutrosophic Computing and Machine Learning. *Revista Asociación Latinoamericana de Ciencias Neutrosóficas*. ISSN 2574-1101, 3(3), 01–78.
[31] Mallik, S., Mohanty, S., & Mishra, B. S. (2023, November). Comparison of neutrosophic logic approach to various deep learning models for predictive analysis in recommender system. In *AIP Conference Proceedings* (Vol. 2878, No. 1). AIP Publishing.
[32] Wajid, M. A., Zafar, A., Bhushan, B., Khanday, A. M. U. D., & Wajid, M. S. (2023). Artificial Intelligence (AI) and Internet of Things (IoT): Application in Detecting and Containing the Spread of COVID-19. In *AI Models for Blockchain-Based Intelligent Networks in IoT Systems: Concepts, Methodologies, Tools, and Applications* (pp. 373–392). Cham: Springer International Publishing.
[33] Sharma, M., Kandasamy, I., & Vasantha, W. B. (2023). Emotion quantification and classification using the neutrosophic approach to deep learning. *Applied Soft Computing*, 148, 110896.

[34] Muhammad Yunus, M.N., Wan Ibrahim, W.M., Mahadi, A.A., Zainalabidin, M.A., Aminuddin, M.F., Pratama, E., Abdul Rahman, M.R., & Yusoff, I. (2022). Machine Learning Vs. *Statistical Models Approach: A Case for Probabilistic AFE Cost in Drilling*. Day 3 Wed, November 02, 2022.
[35] Alshikho, M., Jdid, M., & Broumi, S. (2023). Artificial Intelligence and Neutrosophic Machine learning in the Diagnosis and Detection of COVID 19. *Prospects for Applied Mathematics and Data Analysis*.

Chapter 6

A case study on the use of the fuzzy MOORA approach in university selection

Brajamohan Sahoo and Bijoy Krishna Debnath

6.1 INTRODUCTION

A university is a center for higher learning that offers an extensive selection of academic programs and degrees in a wide variety of subject areas. Universities often consist of several faculties, schools, or colleges that specialize in particular fields, including the humanities and arts, the sciences, the social sciences, engineering, business, law, and medicine.

Universities play a crucial role in the educational sector since they not only offer formal education but also contribute to the overall growth of people, society, and the world at large through research, creativity, critical thinking, and the sharing of knowledge. One of the most important turning points in one's academic development and personal growth is typically considered to be attending university. It provides a platform for people to develop not just specific knowledge but also fundamental abilities, widen their horizons, and create their identities.

Several research questions have emerged in this study, as outlined below

RQ1: What are the main challenges in selecting the right university?
RQ2: What connections exist between these difficulties?
RQ3: Why did we choose the specific fuzzy MOORA approach for this case study?
RQ4: How is this research helpful in selecting a good university?

The subsequent research objectives (RO) will help this book chapter address the aforementioned research questions.

RO1: Identify the key challenges to get a good university.
RO2: Establish the connection between these challenges.
RO3: List the benefits of applying this particular fuzzy MOORA method.
RO4: Describe the implications of this study.

Choosing the correct institution is an important choice that will have a significant impact on a person's academic trajectory and future employment

DOI: 10.1201/9781003606055-6

opportunities. It is crucial to select the right university for your studies. First off, a university's standing and standard of instruction have a big impact on the breadth of knowledge and abilities that can be learned. Access to elite professors, state-of-the-art research facilities, and a diverse student body are all benefits of attending a respected university that promotes learning and personal development. Furthermore, the ideal institution supports a student's academic interests by providing a variety of majors, extracurricular activities, and programs that meet their ambitions and professional objectives. Choosing the correct university also improves networking chances because it links students with industry professionals, alumni networks, and like-minded individuals. These connections can lead to future partnerships, internships, and job placements. Selecting the right university ultimately creates the conditions for a happy learning environment and long-term success in both personal and professional pursuits.

University quality is determined by factors including faculty strength, research funding, job openings, student employment ability, and research publications in higher education. They promote expansion while demonstrating excellence and positive student outcomes. High student-teacher ratios promote learning, researchers raise status, and faculty mold brains. Research funding is what fuels invention, technological leadership, and solutions. Vacancies impact access; balance is important. Employability demands knowledge of the industry and connections therein, enhancing reputation and alumni networks. Quality, worldwide standing, multidisciplinary ties, and funding are measured through research publications. These elements create a diverse perspective at the university, defining identity and influence. Together, they form a holistic ecosystem for learning, research, relevance, and student development. Successful universities embrace these components, producing graduates who are adaptive. The phenomena of education in the future are shaped by these viewpoints.

The key features of our contribution to this manuscript are outlined as follows:

This study identifies the essential criteria for the optimal university selection process.

To convey the uncertainty associated with the university selection process MOORA methodology has been used within a fuzzy environment. The normalized weight is employed by decision-makers to determine the weights normalized decision matrix under a fuzzy environment. In this chapter, the vertex method is employed to determine the overall performance index (Si) for each alternative and rank them according to their performance index value.

To accomplish the goals, the remainder of this chapter is organized as follows: Section 6.2 conducts a literature review. Section 6.3 delves into the details of the fuzzy MOORA method. The case study and result discussion are outlined in Section 6.4. Section 6.5 provides the sensitivity analysis. Section 6.6 addresses the implications of this study. Finally, Section 6.7 represents the conclusion of this chapter.

6.2 LITERATURE REVIEW

The choice of the ideal university has become a crucial one for students in today's quickly changing educational environment. Securing a good education, distinguished teachers, varied learning environments, and plenty of chances for both professional and personal development are all guarantees of choosing the right university. It supports cognitive growth, improves employability possibilities, and is in line with each student's academic objectives. In addition, a good institution fosters networking opportunities, critical thinking, and a friendly environment. By selecting the right university, students can establish the groundwork for a successful future, follow their passions, and realize their full potential. Multi-Criteria Decision-Making (MCDM) approaches have become important tools to help students make decisions that are in line with their academic and personal goals as a result of the complexity of this decision. This paper examines the use of various MCDM techniques for choosing university candidates, providing insights into their efficacy and potential drawbacks. Numerous researchers have delved into various MCDM methods for selecting prestigious institutions such as IIT and NIT in the past. They conducted research aimed at ranking top Indian engineering and technological institutions, notably including all seven older IITs. Notably, these studies consistently found that the older IITs ranked highest on the list, as evidenced by the works of Prathap and Gupta [1], Nishy et al. [2], and Prathap [3, 4]. Biswas, for instance, applied MCDM techniques in the selection process of an IIT [5], while Gardashova utilized the Z-TOPSIS method for university selection [6]. Ighravwe and colleagues employed an MCDM framework to select a simulation tool for renewable energy systems, intended for educational purposes at the university level [7]. Furthermore, Prabadevi and colleagues assessed Asian Higher Education Institutes through the application of an NLP-Based Text Analysis Approach [8] and Das assessed the effectiveness of ITI using a combination of SOWIA-MOORA methodology [9]. Viramontes explored criteria for university math teacher selection [10], while Biswas's innovative MCDM technique was applied to choose commercially available scooters [11]. Additionally, Mugambi and Omuya employed the DEA method to examine how recruitment and selection practices affect employee performance in public universities located in Meru County, Kenya [12]. Tyagi assessed the effectiveness of the performance of 19 academic departments within the Indian Institute of Technology (IIT) located in Roorkee [13]. Sawheny examined the factors motivating international students' intentions and their university selection [14], In contrast, Quigley et al. concentrated on how college students saw graduate employer selection exams [15]. In alignment with this body of research, Malik et al. contributed an article discussing the application of MCDM in education [16].

In this book chapter, we use the fuzzy MOORA method; previously numerous researchers have applied the MOORA technique across diverse fields. For instance, Chakraborty et al. conducted a narrative analysis of MOORA's decision-making approach [17]. Karande utilized the fuzzy MOORA method for selecting Enterprise Resource Planning systems [18]. Additionally, Arslankaya et al. employed fuzzy MOORA and fuzzy Analytic Hierarchy Process techniques to identify environmentally friendly suppliers within the steel door sector [19]. Stanujkic expanded the utilization of the MOORA technique to address decision-making challenges that involve interval data, employing the Extension of the Ratio System Part [20].

Furthermore, Singh et al. conducted an in-depth historical examination and analysis of MOORA and its fuzzy extensions, investigating their utilization in diverse fields [21]. These studies collectively demonstrate the versatility and effectiveness of the MOORA technique in addressing decision-making challenges in different contexts.

6.3 METHODOLOGY

6.3.1 Fuzzy MOORA method

In this chapter, the MCDM model for selecting the best university for graduate studies is being developed using the fuzzy MOORA approach. MCDM uses a modified version of the MOORA technique known as fuzzy (Multi-Objective Optimization on the basis of Ratio Analysis) MOORA. To address uncertainty and ambiguity in decision-making processes, it uses the fuzzy logic notion. Fuzzy MOORA assigns linguistic labels (such as "low, medium, and high") to indicate the importance of certain criteria and the effectiveness of various options. When evaluating alternatives in the face of ambiguous or lacking information, this enables a more flexible and nuanced judgment. The process entails transforming verbal expressions into fuzzy numbers, computing weighted performance scores, and arranging alternatives according to the sum of their fuzzy scores. By taking into consideration the inherent ambiguity that exists in real-world decision-making scenarios, fuzzy MOORA improves on the conventional MOORA approach.

The steps are as follows:

Step 1 The triangle membership function is used to determine each criterion value after creating the fuzzy decision matrix [18].

$$\begin{bmatrix} x_{11}^{l}, x_{11}^{m}, x_{11}^{n} & x_{12}^{l}, x_{12}^{m}, x_{12}^{n} & \cdots & x_{1n}^{l}, x_{12}^{m}, x_{12}^{n} \\ x_{21}^{l}, x_{21}^{m}, x_{21}^{n} & x_{22}^{l}, x_{22}^{m}, x_{22}^{n} & \cdots & x_{22}^{l}, x_{22}^{m}, x_{22}^{n} \\ \cdots & \cdots & \cdots & \cdots \\ x_{m1}^{l}, x_{m1}^{m}, x_{m1}^{n} & x_{m2}^{l}, x_{m2}^{m}, x_{m2}^{n} & \cdots & x_{mn}^{l}, x_{mn}^{m}, x_{mn}^{n} \end{bmatrix} \tag{6.1}$$

Where according to the ith criterion $x_{ij}^{l}, x_{ij}^{m}, x_{ij}^{n}$ stand for the bottom, middle, and top values of a triangular membership function for i, respectively.

Step 2 The vector normalization approach is then used to normalize the fuzzy decision matrix [20].

$$r_{ij}^{l} = \frac{x_{ij}^{l}}{\sqrt{\sum_{i=1}^{m} \left(x_{ij}^{l}\right)^2 + \left(x_{ij}^{m}\right)^2 + \left(x_{ij}^{n}\right)^2}} \tag{6.2}$$

$$r_{ij}^{m} = \frac{x_{ij}^{m}}{\sqrt{\sum_{i=1}^{m} \left(x_{ij}^{l}\right)^2 + \left(x_{ij}^{m}\right)^2 + \left(x_{ij}^{n}\right)^2}} \tag{6.3}$$

$$r_{ij}^{n} = \frac{x_{ij}^{n}}{\sqrt{\sum_{i=1}^{m} \left(x_{ij}^{l}\right)^2 + \left(x_{ij}^{m}\right)^2 + \left(x_{ij}^{n}\right)^2}} \tag{6.4}$$

Step 3 The weighted normalized fuzzy decision matrix is then calculated using the following formula [18]

$$v_{ij}^{l} = w_j \, r_{ij}^{l} \tag{6.5}$$

$$v_{ij}^{m} = w_j \, r_{ij}^{m} \tag{6.6}$$

$$v_{ij}^{n} = w_j \, r_{ij}^{n} \tag{6.7}$$

Here w_j is defined as the normalized weight attribute.
Fuzzy criteria weights can also be employed to create the weighted normalized fuzzy decision matrix.

Step 4 Then, for each choice, we calculate the overall ratings for both beneficial and non-beneficial categories. The triangular membership function's lowest, medium, and upper values are used to calculate an option's overall ratings for the favorable criteria [18].

To determine the total scores of an option based on beneficial criteria, the following procedure is followed:

$$s_i^{+l} = \sum_{j=1}^{n} v_{ij}^{l} \mid j \in J^{\max} \tag{6.8}$$

$$s_i^{+m} = \sum_{j=1}^{n} v_{ij}^{m} \mid j \in J^{\max} \tag{6.9}$$

$$s_i^{+n} = \sum_{j=1}^{n} v_{ij}^{n} \mid j \in J^{\max} \tag{6.10}$$

For non-beneficial criteria, the total scores of an option are determined as follows:

$$s_i^{-l} = \sum_{j=1}^{n} v_{ij}^{l} \mid j \in J^{\min} \tag{6.11}$$

$$s_i^{-m} = \sum_{j=1}^{n} v_{ij}^{m} \mid j \in J^{\min} \tag{6.12}$$

$$s_i^{-n} = \sum_{j=1}^{n} v_{ij}^{n} \mid j \in J^{\min} \tag{6.13}$$

Step 5 The final step is to determine the overall performance index (Si) for each alternative [18, 22, 23].

$$S_i\left(s_i^{+}, s_i^{-}\right) = \sqrt{\frac{1}{3}\left[\left(s_i^{+l} - s_i^{-l}\right)^2 + \left(s_i^{+m} - s_i^{-m}\right)^2 + \left(s_i^{+n} - s_i^{-n}\right)^2\right]} \tag{6.14}$$

Step 6 Based on the values of their overall performance indicators, the alternatives are arranged in descending order; the option with the greatest overall performance index is the best option and the most beneficial option [24].

6.4 CASE STUDY AND RESULTS DISCUSSION

In this chapter of the book, we have employed four decision-makers M_1, M_2, M_3, and M_4 opinion including five alternatives labeled A_1, A_2, A_3, and A_4, A_5, and the evaluation utilizes seven criteria as outlined below.

University fees: University fees are important because they have a direct impact on affordability and financial security for many students. When evaluating their alternatives, students usually consider living expenditures, tuition costs, and the availability of grants or funding. This comprehensive assessment is essential for making informed decisions about their educational journey.

Faculty strength: The competency and expertise of the instructor have a big influence on the students' learning outcomes. To assess the caliber of education provided by an institution, prospective students would find it helpful to look at the research backgrounds, teaching strategies, student-to-faculty ratios, and faculty qualifications.

Research funds: This is an important factor for many students because it has a direct impact on affordability and long-term financial stability. When evaluating, students frequently take into account living costs, tuition costs, and any possible financial assistance or scholarships. Research opportunities are quite important, particularly for those who are seeking postgraduate degrees or want to work in academia. Academic experience can be further enhanced by the abundance of research projects, faculty cooperation possibilities, and internships that universities with significant funding for research offer.

Vacant seat: A prerequisite for admission into any academic institution is the availability of seats in selected programs or courses. Prospective students assess their likelihood of acceptance based on various factors, including acceptance rates, competitiveness, and admission requirements, as they strive to secure a place in their desired university or program.

Student placement: Many students consider a university's reputation for providing opportunities for student job placements as a crucial factor in their decision-making process. Career services, internships, alumni networks, and other similar initiatives play vital roles in helping students transition smoothly from college to the professional world. These resources not only enhance students' academic experiences but also equip them with practical skills and connections necessary for success in their chosen careers.

Research publication: The university's scholarly reputation and contributions to many subjects can be inferred from its research output and publication history. Universities with a strong research culture and publishing history may be given preference by students interested in academic or career-oriented research.

Higher study: Once their undergraduate studies are completed, certain students may seek to pursue graduate or professional programs or further their higher education. In such cases, universities with strong connections to professional programs, graduate schools, and renowned academic institutions may be particularly appealing. These institutions offer opportunities for continued academic advancement and specialization, aligning well with the career aspirations and academic goals of ambitious students.

Figure 6.1 shows that to select the best university the decision-maker entails the evaluation of seven criteria, each having five alternative alternatives.

In applying the fuzzy MOORA approach to address the identical problem, we start by normalizing the fuzzy decision matrix found in Table 6.2. This matrix contains criteria values represented in triangular fuzzy numbers, as displayed in Table 6.1, and is normalized using equations (6.2)–(6.4).

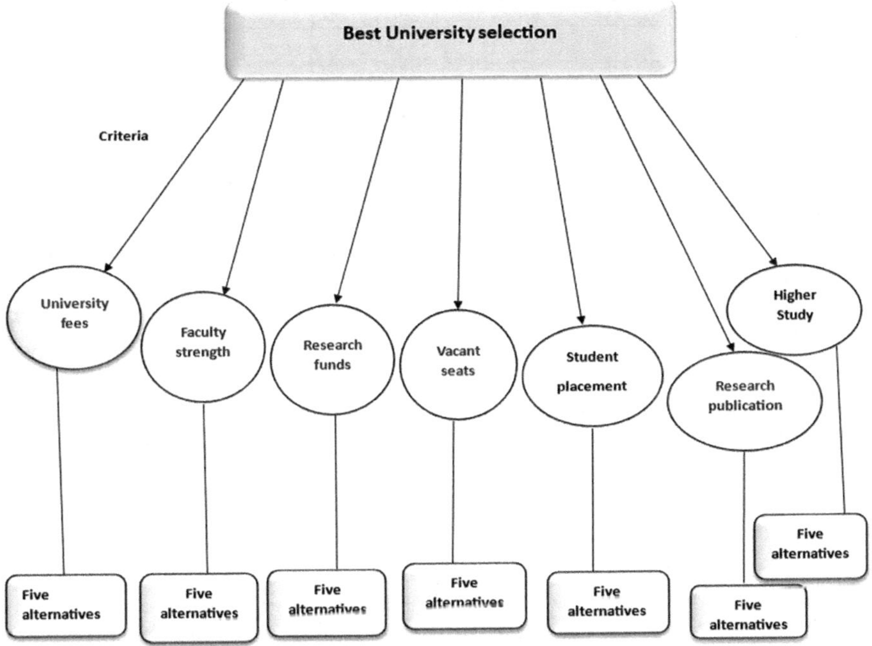

Figure 6.1 Best university selection.

Table 6.1 Linguistic terms and the corresponding triangular fuzzy numbers

Linguistic variable	*Triangular fuzzy number*
Very low	(1,2,2)
Low	(2,3,4)
Moderate low	(3,4,5)
Moderate	(4,4,6)
Moderately high	(5,6,6)
High	(6,7,8)
Very high	(8,8,9)

Table 6.2 Fuzzy decision matrix

Alternative	*U.F*	*F.S*	*R.F*	*V.S*	*S.E*	*R.P*	*H.S*
A_1	(3,4,5)	(6,7,8)	(8,8,9)	(2,3,4)	(8,8,9)	(6,7,8)	(5,6,6)
A_2	(2,3,4)	(8,8,9)	(6,7,8)	(1,2,2)	(8,8,9)	(6,7,8)	(8,8,9)
A_3	(5,6,6)	(3,4,5)	(6,7,8)	(2,3,4)	(5,6,6)	(8,8,9)	(3,4,5)
A_4	(5,6,6)	(2,3,4)	(3,4,5)	(3,4,5)	(6,7,8)	(2,3,4)	(8,8,9)
A_5	(2,3,4)	(5,6,6)	(3,4,5)	(1,2,2)	(6,7,8)	(8,8,9)	(5,6,6)

This initial step ensures that the components of the decision matrix are dimensionless and comparable, as stipulated by an MCDM technique. The resulting normalized fuzzy decision matrix is illustrated in Table 6.3.

Next, the normalized fuzzy criteria values are multiplied by the corresponding crisp criteria weights, as determined by equations (6.5)–(6.7), this process yields the weighted normalized fuzzy decision matrix, which is presented in Table 6.4.

Subsequently, employing equations (6.8)–(6.10), the overall ratings of the beneficial criteria for the choices under scrutiny are computed. Conversely, equations (6.11)–(6.13) facilitate the calculation of the overall ratings of the non-beneficial criteria for the alternatives. The culmination of these computations is the presentation of the total ratings for both favorable and unfavorable traits associated with the four distinct firm selection methods, as illustrated in Table 6.5.

To ascertain the values of the overall performance index for all choices, we employ the vertex technique, utilizing equations (6.14) to defuzzify the overall ratings for both beneficial and non-beneficial criteria.

Addressing the challenge at hand involves reducing university fees and decreasing the number of vacant seats, while simultaneously requiring an increase in faculty, research funds, student placement opportunities, and research publications, as well as support for higher education studies. Consequently, decision-makers classify university fees and vacant seats as non-beneficial, while considering the rest to be beneficial across all scenarios. The highest value of the overall performance index is recorded for alternative 2.

In this case study, decision-makers use the following fuzzy weight attribute of the decision criterion

$$w_{uf} = 0.292, w_{fs} = 0.210, w_{rf} = 0.180, w_{vs} = 0.125, w_{sp} = 0.108,$$
$$w_{rp} = 0.050, w_{hs} = 0.035.$$

The decision-makers compare the criteria or alternatives using the linguistic terminology presented in Table 6.1.

Table 6.3 Normalized fuzzy decision matrix

Alt.	*U.F*	*F.S*	*R.F*	*V.S*	*S.P*	*R.P*	*H.S*
A1	(0.173, 0.230, 0.288)	(0.260, 0.303, 0.346)	(0.325, 0.325, 0.365)	(0.178, 0.267, 0.356)	(0.280, 0.280, 0.316)	(0.219, 0.256, 0.293)	(0.194, 0.233, 0.233)
A2	(0.115, 0.173, 0.230)	(0.346, 0.346, 0.389)	(0.244, 0.284, 0.324)	(0.089, 0.178, 0.178)	(0.280, 0.280, 0.316)	(0.219, 0.256, 0.293)	(0.310, 0.310, 0.350)
A3	(0.288, 0.345, 0.345)	(0.130, 0.173, 0.216)	(0.244, 0.284, 0.324)	(0.178, 0.267, 0.356)	(0.175, 0.210, 0.210)	(0.293, 0.293, 0.329)	(0.117, 0.155, 0.194)
A4	(0.288, 0.345, 0.345)	(0.087, 0.130, 0.173)	(0.122, 0.162, 0.203)	(0.267, 0.356, 0.445)	(0.210, 0.245, 0.280)	(0.073, 0.109, 0.147)	(0.310, 0.310, 0.350)
A5	(0.115, 0.173, 0.230)	(0.216, 0.260, 0.260)	(0.122, 0.162, 0.203)	(0.089, 0.178, 0.178)	(0.210, 0.245, 0.280)	(0.293, 0.293, 0.329)	(0.194, 0.233, 0.233)

Table 6.4 Weighted normalized fuzzy decision matrix

Alt.	*U.F*	*F.S*	*R.F*	*V.S*	*S.P*	*R.P*	*H.S*
A_1	(0.050, 0.067, 0.084)	(0.054, 0.063, 0.072)	(0.058, 0.058, 0.066)	(0.022, 0.033, 0.044)	(0.030, 0.030, 0.034)	(0.010, 0.012, 0.014)	(0.007, 0.008, 0.008)
A_2	(0.034, 0.050, 0.067)	(0.072, 0.072, 0.082)	(0.044, 0.051, 0.058)	(0.011, 0.022, 0.022)	(0.030, 0.030, 0.034)	(0.010, 0.012, 0.014)	(0.010, 0.010, 0.012)
A_3	(0.084, 0.100, 0.100)	(0.0273, 0.036, 0.045)	(0.044, 0.051, 0.058)	(0.022, 0.033, 0.044)	(0.018, 0.023, 0.023)	(0.014,0 .014, 0.016)	(0.004, 0.005, 0.007)
A_4	(0.084, 0.100, 0.100)	(0.018, 0.027, 0.036)	(0.022, 0.029, 0.037)	(0.033, 0.044, 0.055)	(0.022, 0.026, 0.030)	(0.003, 0.005, 0.007)	(0.010, 0.010, 0.012)
A_5	(0.034, 0.050, 0.067)	(0.045, 0.054, 0.054)	(0.022, 0.029, 0.037)	(0.011, 0.022, 0.022)	(0.022, 0.026, 0.030)	(0.014, 0.014, 0.016)	(0.007, 0.008, 0.008)

Table 6.5 Ranking of the alternatives

	S^+			S^-			S	
Alternative	*l*	*m*	*n*	*l*	*m*	*n*	*Crisp value*	*Rank*
A_1	0.159	0.171	0.194	0.072	0.100	0.128	0.075	2
A_2	0.166	0.175	0.200	0.045	0.072	0.089	0.111	1
A_3	0.107	0.129	0.149	0.106	0.133	0.144	0.003	5
A_4	0.075	0.097	0.122	0.117	0.144	0.155	0.041	4
A_5	0.110	0.131	0.145	0.045	0.072	0.089	0.060	3

According to the data presented in Table 6.5, it is evident that A_2 University stands out as the most preferred option for furthering one's education.

Note: U.F stands for university fees, F.S stands for Faculty strength, R.F stands for Research funds, V.S stands for Vacant seat, S.P stands for Student placement, R.P stands for Research publication, and H.S stands for Higher study.

6.5 SENSITIVITY ANALYSIS

Sensitivity analysis must be conducted to verify the accuracy of the data being utilized. This analysis proves valuable in scenarios where models need validation. The method aids in detecting even subtle variations in the importance of criteria that could potentially impact the ranking of options. Decision-makers can make more informed choices and reach better conclusions when they understand the significance or sensitivity of each consideration while ranking possibilities.

After normalization, the values of the alternatives are as follows: A_1 = 0.25861, A_2 = 0.382759, A_3 = 0.010345, A_4 = 0.141379, A_5 = 0.206897. These values serve as the weights of the alternatives (Figure 6.2). Table 6.6 displays the values for each step of sensitivity analysis, along with a graphical and radar diagram depicting the analysis. It is observed from the 10th iteration, where the value is 0.33, that the ranking order of the alternatives remained unchanged. Furthermore, following the 10th step, alternative A_2 continued to maintain its top position in the ranking.

Sensitivity analysis thus plays a crucial role in ensuring the robustness and reliability of decision-making processes, particularly in complex situations such as university selection where multiple factors influence the final outcome (Figure 6.3).

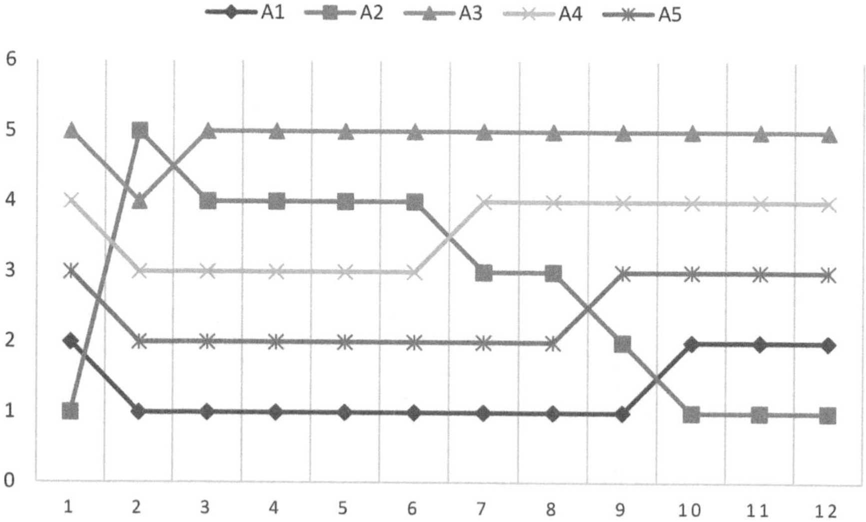

Figure 6.2 Graphical representation of sensitivity analysis.

Table 6.6 Sensitivity analysis matrix

Alt.	*Original*	*0.01*	*0.05*	*0.09*	*0.13*	*0.17*	*0.21*	*0.25*	*0.29*	*0.33*	*0.37*	*0.41*
A1	0.258621	0.414804	0.398045	0.381285	0.364525	0.347765	0.331006	0.314246	0.297486	0.280726	0.263966	0.247207
A2	0.382759	0.01	0.05	0.09	0.13	0.17	0.21	0.25	0.29	0.33	0.37	0.41
A3	0.010345	0.016592	0.015922	0.015251	0.014581	0.013911	0.01324	0.01257	0.011899	0.011229	0.010559	0.009888
A4	0.141379	0.22676	0.217598	0.208436	0.199274	0.190112	0.18095	0.171788	0.162626	0.153464	0.144302	0.13514
A5	0.206897	0.331844	0.318436	0.305028	0.29162	0.278212	0.264804	0.251397	0.237989	0.224581	0.211173	0.197765
Total	1	1	1	1	1	1	1	1	1	1	1	1

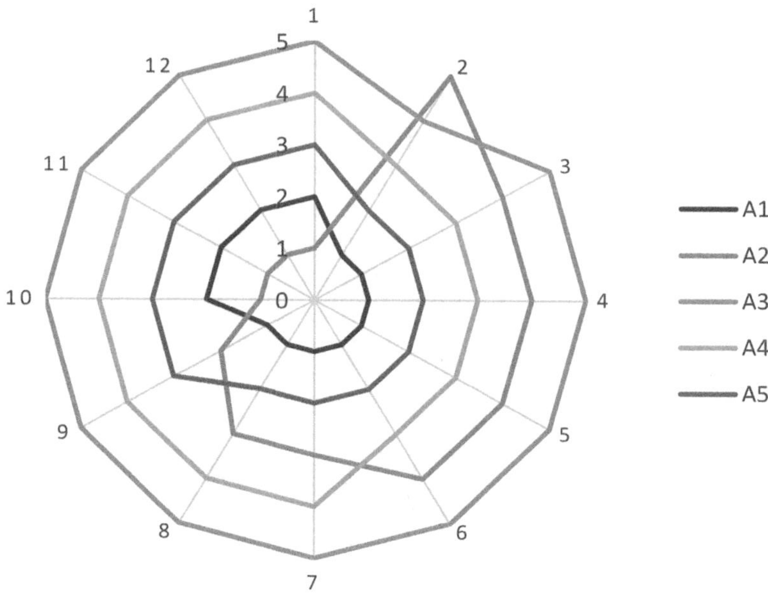

Figure 6.3 Radar diagram of sensitivity analysis.

6.6 IMPLICATION OF THIS STUDY

6.6.1 Educational implications

There are numerous educational implications from the case study on the fuzzy MOORA approach to university selection, particularly in the domains of operations research, decision science, and higher education administration. These are some major consequences for education.

Introduction to Decision-Making Techniques: Students are exposed to decision-making strategies that go beyond conventional approaches through the case study. Their knowledge of decision science is expanded as a result of being introduced to fuzzy logic and MCDM techniques like MOORA.

Understanding Fuzzy Logic: A key idea in making decisions in the face of ambiguity is fuzzy logic. Pupils can gain knowledge of how fuzzy logic enables the representation of imprecise and vague information, which is frequently encountered in real-world decision-making situations.

Application of MOORA: The case study provides students with insights into applying the MOORA approach to intricate decision challenges that involve numerous conflicting criteria. They grasp the procedure of setting objectives, defining criteria, allotting weights, and evaluating alternatives.

Interdisciplinary Learning: The case study serves as a connection between multiple academic areas, including mathematics, computer science, and management. It enables students to recognize the interdisciplinary aspect of decision science and how it is relevant across a wide range of fields.

Critical Thinking and Analysis: Interdisciplinary education encourages students to cultivate critical thinking abilities through the evaluation of both the advantages and constraints of the fuzzy MOORA technique. As they engage with this methodology, they gain the capacity to gauge its relevance across various scenarios and juxtapose it with alternative approaches to decision-making.

Practical Application in Higher Education Administration: The case study offers valuable perspectives on how decision-making techniques can be practically applied in the administration of higher education. It helps students grasp how universities can utilize quantitative approaches to refine their selection procedures and improve the outcomes of their decisions.

Ethical Considerations: The case study might prompt ethical reflections concerning the criteria used in university selection, as well as issues of fairness and transparency. It offers students the opportunity to examine ethical challenges and find a harmonious balance between quantitative analysis and ethical values in the decision-making process.

Optimization of Resources: Students can make the most of their educational experience and resources by choosing universities that best fit their preferences and ambitions. Universities can also draw in students who best fit their curricula and culture, which increases student happiness and retention.

Research Opportunities: The case study has the potential to motivate students to delve into research avenues within decision science, optimization, and the management of higher education. They have the opportunity to explore various possibilities, such as extending the methods used, making modifications, or conducting comparative analyses with alternative decision-making strategies.

6.6.2 Business and human resources implications

The case study's conclusions have important ramifications for companies and HR divisions. By using the fuzzy MOORA technique in the candidate screening process, recruiting decisions can become more accurate and well-informed, improving the caliber of the workforce as a whole. Organizations may increase employee retention rates, optimize recruitment procedures, and more effectively discover top talent by utilizing advanced decision-making methodologies. Adopting such cutting-edge strategies also shows a dedication to maintaining competitiveness in a changing market, making sure that companies maintain their adaptability and responsiveness to changing industry demands while making the best use of their human capital.

6.6.3 Technological implications

The research outcomes have the capacity to stimulate the advancement and extensive use of technological remedies that integrate fuzzy logic and MCDM techniques in various fields. Beyond college selection, these cutting-edge methods have the potential to transform industries like product design and make it possible to create more efficient and customized goods that satisfy the wants of customers. Furthermore, by taking into account ambiguous and inaccurate information, fuzzy logic integration in project management could improve decision-making processes and result in more precise resource allocation and planning. Moreover, the implementation of these methodologies could aid enterprises facing difficulties with resource management by optimizing resource allocation strategies for increased sustainability and efficiency.

6.6.4 Policy implications

Policymakers in the education and labor sectors hold a critical role in shaping the trajectory of higher education, workforce development, and recruitment practices. By recognizing the implications of employing advanced decision-making techniques such as fuzzy MOORA in university selection processes, policymakers can advocate for policies that foster fairness, transparency, and efficiency. These policies can not only enhance the quality of higher education but also contribute to the cultivation of a skilled and competitive workforce. Additionally, by incorporating such methodologies into policy frameworks, policymakers can stay abreast of technological advancements and global trends, ensuring that educational and labor policies remain relevant and effective in an ever-evolving landscape of academia and employment.

6.6.5 Globalization implications

In an age where universities and businesses are significantly impacted by global factors, adopting advanced decision-making methods like fuzzy MOORA becomes essential. Utilizing these sophisticated techniques allows organizations to enhance their competitiveness and adaptability in the ever-changing global environment. Fuzzy MOORA provides a nuanced strategy that addresses the intricacies and uncertainties present in international settings, enabling more informed and strategic decision-making processes. This increased level of sophistication not only improves operational effectiveness but also encourages innovation and resilience when confronted with various challenges. Ultimately, embracing such methodologies is crucial for organizations seeking success in a progressively interconnected and dynamic global economy.

To conclude, the case study delving into the application of the fuzzy MOORA approach for university selection provides significant educational value by shedding light on decision-making methods, fostering interdisciplinary understanding, illustrating real-world applications in higher education, and prompting ethical reflection. It prompts students to engage in critical thinking, tackle intricate challenges, and delve into research prospects within decision science and associated disciplines.

6.7 CONCLUSION

Considering the changing overall performance index values, A_2 University emerges as the top choice within the educational sector, as evidenced in Table 6.5. This chapter delineates a robust methodology for university selection processes by leveraging the fuzzy MOORA approach effectively. Importantly, this approach transcends academia; its applicability extends to diverse socioeconomic domains grappling with selection and ranking challenges. Stakeholders across various sectors stand to enhance efficiency and accountability by embracing such systematic frameworks, empowering them to make well-informed decisions.

Navigating intricate decision landscapes and fostering progress within our communities necessitates a steadfast commitment to employing rigorous methodologies like fuzzy MOORA. As we continuously refine and adapt these strategies, our collective capacity to address complex issues and optimize outcomes for societal benefit strengthens. This underscores the imperative of ongoing improvement and innovation in decision-making processes to drive meaningful advancement.

ACKNOWLEDGMENTS

The authors are grateful to the editors and reviewers for their invaluable feedback, which greatly improved this chapter.

REFERENCES

[1] Prathap G, Gupta BM. Ranking of Indian engineering and technological institutes for their research performance during 1999–2008. *Current Science*. 2009 Aug 10;97(3):304–6. https://www.jstor.org/sable/24111992

[2] Nishy P, Panwar Y, Prasad S, Mandal GK, Prathap G. An impact-citations-exergy (iCX) trajectory analysis of leading research institutions in India. *Scientometrics* 2012 Apr 1;91(1):245–51. https://doi.org/10.1007/s11192-011-0594-4

[3] Prathap G. Benchmarking research performance of the IITs using "Web of Science" and "Scopus" bibliometric databases. *Current Science* 2013 Oct 25:1134–8. https://www.jstor.org/stable/24098220
[4] Prathap G The performance of research-intensive higher educational institutions in India. *Current Science* 2014 Aug 10:389–96. https://www.jstor.org/stable/24103492
[5] Biswas TK, Chaki S, Das MC. MCDM technique application to the selection of an Indian institute of technology. *Operational Research in Engineering Sciences: Theory and Applications* 2019 Dec 16; 2(3):65–76. https://doi.org/10.31181/10.31181/oresta1903065b
[6] Gardashova LA. University Selection by Using Z-TOPSIS Methodology. In *World Conference Intelligent System for Industrial Automation* 2022 Nov 25 (pp. 11–21). Cham: Springer Nature Switzerland. http://doi.org/10.1007/978-3-031-51521-74
[7] Ighravwe DE, Babatunde MO, Mosetlhe TC, Aikhuele D, Akinyele D. A MCDM-based framework for the selection of renewable energy system simulation tool for teaching and learning at university level. *Environment, Development and Sustainability* 2022 Nov 1:1–22. https://doi.org/10.1007/s10668-021-01981-1
[8] Prabadevi B, Deepa N, Ganesan K, Srivastava G. A decision model for ranking Asian Higher Education Institutes using an NLP-based text analysis approach. *ACM Transactions on Asian and Low-Resource Language Information Processing* 2023 Mar 10;22(3):1–20. http://doi.org/10.1145/3534562
[9] Das MC, Sarkar B, Ray S. On the performance of Indian technical institutions: a combined SOWIA-MOORA approach. *Opsearch* 2013 Sep;50:319–33. http://doi.org/10.1007/s12597-012-0116-z
[10] Viramontes-Miranda JD, Sánchez Aguilar M. Exploring criteria for university mathematics teachers' selection of calculus textbooks. *Asian Journal for Mathematics Education* 2023 Dec;2(4):430–44. https://doi.org/10.1177/27527263231217575
[11] Biswas T, Saha P. Selection of commercially available scooters by new MCDM method. *International Journal of Data and Network Science* 2019;3(2):137–44. http://doi.org/10.5267/j.ijdns.2018.12.002
[12] Wajid MS, Wajid MA. (2021). The importance of indeterminate and unknown factors in nourishing crime: a case study of South Africa using neutrosophy. *Neutrosophic Sets and Systems*, 41(2021), 15.
[13] Tyagi P, Yadav SP, Singh SP. Relative performance of academic departments using DEA with sensitivity analysis. *Evaluation and Program Planning* 2009 May 1;32(2):168–77. https://doi.org/10.1016/j.evalprogplan.2008.10.002
[14] Sawheny Y Factors Motivating International Student Intentions and Selection of University: A Critical Analysis of a Private University in Thailand. *Eurasian Journal of Educational Research* 2023 Oct 18;105(105):256–76. http://doi.org/10.14689/ejer.2023.105.015
[15] Quigley M, Smith C, Stocker E, Bradley A. University students' perceptions of graduate employer selection tests. *Education Training* 2024 Feb 19;66(1):1–6. http://doi.org/10.1108/ET-05-2022-0188
[16] Wajid MS, Terashima-Marin H, Najafirad P., Pablos SEC, Wajid MA (2024). DTwin-TEC: An AI-based TEC district digital twin and emulating

security events by leveraging knowledge graph. *Journal of Open Innovation: Technology, Market, and Complexity*, 10(2), 100297.

[17] Chakraborty S, Datta HN, Kalita K, Chakraborty S. A narrative review of multi-objective optimization on the basis of ratio analysis (MOORA) method in decision making. *OPSEARCH* 2023 Dec;60(4):1844–87. http://doi.org/10.1007/s12597-023-00676-7

[18] Karande P, Chakraborty S. A Fuzzy-MOORA approach for ERP system selection. *Decision Science Letters* 2012;1(1):11–21. http://doi.org/10.5267/j.dsl.2012.07.001

[19] Arslankaya S, Çelik MT. Green supplier selection in steel door industry using fuzzy AHP and fuzzy Moora methods. *Emerging Materials Research* 2021 Jul 13;10(4):357–69. http://doi.org/10.1680/jemmr.21.00011

[20] Stanujkic D, Magdalinovic N, Stojanovic S, Jovanovic R. Extension of ratio system part of MOORA method for solving decision-making problems with interval data. *Informatica* 2012 Jan 1;23(1):141–54. http://doi.org/10.15388/Informatica.2012.353

[21] Singh R, Pathak VK, Kumar R, Dikshit M, Aherwar A, Singh V, Singh T. A historical review and analysis on MOORA and its fuzzy extensions for different applications. *Heliyon* 2024 Feb 1. http://doi.org/10.1016/j.heliyon.2024.e25453

[22] Huiqun H, Guang S. ERP software selection using the rough set and TPOSIS methods under fuzzy environment. *Advances in Information Sciences and Service Sciences* 2012; 4(3):111–8. http://doi.org/10.4156/AISS.vol4.issue3.15

[23] Wajid MA, Zafar A (2022). Neutrosophic image segmentation: An approach for the treatment of uncertainty in multimodal information systems. *International Journal of Neutrosophic Science (IJNS)*, 19(1), p. 217, doi. http://doi.org/10.54216/IJNS.190117

[24] Wajid MA, Zafar A (2022). A multimodal approach of information access and retrieval using neutrosophic sets. In *Emerging Trends in IoT and Computing Technologies* (pp. 382–8). Routledge.

Chapter 7

Multi-criteria feature selection evaluation for multi-label text classification

Hesitant fuzzy geometric approach

Mohsen Miri, Mohammad Bagher Dowlatshahi, and Amin Hashemi

7.1 INTRODUCTION

With the advancement of new technologies and the rise of social media, the amount of textual information is growing exponentially. Text classification has become an important task due to the growth of digital text data. The recognition of text data labels is predictable depending on the content (e.g., spam detection). Nowadays, not only the volume of text data has increased, but also its scope. The number of labels related to the content of the text grows with the dimension of the text (each text can be assigned to one or more different titles). For example, a document can have a sports and a business content at the same time [1–3]. Text classification requires feature selection (FS) because redundant and unrelated features in each data set either have no effect on the result or have a detrimental effect. FS is used as a dimension-reduction technique to eliminate these features. It can also reduce computational cost, complexity, learning time and, most importantly, improve classification accuracy [4–6].

FS is often divided into the filter, wrapper, and embedded approaches, where at least the relevant text features are selected to optimize the classification. The filtering methods are fast for high-dimensional data as each feature is evaluated using static experiments and key data. Instead, each feature is evaluated by a specific classifier when learning in the wrapper techniques. A wrapper approach uses a learning algorithm to select a subset of features with the best performance according to certain criteria. Finally, the embedded technique utilizes the advantages of the filter and wrapper methods in a complementary way [7, 8, 9].

Recently, many significant problems have arisen for researchers where decisions or choices are unclear or tentative, such as the choice of an FS for classification. In 1965, Zadeh [10] proposed fuzzy sets to solve these problems, which attracted much attention and were very popular among

DOI: 10.1201/9781003606055-7

researchers, and many extensions of them have been introduced until today. In these methods, the solutions are evaluated by an expert based on some criteria. In 2010, Torra [11] proposed an extension of fuzzy sets, which he called Hesitant Fuzzy Sets, in which each problem is evaluated based on different measures by different experts. The biggest advantage of Torra's method over others is that it allows the fuzzy and inconclusive problems to be evaluated by several experts except one. Thus, the researcher can consider and evaluate different aspects of uncertainty in an experiment to arrive at a good solution [12, 13].

In this chapter, the hesitant fuzzy approach is used for the first time to solve the problem of selecting MLFS – an appropriate FS method for MTC. Ten MLFS methods were run on 18 MTDs Based on six criteria for evaluating the methods according to the difference and nonlinearity of their relationship with each other, by which they can comprehensively investigate the methods. We modeled this procedure as a hesitant fuzzy multi-criteria decision process in which each MLFS is considered as an alternative. These methods are run on 18 MTDs and their results are evaluated by six performance experts using ten equivalent validations. In each dataset, the MLFS methods evaluate and sort based on Hesitant Fuzzy Geometric (HFG) calculations. Finally, the results of all datasets are aggregated using the order statistics strategy to produce the final ranking of methods.

This chapter consists of five sections, and after the introduction, Section 7.2 sets out the basic concepts required for the proposed method. Section 7.3 presents the proposed method in detail. The data required to perform the experiments and their results are described and explained in Section 7.4. And finally, we draw a general conclusion to conclude our chapter.

7.2 RELATED WORKS AND FUNDAMENTALS

7.2.1 Related works

In contrast to the importance of MLFS in MTC, many studies in the field of text classification use only one criterion or even sometimes have not used any criterion at all for evaluating different methods for selecting a preprocessor. However, the performance of MTC can be evaluated using three different sets of measures, which are not interchangeable. Since the strengths, weaknesses, and especially the structure of MLFS methods are different, choosing an MLFS is a tentative choice. Therefore, choosing an MLFS based on a specific criterion or two criteria separately may not perform well. For example, if you have good classification accuracy but need too much time to capture all labels. So which criterion is better? Isn't it better to find an MLFS that performs well based on a group of different criteria? Some of them are briefly introduced below [14].

Shawkat et al. [15] conducted a selection algorithm using a rule-based method. They evaluated eight algorithms based on the two criteria of accuracy and complexity. In one of the recent researches, Gang Kou et al. [16] presented a method for evaluating algorithms using MCDM methods. They performed this evaluation on ten small text datasets. Gang Kou et al. [17] also proposed an algorithm in 2012 to evaluate classification algorithms using a multi-criteria strategy by considering Spearman's correlation rank. In addition, Kashef et al. [18] presented a comprehensive review of MLFS in which some methods were evaluated using multi-label classification criteria. All these authors evaluated several methods separately according to specific criteria and finally selected one approach as the best. However, the overall performance of an MLFS for classification is not specified by one criterion.

7.2.2 Multi-label learning

The structures of the two categories of single-label and multi-label data used for machine learning are shown in Table 7.1. In multi-label datasets, two or more labels are present. Each instance in this data has a vector of features as well as labels as $X_i = (x_{i1}, x_{i2}, \ldots\ldots, x_{iM})$, and $Y_i = (y_{i1}, y_{i2}, \ldots\ldots, y_{iL})$. The letters M and L stand for the number of features and labels respectively. Each label in multi-label data has a binary value. In contrast, different values, such as real, nominal, ordinal or binary values, can be specified for single-label data. Multi-label learning creates a model that can predict the label for future samples based on the N data samples, sometimes referred to as training data [19, 20].

Table 7.1 Multi-label data structure

X_1	X_2	$\cdots$	X_M	Y
X_{11}	X_{12}	$\cdots$	X_{1M}	y_1
X_{21}	X_{22}	$\cdots$	X_{2M}	y_2
$\vdots$	$\vdots$	$\ddots$	$\vdots$	$\cdots$
X_{N1}	X_{N2}	$\cdots$	X_{NM}	y_N

X_1	X_1	$\cdots$	X_M	Y_1	Y_2	$\cdots$	Y_L
X_{11}	X_{12}	$\cdots$	X_{1M}	0	1	$\cdots$	0
X_{21}	X_{22}	$\cdots$	X_{2M}	1	0	$\cdots$	0
$\vdots$	$\vdots$	$\ddots$	$\vdots$	$\vdots$	$\vdots$	$\ddots$	$\vdots$
X_{N1}	X_{N2}	$\cdots$	X_{NM}	0	1	$\cdots$	1

7.2.3 Multi-label feature selection methods (MLFS)

FS stands out as one of the most crucial aspects of any classification task. In this paper, ten alternative MLFS algorithms were evaluated to determine the most suitable one for MTC. As illustrated in Table 7.2, eight of the selected methods belong to the category of filter-based approaches. Additionally, within the embedded category, we find the MCLS and MDFS approaches. The eight filter-based methods employed in this experiment are classified under the global category of filter-based methods, which can further be subdivided into local and global categories based on their scoring mechanisms [21–24].

A brief description of the MLFS methods is as follows:

7.2.3.1 MLACO

Presented in 2020 by Paniri et al. [25]. The MLACO algorithm is based on the Ant Colony Optimization (ACO) method for MLFS and combines correlation with class labels and feature redundancy. The features were evaluated in this method using a combination of supervised and unsupervised heuristics. The starting values of the pheromones are determined by the cosine similarity between the features and labels.

7.2.3.2 Ant-TD

This method is similar to the MLACO algorithm; the difference is that it uses reinforcement learning [26]. In the MLACO algorithm, according to Eq 7.1, the correlation was used to route the ants as a fixed learning heuristic for routing. However, Ant-TD uses temporal difference (TD) reinforcement learning instead to deal with fixed heuristics during the learning process. This algorithm describes the FS search space as a Markovian Decision Process, transforming the ACO problem into a reinforcement learning problem (MDP). States (s) in the new model define features, whereas actions (a) represent the ants' choice of unexplored features.

Table 7.2 Categories of MLFS methods

Categories	*MLFS methods*
Filter based	MGFS, MLACO, BMFS, Ant-TD, MFS-MCDM, Pareto Cluster, LRFS, PPT-ReliefF,
Embedded	MCLS, MDFS

7.2.3.3 LRFS

A new technique for selecting a multi-label feature, the so-called LRFS method [27], uses a new correlation criterion that takes the redundancy of the labels into account when selecting the feature. In this approach, the labels are first divided into independent and dependent labels. The conditional mutual information between each label and the candidate features was estimated in this new criterion assuming other labels. Finally, the redundancy of the features was aggregated with this new criterion.

7.2.3.4 Cluster pareto

The idea of Pareto dominance and clustering are used in this approach [28] to tackle the MLFS problem. First, the following equation is used to calculate the symmetric uncertainty (SU) of each feature with each label.

$$Su(X_i, Y_j) = 2 \times \left(\frac{MI(X_i, Y_j)}{H(X_i) + H(Y_j)} \right) \tag{7.1}$$

In this context, $H(X_i)$ and $H(Y_j)$ are the entropy of the j-th label and the i-th feature according to Eq. 7.13 and $MI(X_i, Y_j)$ refers to the mutual information between them according to Eq. 7.12. Then, the Su matrix is created. The rows represent the features and the columns represent the labels. Using the values of this matrix, we arrange the features in a space of L dimensions corresponding to the number of labels and rank the features based on the Pareto front. Since in this case, several features receive the same rank, a second pass must be used to distinguish these features. The second criterion is clustering, which is performed using the hierarchical clustering method and the full-link communication mode. The Pareto front set is a set of features that is not dominated by all other features and is called a non-dominant feature set.

7.2.3.5 MCLS

The MCLS technique [29] is a three-step approach based on the Laplacian score and manifold structure.

In the first stage, manifold learning is used to convert discrete (binary) labels into continuous values. The following equation is minimized to convert binary labels into constant values.

In the second step, three similarity criteria are applied based on the association between labels and local data. The first criterion is the constraint score technique, where the similarity between continuous labels is used as constraints for two instances.

Finally, a final score is assigned for each feature based on the combination of the three factors.

7.2.3.6 MDFS

This strategy is proposed by Zhang et al. [30] to select multi-label features, and it uses an embedded method that locally determines the correlation between features and labels. The main feature space is first transformed into a low-dimensional embedded space based on the manifold structure. If two features are very similar in this new space, it can be concluded that they are also close to each other in the primary space. At the local level, the FS procedure uses the correlation of features in the embedded space. This criterion takes into account the redundancy of the features. The association between the features and the labels is then discovered using the manifold setting. Ultimately, these two factors were taken into account when selecting the features.

7.2.3.7 PPT-Relief

The Pruned Problem Transformation (PPT) method, which selectively converts labels occurring beyond a certain threshold, is among the strategies employed in binary transformation methods to reduce computational complexity and eliminate single-label problems [31]. For example, a label that occurs only once in a dataset of 1,000 informative samples remains constant. The PPT-Relief method, which incorporates this strategy, is one of the most widely utilized MLFS techniques. This method initially partitions multi-label data into individual single-label datasets based on the number of labels. Labels occurring fewer than six times in total are then pruned or eliminated. Finally, a weight vector, W, is computed for the features within the range of [0,1]. Features with higher weights are deemed more significant.

7.2.3.8 BMFS

For the first time, Hashemi et al. [32] introduced the MLFS problem as a bipartite graph in their paper. In this representation, the features constitute the first section, while the class labels form the second section. A bipartite graph is a model wherein all nodes on the right side can connect to all nodes on the left, with nodes within each section being disconnected from one another. They utilized the renowned graph algorithm technique known as the Hungarian algorithm to identify the optimal features concerning labels.

7.2.3.9 MGFS

In 2020, Hashemi et al. introduced the MGFS technique [33], which models the MLFS problem using a full graph. In this network, nodes represent features, and each edge signifies the Euclidean distance between two features, reflecting their relationship with the labels. Initially, a correlation distance matrix is computed to establish these values. The correlation distance remains consistent for each layer of this matrix between the two characteristics in the row and column. Subsequently, using the previously determined correlation distances to represent edges between nodes, the Euclidean distance between features is calculated. Finally, the features are ranked by executing the PageRank algorithm on the resulting graph.

7.2.3.10 MFS-MCDM

The MFS-MCDM method [34] is an MLFS technique that integrates the FS process into a Multi-Criteria Decision Making (MCDM) framework. In essence, MCDM involves selecting a candidate based on various factors. In the context of FS, features serve as candidates and are evaluated against several criteria, which are designated as labels here. Initially, this method employs the Ridge Regression algorithm to derive a coefficient matrix W, which indicates the importance of each feature. The correlation matrix between features and labels serves as input to the TOPSIS method, an MCDM algorithm that generates a ranking vector.

7.2.4 Performance measures

In this study, we employed four classification-based criteria: coverage, hamming loss (HL), average precision (AP), and accuracy. Additionally, we utilized two ranking-based metrics: ranking loss (RL) and one-error (OE). Notably, we assigned equal weight to these parameters, considering both their impact on classification performance in Multi-Label Text Classification (MTC) and the importance of selecting the most relevant label for assignment [18, 35, 38].

Assume that T is a collection of test data ($T = \{(x_i, Y_i) : i = 1,\ldots,n\}$), L is a set of all labels, and Y_i is a subset of L ($Y_i \subseteq L$). Here are the necessary definitions for each of the evaluation criteria:

7.2.4.1 Accuracy

Classification algorithms are one of the most used methods of evaluation. The percentage of documents that the classifier accurately predicted is known as accuracy. Because accuracy is a gauge of profit, the higher the better.

$$\text{Accuracy}(h,T) = \frac{1}{n}\sum_{i=1}^{n}\frac{|Y_i \Delta Z_i|}{|L|} \tag{7.2}$$

The labels that the classifier predicted are indicated as Z_i in accordance with sample x_i.

7.2.4.2 *Ave-precision*

For ranked retrieval outcomes, AP is a measure that combines recall and precision. The average precision for one information demand is the average of the precision scores following the retrieval of each pertinent document.

$$\text{Ave\_pre}(f) = \frac{1}{n}\sum_{i=1}^{n}\frac{1}{|Y_i|}\sum_{y\in Y_i}\frac{\{y' \mid \text{rank}_f(x,y') \le \text{rank}_f(x_i,y), y' \in Y_i\}}{\text{rank}_f(x_i,y)} \tag{7.3}$$

A function produced by learning systems is called $f()$.

7.2.4.3 *Ranking loss*

The purpose of RL is to predict the relative distance between inputs. An irrelevant projected label placed higher than a corresponding label.

$$\text{Rank\_loss}(f) = \frac{1}{n}\sum_{i=1}^{n}\frac{1}{|Y_i||\overline{Y_i}|} \mid \{(y',y) \mid f(x_i,y') \le f(x,y), (y',y'') \in Y_i \times \overline{Y_i}\} \mid \tag{7.4}$$

7.2.4.4 *Hamming loss*

The Hamming distance between y_{true} and $y_{predicted}$ is connected to the HL in multi-label classification. If $y_{predicted}$ does not exactly match the actual set of labels, HL merely penalizes the individual labels and does not consider all of the tags in a sample to be false.

$$\text{Hamming\_loss}(h,T) = \frac{1}{n}\sum_{i=1}^{n}\frac{|Y_i \Delta Z_i|}{|L|} \tag{7.5}$$

where the symmetric difference between two sets is represented byΔ.

7.2.4.5 *Coverage*

The number of top-scoring final prediction labels that must be included without omitting any actual labels is indicated by the coverage error. This information is helpful if we want to estimate the typical number of top-scored predictions needed to make a forecast without overlooking any genuine labels.

$$\text{Coverage}(f) = \frac{1}{n}\sum_{i=1}^{n} \max_{y \in Y_i} \text{rank}_f(x_i, y) - 1 \tag{7.6}$$

7.2.4.6 One-error

The rejection of an actual null hypothesis is OE. It counts the number of times the ranking label differs from the instance labels.

$$\text{One\_error}(f) = \frac{1}{n}\sum_{i=1}^{n} \left[\!\left[\left[\arg \arg \max_{y \in Y_i} f(x_i, y) \right] \notin Y_i \right]\!\right] \tag{7.7}$$

7.2.5 Fuzzy methods

Fuzzy methods are indispensable for addressing problems characterized by uncertainty, allowing researchers to make informed decisions and selections. These methods facilitate the evaluation of issues from multiple perspectives or aspects. Each fuzzy method considers a potential membership degree of an element to a set, with these degrees, provided by one or more experts, not necessarily carrying equal weight [12, 36]. Over time, the concept of fuzzy sets has evolved into various types, some of which are mentioned below:

Torra and Narukawa [37] introduced the hesitant fuzzy set as a generalization of hesitant fuzzy sets, where the degree of membership is a set of multiple values rather than a single one, a concept that has seen various developments in subsequent years. They defined the fuzzy set as a reference set known as the universe of discourse. In this definition, if the global set is represented as $U = \{u_1, u_2, u_3, \ldots, u_N\}$, then the fuzzy set A in U is defined as a set of ordered pairs as follows:

$$A = \{(u_i, \mu_A(u_i))\} \tag{7.8}$$

where $u_i \in U$, μ_A is a membership function that maps the values of U in ranges $[0,1]$, and $\mu_A(u_i) \in [0,1]$ is the degree of membership of u in A.

For example, considering the universal set $U = \{fs_1, fs_2, fs_3, fs_4\}$, where fs_1, fs_2, fs_3, fs_4 represent the four FS methods. Then fuzzy set A, which refers to the concept of a better method, is as follows:

$$A = \{(fs_1, 0.2), (fs_2, 0), (fs_3, 1), (fs_4, 0.7)\} \tag{7.9}$$

Notice that the higher the membership degree of a method, the method is better.

In later years, Torra [11] introduced the Hesitant Fuzzy Series (HFSe). U is a universal set, and the hesitant fuzzy set A is represented as follows:

$$A = \{(u_i, h(u_i)) : u_i \epsilon U\} \tag{7.10}$$

where $h(u_i)$ refers to a hesitant fuzzy element and is a set of values between 0 and 1 that indicates the possible degree of membership of an element u_i to the set A.

For example, considering the universal set $U = \{fs_1, fs_2, fs_3, fs_4\}$, where represent the four FS methods. Then the hesitant fuzzy set A, which presents the concept of a better method, is as follows:

$$A = \{(fs_1, \{0.2, 0.6\}), (fs_2, \{0.1, 0.15\}), (fs_3, \{0.8.1\}), (fs_4, \{0.7, 0.9\})\}$$

$fs_1, \{0.2, 0.6\}$ are scores given by different experts to the fs_1 method based on one criterion. That is, the first expert gave fs_1 0.2 score, and the second also gave fs_1 0.6 score. It means that we are evaluating an alternative based on the judgment of several experts.

Xia and Xu also introduced the concept of Hesitant Fuzzy-Weighted Geometric (HFWG), assuming $h = \{h_1, h_2, \dots, h_n\}$ are a set of HFEs. HFWG maps H^n to H such that

$$HFWG(h_1, h_2, \dots, h_n) = \oplus \left(h_i^{w_i}\right) = \bigcup_{\gamma_1 \in h_1, \dots, \gamma_n \in h_n} \left\{\prod_{i=1}^{n} (\gamma_i)^{w_1}\right\} \tag{7.11}$$

where $w - (w_1, w_2, \dots, w_n)^T$ is a weighting vector of h. $w_i \in [0,1]$, and the sum of all weights is 1. If all weights w are equal ($w = \left(\frac{1}{w_1}, \frac{1}{w_2}, \dots, \frac{1}{w_n}\right)^T$), then HFWG becomes the HFG.

$$\text{HFG}(h_1, h_2, \dots, h_n) = \oplus \left(h_i^{\frac{1}{n}}\right) = \bigcup_{\gamma_1 \in h_1, \dots, \gamma_n \in h_n} \left\{\prod_{i=1}^{n} (\gamma_i)^{\frac{1}{n}}\right\} \tag{7.12}$$

Finally, having HFEs and mapping them should determine a final score for each member of the U global set. The mean score functions, minimum score, and maximum score used to calculate the final score. Assuming $h = \{h_1, h_2, \dots, h_n\}$, the score functions are as follows.

$$S_{mean}(h) = \frac{1}{n} \sum_{i=1}^{n} h_i \tag{7.13}$$

$$S_{\min}(h) = \min\{h_1, h_2, \dots, h_n\} \tag{7.14}$$

$$S_{\max}(h) = \max\{h_1, h_2, \ldots, h_n\} \tag{7.15}$$

Notice that the length of elements of h_1, and h_2 must be equal. If the elements of h_1 are less than h_2, then to calculate the mean score, the length of h_1 must be equal to h_2, so that the most significant score in h_1 is repeated several times to reach the length of h_2.

7.3 PROPOSED METHOD

This chapter not only introduces ten MLFS methods but also evaluates them using the HFG approach to determine the best preprocessing algorithm for MTC. In this method, all MLFSs are applied to eighteen MTDs, assessed by six performance experts, and subjected to ten-fold cross-validation on each dataset. The proposed method is outlined in Figure 7.1 and elaborated step by step in Algorithm 7.1.

As depicted in Figure 7.1, the ten MLFS algorithms are sequentially applied to each of the eighteen MTDs. Each MLFS method is evaluated by six different performance experts using ten-fold cross-validation – wherein ten subsets of top-selected features are determined for each MLFS and performed separately on all datasets. These evaluations result in the scoring of MLFSs, yielding 60 scores for each method within a specific MTD. This implies that each MLFS is assigned 60 different scores, as each expert evaluates all MLFSs based on ten validations. After obtaining the fuzzy results,

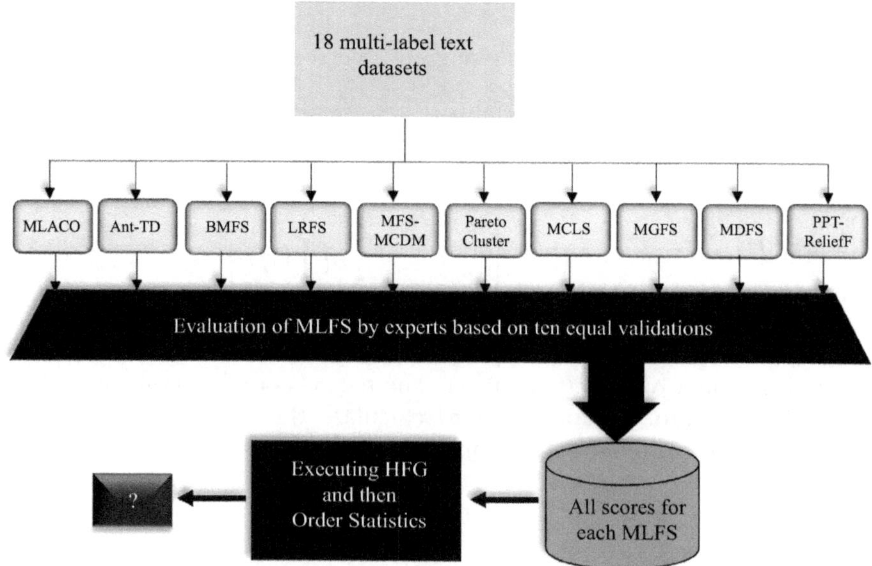

Figure 7.1 Graphical abstract of the proposed method.

we employ the order statistic approach to identify the best MLFS across all datasets for text classification preprocessing. This involves fusing the obtained results of MLFSs from all datasets.

In this chapter, we utilized the HFG approach to address the MLFS selection problem in choosing a preprocessing method for MTC. This is considered a hesitancy issue because a selected MLFS may not be suitable if it hasn't been evaluated by different experts simultaneously. In this method, six experts were employed to evaluate various MLFS on all MTDs individually. Given that all experts are considered equally valuable to us, we assigned equal weight to their evaluations. Four classification-based and two ranking-based measures were utilized by our experts, with MLFS methods serving as alternatives. Furthermore, our aim was to evaluate the methods based on the judgments of all six experts, considering ten-fold cross-validation, to determine the best performance among them.

Firstly, all alternatives are executed on datasets. Then, experts assess the alternatives to score them based on each validation. These scores, ranging from zero to one and represented as fuzzy values, collectively form elements of the HFEs. Subsequently, the fuzzy operation of HFG (Eq. 7.12) is applied to the HFEs to compute the HFG scores. A max score function is then employed to determine the best and final score for alternatives. Following the fuzzy operations, the alternatives are arranged based on the obtained HFG scores. Ultimately, 18 different scores are obtained for MLFSs, and utilizing the order statistic approach, we calculate the overall score for each MLFS by merging their scores.

Furthermore, as outlined in Algorithm 7.1, the method's output is a vector of algorithms arranged in descending order. In line 3, the universal set U is defined as the ten MLFS algorithms utilized in this method. Lines 7–9 depict the execution of MLFSs on one of the MTDs. Subsequently, in line 11, the score function is invoked, where experts evaluate the alternatives and assign scores based on all ten validations. In this approach, we considered six performance measures as the experts, namely Accuracy, AP, HL, RL, OE, and Coverage. Following this, the HFG fuzzy operations are performed on the HFEs to obtain fuzzy scores for each MLFS. Finally, a set of fuzzy scores is obtained for each MLFS, and we utilize the max score function (Eq. 7.15) in line 14 to derive the final score for each alternative. After performing these computations for all alternatives in each dataset, they are arranged in descending order based on their scores.

For example, suppose $U = \{Fs_1, Fs_2, Fs_3\}$, is the universal set, *s= {expert1, expert2, expert3}* is the experts, and *Cr = {cr1, cr2}* is the criteria, and the required information to calculate HFG is given in Table 7.3.

$$\begin{aligned} A = \{&(Fs_{1,}(\{0.78, 0.1, 0.12\}, \{0.8, 0.09, 0.1\}),\\ &(Fs_{2,}(\{0.85, 0.08, 0.09\}, \{0.89, 0.06, 0.07\}),\\ &(Fs_{3,}(\{0.70, 0.12, 0.2\}, \{0.73, 0.11, 0.17\})\}\end{aligned}$$

Algorithm 7.1 Methodology of evaluation and algorithm selection

```
Input: 18 multi label text datasets
Output: Algorithms ranking vector Algo_ranks;

 1. Begin
 2. A = ∅;
 3. U = [MLACO, AntTD, BMFS, PPTReliefF, MFS_MCDM,
         MGFS, MDFS, MCLS, LRFS, Pareto];
 4. % Apply ten MLFS on eighteen datasets.
 5. While exist dataset in
 6.     For i=1:size(U) % number of alternatives.
 7.         For j=1:10 % number of validations.
 8.             Cr(j)= execute the i_th MLFS on the desired
                dataset.
 9.         End
10.         % evaluation and scoring by six experts
11.         HFE(i)= scoring(U_i , Experts ,Cr);
12.         % The i_th HFE is a matrix 10*6 that consist of
            experts scores based on different validations
            for _ith alternative.
13.         HFG(i) = ⋃_{γ1 ∈ HFE1, ..., γ10 ∈ HFE10} {1 - ∏_{i=1}^{10} (1 - γ_i)^(1/6)};
14.         AScore(i) = max(HFG(i)); %based on equation 38
15.     End
16. End
17. % Ascore is a ranking matrix 10*18 which represented
    num of alternatives and datasets respectively
18. A=sort(AScore, 'Descending');
19. Alg_ranks = orderstatistic(A); %based on algorithm 7.3
20. End
```

$$= \{(Fs_1, \{0.78^{0.33} \times 0.8^{0.33}, 0.78^{0.33} \times 0.09^{0.33}, 0.78^{0.33} \times 0.1^{0.33}, 0.1^{0.33} \times 0.8^{0.33}, 0.1^{0.33} \times 0.09^{0.33}, 0.1^{0.33} \times 0.1^{0.33}, 0.12^{0.33} \times 0.8^{0.33}, 0.12^{0.33} \times 0.09^{0.33}, 0.12^{0.33} \times 0.1^{0.33}\}),$$

$$(Fs_2, \{0.85^{0.33} \times 0.89^{0.33}, 0.85^{0.33} \times 0.06^{0.33}, 0.85^{0.33} \times 0.07^{0.33}, 0.08^{0.33} \times 0.89^{0.33}, 0.08^{0.33} \times 0.06^{0.33}, 0.08^{0.33} \times 0.07^{0.33}, 0.09^{0.33} \times 0.89^{0.33}, 0.09^{0.33} \times 0.06^{0.33}, 0.09^{0.33} \times 0.07^{0.33}\}),$$

$$(Fs_2, \{0.70^{0.33} \times 0.73^{0.33}, 0.70^{0.33} \times 0.11^{0.33}, 0.70^{0.33} \times 0.17^{0.33}, 0.12^{0.33} \times 0.73^{0.33}, 0.12^{0.33} \times 0.11^{0.33}, 0.12^{0.33} \times 0.17^{0.33}, 0.2^{0.33} \times 0.73^{0.33}, 0.2^{0.33} \times 0.11^{0.33}, 0.2^{0.33} \times 0.17^{0.33}\})\}$$

Table 7.3 Required calculations for preparing HFEs

U	*Criteria*	*Expert1*	*Expert2*	*Expert3*	*HFEs*
Fs_1	Cr1	0.78	0.1	0.12	$h_1(Fs_1) = \{0.78, 0.1, 0.12\}$
	Cr2	0.8	0.09	0.1	$h_2(Fs_1) = \{0.8, 0.09, 0.1\}$
Fs_2	Cr1	0.85	0.08	0.09	$h_1(Fs_2) = \{0.85, 0.08, 0.09\}$
	Cr2	0.89	0.06	0.07	$h_2(Fs_2) = \{0.89, 0.06, 0.07\}$
Fs_3	Cr1	0.70	0.12	0.2	$h_1(Fs_3) = \{0.7, 0.12, 0.2\}$
	Cr2	0.73	0.11	0.17	$h_2(Fs_3) = \{0.73, 0.11, 0.17\}$

$$= \{(Fs_1, \max\{0.85587, 0.41618, 0.43091, 0.43452, 0.2113, 0.21877, 0.46147, 0.2244, 0.23234\}), (Fs_2, \max\{0.9120, 0.37453, 0.39408, 0.41813, 0.17171, 0.18067, 0.4347, 0.17852, 0.18783\}), (Fs_3, \max\{0.80127, 0.42908, 0.49537, 0.44774, 0.23976, 0.2768, 0.52995, 0.28379, 0.32763\})$$

$$= \{(Fs_1, 0.85587), (Fs_2, 0.9120), (Fs_3, 0.80127)\}$$

Note that the proposed method employs six performance measure experts, of which scores for two measures should be maximized, while the scores for the other four should be minimized. This is because accuracy and AP measure beneficial performance, whereas the other four criteria aim to minimize certain aspects of performance. To utilize the Max Score function effectively, the scores for these four criteria should be subtracted from one. By making these adjustments, the scores obtained for all performance criteria can be directly compared.

Algorithm 7.2 Order statistic method

```
Input: N × K Score matrix A
Output: 1 × N Algorithms score vector O _ scores

1.O_score = ∅;
2.For i = 1 : N
3.  For j=1 : K
4.    Ones = (−1) ^ (i − 1);
5.    V (j + 1) = ones × (V_{j-i} / i !) A^i_{K-j-1}; % where V_0=1
6.  End
7.  O_score (i) = i !× V_i;
8.End
```

Finally, after obtaining the final scores of all MLFS algorithms based on all MTDs, it is time to aggregate the overall score of the algorithms to determine the best one in line 19. The order statistic is an aggregation method utilized to aggregate scores of MLFS algorithms.

7.4 EXPERIMENTS AND RESULTS

In this section, experiments and their results are documented. Initially, the MTDs used in the experiments are introduced and explained. Following this, the results of the experiments are displayed and described in detail, along with a discussion of their implications.

7.4.1 Datasets

In this empirical study, 18 MTDs were utilized in the experiments. A brief synopsis of these datasets is provided in Table 7.4. The MTCs present various challenges across different dimensions. For example, the Arts dataset exhibits very high dimensionality, comprising 7,484 samples with 23,150 characteristics and 26 labels. Similarly, the cooking dataset, consisting of 10,490 samples, presents 577 features and 400 labels. Handling such high-dimensional and diverse datasets poses a significant challenge in MTC.

Table 7.4 Datasets descriptions

Num	*Dataset*	*Samples*	*Features*	*Labels*
1	20NG	19300	1006	20
2	Yelp	10810	671	5
3	Cooking	10490	577	400
4	Rcv1subset5	6000	1045	101
5	TMC2007	28600	49060	22
6	Ohsumed	13930	1002	23
7	Recreation	12830	30320	22
8	Chemistry	6961	715	175
9	Chess	1675	585	227
10	Arts	7484	23150	26
11	Bibtex	7395	1836	159
12	Cs	9270	635	274
13	Philosophy	3971	842	233
14	Medical	978	1449	45
15	Enron	1702	1001	53
16	Slashdot	3782	1079	22
17	Science	6428	37190	40
18	Language_log	1460	1004	75

7.4.2 Results

In this section, we provide detailed insights into the experiments and their results. The evaluation process encompasses several steps, as outlined below:

1. Evaluation of MLFS by Performance Experts:
   - Ten MLFS were evaluated by six performance experts based on ten equal validations across 18 MTDs.
   - The multi-label KNN algorithm was employed for MTC in this study.
   - The datasets were split into training and test subsets in a 60:40 ratio, with all results recorded as averages over ten separate executions.
2. Visualization of Performance Scores:
   - Figures 7.2–7.4 illustrate the performance scores assigned by experts to all MLFS across ten validations for three datasets: Arts, Bibtex, and Chemistry.
   - Figure 7.2 displays the performance scores in the Art MTD, highlighting the superior performance of MGFS, Ant-TD, and BMFS among the methods evaluated.
   - Figure 7.3 depicts the evaluation performance in the Bibtex dataset, where MCLS, MDFS, Ant-TD, MGFS, and BMFS exhibit strong scores across multiple experts.
   - Figure 7.4 showcases the evaluation results in the Chemistry dataset, with Ant-TD demonstrating higher accuracy and lower loss in OE and HL. MFS-MCDM outperforms others based on RL, while BMFS and MDFS excel in AP and Coverage, respectively.

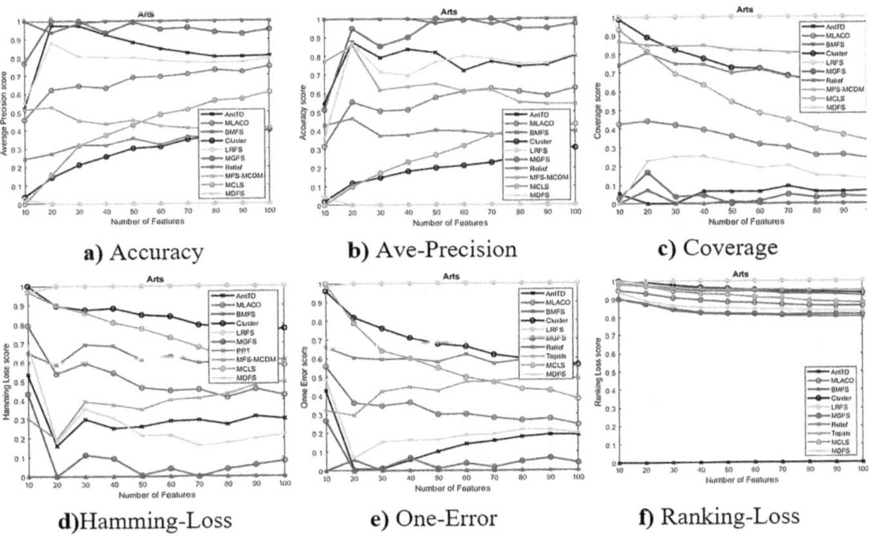

Figure 7.2 Performance of algorithms on Arts dataset.

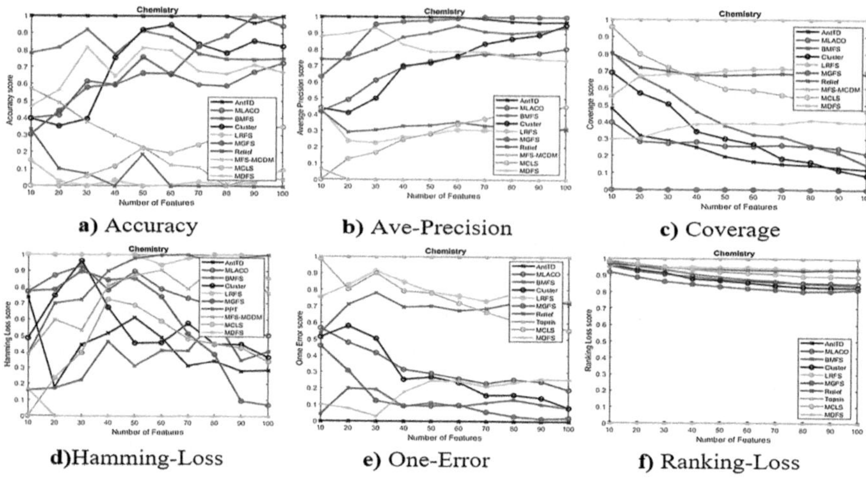

Figure 7.3 Performance of algorithms on Bibtex dataset.

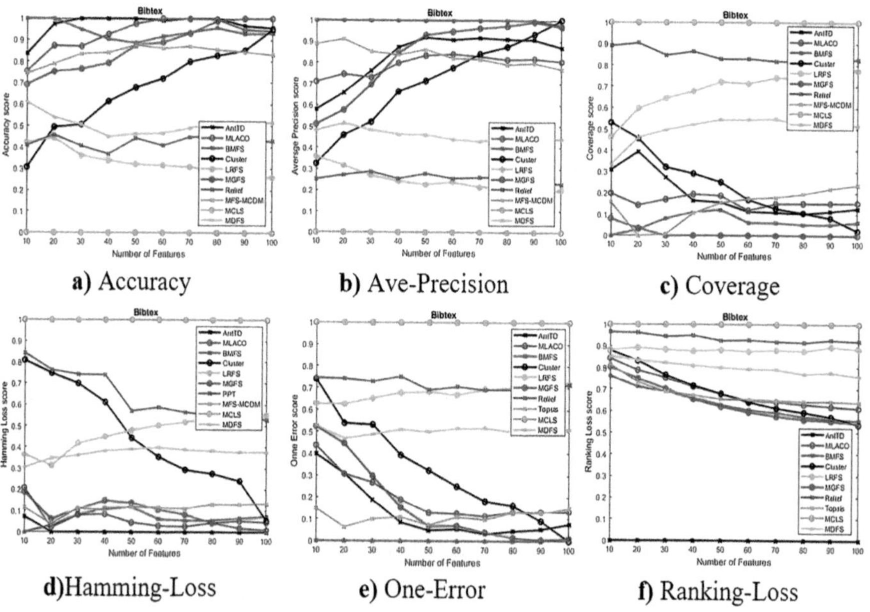

Figure 7.4 Performance of algorithms on Chemistry dataset.

3. Tabulation of Mean Expert Scores:
   - Tables 7.5–7.9 present the mean scores assigned by experts across all validations for each MLFS in the entire dataset.
   - Each table illustrates the mean scores of all methods across all datasets for a specific expert, with rows representing the scores for each MLFS in a particular dataset.

Table 7.5 Accuracy expert

Dataset	*MLACO*	*Ant-TD*	*BMFS*	*Pareto Cluster*	*LRFS*	*MGFS*	*PPT-ReliefF*	*MFS-MCDM*	*MCLS*	*MDFS*
20NG	0.1386	0.1483	**0.1513**	0.0997	0.0483	0.1443	0.0444	0.1462	0.0084	0.0705
Arts	0.1003	0.1361	**0.1761**	0.0416	0.0110	0.1587	0.0766	0.1173	0.0531	0.1292
Bibtex	0.1375	**0.1440**	0.1404	0.1014	0.0584	0.1290	0.0700	0.1254	0.0123	0.0815
Chemistry	0.1117	**0.1898**	0.1629	0.1315	0.0252	0.1644	0.0660	0.0655	0.0417	0.0413
Chess	0.0829	**0.1675**	0.1607	0.0812	0.0801	0.0833	0.0335	0.0898	0.0800	0.1360
Cooking	**0.1549**	0.1010	0.1368	0.1171	0.0221	0.0202	0.0346	0.1402	0.0522	0.0512
Cs	0.1098	**0.1905**	0.1447	0.1064	0.0512	0.1595	0.0467	0.0414	0.0309	0.1190
Enron	0.1180	0.0749	**0.1231**	0.1046	0.0795	0.1194	0.1078	0.1165	0.0978	0.0584
Language log	0.0829	0.1007	**0.1096**	0.1068	0.1013	0.1091	0.0829	0.1059	0.0996	0.1013
Medical	0.0462	0.1388	**0.1426**	0.0772	0.0547	0.1180	0.1177	0.1166	0.0808	0.1074
Ohsumed	0.1112	**0.1435**	0.1342	0.1084	0.0721	0.1365	0.0374	0.1197	0.0436	0.0935
Philosophy	0.0796	**0.1953**	0.1663	0.0789	0.0710	0.1354	0.0348	0.0474	0.0476	0.1435
Rcv1subset5	0.0660	**0.1430**	0.1266	0.1293	0.1023	0.1260	0.1012	0.0840	0.0175	0.1041
Recreation	0.1428	0.1607	0.2045	0.0319	0.0036	**0.2112**	0.0276	0.0653	0.0173	0.1351
Science	0.1110	0.1625	**0.1801**	0.0271	0.0091	0.1693	0.0671	0.1097	0.0236	0.1406
Slashdot	0.1001	0.1012	0.0989	0.0977	0.0986	0.0995	0.0980	0.1025	0.0989	**0.1046**
TMC2007	0.0954	0.1128	**0.1144**	0.1009	0.0909	0.1144	0.1005	0.0815	0.0780	0.1112
Yelp	0.1264	0.1292	0.1267	0.1175	0.1076	**0.1308**	0.1180	0.1198	0.0111	0.0125

Note: Bold value shows the best mean expert scores on different datasets.

Table 7.6 Average-Precision expert

Dataset	*MLACO*	*Ant-TD*	*BMFS*	*Pareto Cluster*	*LRFS*	*MGFS*	*PPT-ReliefF*	*MFS-MCDM*	*MCLS*	*MDFS*
20NG	0.1117	0.1170	0.1167	0.1006	0.0798	**0.1176**	0.0852	0.1143	0.0626	0.0945
Arts	0.1022	0.1057	**0.1090**	0.0940	0.0877	0.1080	0.0950	0.0971	0.0969	0.1044
Bibtex	0.1129	0.1156	**0.1263**	0.1079	0.0767	0.1164	0.0776	0.1155	0.0604	0.0907
Chemistry	0.1028	**0.1106**	0.1076	0.1037	0.0917	0.1095	0.0927	0.0839	0.0920	0.1056
Chess	0.1017	**0.1171**	0.1137	0.1016	0.0870	0.1044	0.0859	0.0864	0.0982	0.1041
Cooking	0.1038	**0.1306**	0.1205	0.1129	0.0624	0.1222	0.0785	0.0712	0.0771	0.1206
Cs	0.1024	**0.1232**	0.1141	0.1029	0.0893	0.1194	0.0822	0.746	0.0825	0.1094
Enron	0.1041	0.0940	**0.1057**	0.1014	0.0970	0.1030	0.1014	0.1038	0.0991	0.0904
Language log	0.0963	**0.1042**	0.1005	0.1006	0.1008	0.1005	0.0963	0.0998	0.0979	0.1013
Medical	0.0803	0.1143	**0.1156**	0.0937	0.0776	0.1059	0.1063	0.1051	0.0981	0.1030
Ohsumed	0.1022	0.1057	**0.1063**	0.1024	0.0949	0.1062	0.0943	0.0973	0.0932	0.0975
Philosophy	0.1024	**0.1185**	0.1131	0.1022	0.0845	0.1128	0.0857	0.0820	0.0942	0.1045
Rcv1subset5	0.0916	**0.1120**	0.1064	0.1070	0.0971	0.1112	0.0995	0.0898	0.0810	0.1044
Recreation	0.1084	0.1090	0.1165	0.0899	0.0813	**0.1185**	0.0915	0.0900	0.0881	0.1069
Science	0.1031	0.1087	0.1097	0.0919	0.0889	**0.1113**	0.0942	0.0926	0.0929	0.1067
Slashdot	0.0999	0.1012	0.0989	0.0977	0.0986	0.0995	0.0980	0.1025	0.0989	**0.1046**
TMC2007	0.0980	0.1058	0.1061	0.1009	0.0972	**0.1067**	0.0996	0.0922	0.0882	0.1053
Yelp	0.1234	0.1244	0.1236	0.1208	0.1176	**0.1251**	0.1215	0.1194	0.0119	0.0123

Note: Bold value shows the best mean expert scores on different datasets.

Table 7.7 Coverage expert

Dataset	*MLACO*	*Ant-TD*	*BMFS*	*Pareto Cluster*	*LRFS*	*MGFS*	*PPT-ReliefF*	*MFS-MCDM*	*MCLS*	*MDFS*
20NG	0.0866	0.0786	0.0849	0.0974	0.1265	**0.0773**	0.1109	0.0845	0.1450	0.1036
Arts	0.0983	0.0941	0.0933	0.1044	0.1087	**0.0939**	0.1040	0.1059	0.1013	0.0961
Bibtex	0.0908	0.0910	0.0869	0.0923	0.1136	**0.0844**	0.1191	0.0907	0.1258	0.1055
Chemistry	0.0975	0.0966	0.0989	0.0976	0.1033	**0.0940**	0.1033	0.1076	0.1021	0.0991
Chess	0.0987	**0.0969**	0.0980	0.0977	0.1036	0.0970	0.1034	0.1040	0.1004	0.1004
Cooking	0.1019	0.1015	0.1061	0.1038	0.1152	0.0978	0.1126	0.1177	0.1065	**0.0367**
Cs	0.0973	0.0944	0.0980	0.0979	0.1043	**0.0923**	0.1057	0.1092	0.1041	0.0967
Enron	0.0985	0.1025	**0.0954**	0.0974	0.1011	0.1003	0.1005	0.0983	0.1022	0.1039
Language log	0.1000	**0.0974**	0.1021	0.1011	0.0995	0.1015	0.1000	0.1006	0.1001	0.0976
Medical	0.1212	0.0788	**0.0748**	0.1022	0.1491	0.0858	0.0962	0.0973	0.0949	0.0994
Ohsumed	0.0983	0.0939	0.0930	0.0959	0.1036	**0.0927**	0.1060	0.1053	0.1067	0.1046
Philosophy	0.0971	0.0964	0.0989	0.0986	0.1038	**0.0950**	0.1047	0.1051	0.1000	0.1003
Rcv1subset5	0.1055	0.0904	0.0920	0.0913	0.1058	**0.0865**	0.1023	0.1114	0.1188	0.0958
Recreation	0.0944	0.0963	0.0894	0.1050	0.1119	**0.0885**	0.1045	0.1077	0.1058	0.0962
Science	0.0978	0.0943	0.0928	0.1048	0.1076	**0.0920**	0.1040	0.1063	0.1043	0.0961
Slashdot	0.0980	0.0943	**0.0948**	0.1005	0.1022	0.0954	0.1010	0.1021	0.1073	0.1046
TMC2007	0.1029	0.0862	0.0869	0.0945	0.1047	**0.0826**	0.1034	0.1183	0.1329	0.0875
Yelp	0.1186	0.1180	0.1181	0.1237	0.1305	0.1155	0.1224	0.1282	0.0129	**0.0122**

Note: Bold value shows the best mean expert scores on different datasets.

Table 7.8 Hamming loss expert

Dataset	*MLACO*	*Ant-TD*	*BMFS*	*Pareto Cluster*	*LRFS*	*MGFS*	*PPT-ReliefF*	*MFS-MCDM*	*MCLS*	*MDFS*
20NG	0.0912	0.0894	**0.0881**	0.0983	0.1125	0.0899	0.1132	0.0898	0.1202	0.1073
Arts	0.1003	0.0984	**0.0957**	0.1033	0.1047	0.0964	0.1014	0.0993	0.1024	0.0981
Bibtex	0.0965	**0.0958**	0.0966	0.1013	0.1020	0.0969	0.1040	0.0972	0.1089	0.1007
Chemistry	0.1002	0.0998	0.0998	0.1000	0.1005	0.1000	0.1004	0.1003	0.0999	0.**0993**
Chess	0.1006	**0.0976**	0.0977	0.1009	0.1011	0.1006	0.1032	0.1000	0.1002	0.0978
Cooking	0.0885	**0.0858**	0.0867	0.0877	0.0922	0.0864	0.0906	0.0908	0.0923	0.1989
Cs	0.0996	**0.0977**	0.0987	0.0997	0.1017	0.0984	0.1015	0.1014	0.1017	0.0996
Enron	0.0955	0.1062	**0.0948**	0.1001	0.1042	0.0968	0.0973	0.0962	0.1001	0.1089
Language log	0.1076	0.1050	**0.0924**	0.0953	0.1007	0.0955	0.1076	0.0948	0.0964	0.1048
Medical	0.1276	0.1276	**0.0763**	0.1084	0.1303	0.0867	0.0915	0.0937	0.1093	0.0975
Ohsumed	0.0989	0.1009	0.1007	0.1015	0.1003	0.1005	**0.0977**	0.1001	0.0997	0.0997
Philosophy	0.1003	0.0970	**0.0976**	0.1005	0.1012	0.0983	0.1020	0.1020	0.1023	0.0988
Rcv1subset5	0.1023	**0.0973**	0.0989	0.0983	0.0988	0.0991	0.0996	0.1002	0.1059	0.0995
Recreation	0.0970	0.0964	**0.0931**	0.1045	0.1060	0.0928	0.1050	0.1013	0.1059	0.0975
Science	0.1000	0.0977	**0.0974**	0.1026	0.1031	0.0975	0.1009	0.1001	0.1025	0.0982
Slashdot	0.0995	0.0967	0.1026	0.1053	0.1032	0.1010	0.1046	0.0938	0.1034	**0.0900**
TMC2007	0.1028	0.0931	0.0929	0.1001	0.1062	**0.0922**	0.0991	0.1070	0.1130	0.0936
Yelp	0.1169	0.1135	0.1200	0.1261	0.1348	0.1110	0.1270	0.1254	0.0133	**0.0120**

Note: Bold value shows the best mean expert scores on different datasets.

Table 7.9 One error expert

Dataset	*MLACO*	*Ant-TD*	*BMFS*	*Pareto Cluster*	*LRFS*	*MGFS*	*PPT-ReliefF*	*MFS-MCDM*	*MCLS*	*MDFS*
20NG	0.0833	0.0847	**0.0828**	0.0969	0.1199	0.0845	0.1164	0.0860	0.1340	0.1066
Arts	0.0979	0.0934	**0.0898**	0.1072	0.1157	0.0910	0.1051	0.1013	0.1038	0.0948
Bibtex	0.0915	0.0883	**0.0836**	0.0960	0.1138	0.0893	0.1152	0.0855	0.1279	0.1060
Chemistry	0.0984	**0.0943**	0.0959	0.0978	0.1050	0.0958	0.1039	0.1079	0.1041	0.0970
Chess	0.0993	**0.0873**	0.0895	0.0989	0.1084	0.0982	0.1101	0.1095	0.1023	0.0965
Cooking	0.0960	0.0974	**0.0841**	0.0899	0.1225	0.0856	0.1124	0.1143	0.1118	0.1041
Cs	0.0977	**0.0873**	0.0919	0.0979	0.1072	0.0906	0.1111	0.1129	0.1077	0.0957
Enron	**0.0853**	0.1172	0.0863	0.1018	0.1121	0.0888	0.0921	0.0885	0.0993	0.1287
Language log	0.1198	**0.0785**	0.0962	0.0971	0.0952	0.0965	0.1198	0.0992	0.1148	0.0829
Medical	0.1509	0.0650	**0.0629**	0.1098	0.1618	0.0806	0.848	0.0888	0.1021	0.0933
Ohsumed	0.0977	**0.0935**	0.0938	0.0978	0.1064	0.0938	0.1068	0.1016	0.1072	0.1016
Philosophy	0.0988	**0.0861**	0.0903	0.0991	0.1117	0.0907	0.1093	0.1128	0.1055	0.0955
Rcv1subset5	0.1106	0.0918	0.0976	0.0965	**0.0873**	0.0948	0.0908	0.1089	0.1215	0.1002
Recreation	0.0925	0.0915	**0.0841**	0.1096	0.1173	0.0829	0.1085	0.1087	0.1113	0.0937
Science	0.0973	0.0925	0.0915	0.1074	0.1112	**0.0904**	0.1050	0.1043	0.1063	0.0941
Slashdot	0.0991	0.1009	**0.0986**	0.0998	0.1007	0.0993	0.1003	0.1006	0.1001	0.1007
TMC2007	0.1034	0.0877	**0.0869**	0.0992	0.1097	0.0865	0.0973	0.1182	0.1220	0.0889
Yelp	0.1165	0.1116	0.1161	0.1263	0.1392	0.1083	0.1260	0.1308	0.0136	**0.0117**

Note: Bold value shows the best mean expert scores on different datasets.

- Tables 7.7–7.10 focus on cost-based criteria, where lower scores are more significant, while Tables 7.5 and 7.6 prioritize benefit-based criteria, with higher scores being more significant.

Table 7.10 highlights the significant performance of the Ant-TD algorithm based on RL expert perspective, exhibiting its strength in ranking-based performance. Additionally, BMFS received higher acceptance according to the OE expert perspective. However, Ant-TD consistently demonstrated strong performance across various datasets, emerging as the top performer in six datasets and ranking among the top three MLFS in other executions.

In addition to the evaluations by the Accuracy expert, methods have also been rigorously assessed by the AP expert, which measures methods based on classification-based performance similar to HL and Coverage experts. Table 7.6 presents the mean scores given by the AP experts for all MLFS algorithms, highlighting Ant-TD, BMFS, MGFS, and MDFS as the top methods. Ant-TD emerges as the best MLFS according to the AP expert's perspective. These four methods also stand out as the best based on the given scores by the Coverage expert in Table 7.7, where, contrary to previous measurements, MGFS emerges as the top method. Furthermore, Table 7.8 illustrates the mean performance of methods based on all validations in HL, where BMFS, Ant-TD, MDFS, and MGFS are identified as the top four methods, respectively.

The next step in this evaluation is the computation of HFG scores, the output of which, after sorting and ranking, is recorded in Table 7.11. Once the scores of each method are obtained and fully listed and sorted in the aforementioned table, these rankings must be combined. One of the most important statistical methods in nonparametric composition is the order statistic method presented in Algorithm 7.1. The input of this method is a matrix of information similar to that in Table 7.11. Each row of this matrix represents an algorithm, and each column represents an MTD, where each cell (e.g., Art, MLACO) shows the MLACO ranking in the Art dataset. A Q statistic is then calculated from all rank ratios using an N-dimensional order statistic.

$$Q(r_1, r_2, r_3, \dots, r_N) = N! V_N \tag{7.16}$$

$$V_k = \sum_{i=1}^{k} (-1)^{i-1} \frac{V_{k-i}}{i!} r_{N-k-1}^{i} \tag{7.17}$$

where N is the number of algorithms, K is the number of dimensions, r_i is the rank ratio for algorithm i, $r_0 = 0$, and $V_0 = 1$.

Table 7.10 Ranking loss expert

Dataset	*MLACO*	*Ant-TD*	*BMFS*	*Pareto Cluster*	*LRFS*	*MGFS*	*PPT-ReliefF*	*MFS-MCDM*	*MCLS*	*MDFS*
20NG	0.0967	**0.0043**	0.0920	0.1124	0.1368	0.0835	0.1102	0.0894	0.1627	0.1120
Arts	0.1085	**0.0014**	0.1011	0.1168	0.1216	0.1018	0.1152	0.1166	0.1122	0.1048
Bibtex	0.0994	**0.0001**	0.0910	0.0977	0.1277	0.0911	0.1352	0.0977	0.1445	0.1156
Chemistry	0.1209	0.0001	0.1223	0.1195	0.1296	0.1150	0.1288	0.1370	0.1267	**0.0000**
Chess	0.1087	**0.0001**	0.1072	0.1064	0.1162	0.1071	0.1175	0.1180	0.1097	0.1091
Cooking	0.1166	**0.0000**	0.1214	0.1181	0.1356	0.1108	0.1321	0.1392	0.1260	**0.0000**
Cs	0.1190	**0.0001**	0.1185	0.1192	0.1305	0.1101	0.1327	0.1385	0.1313	**0.0001**
Enron	0.1083	**0.0003**	0.1049	0.1079	0.1141	0.1115	0.1128	0.1092	0.1142	0.1168
Language log	0.1114	**0.0000**	0.1115	0.1112	0.1112	0.1107	0.1114	0.1134	0.1124	0.1067
Medical	0.1503	**0.0009**	0.0776	0.1112	0.1549	0.0928	0.0994	0.0972	0.1114	0.1043
Ohsumed	0.1072	**0.0023**	0.1030	0.1048	0.1138	0.1003	0.1137	0.1204	0.1177	0.1168
Philosophy	0.1052	**0.0001**	0.1085	0.1071	0.1160	0.1025	0.1173	0.1176	0.1121	0.1134
Rcv1subset5	0.1172	**0.0001**	0.0997	0.0987	0.1178	0.0939	0.1133	0.1217	0.1322	0.1055
Recreation	0.1023	**0.0019**	0.0950	0.1170	0.1258	0.0941	0.1168	0.1208	0.1205	0.1059
Science	0.1077	**0.0010**	0.1020	0.1164	0.1202	0.1012	0.1130	0.1181	0.1151	0.1053
Slashdot	0.1076	**0.0026**	0.1073	0.1115	0.0119	0.1073	0.1112	0.1118	0.1155	0.1132
TMC2007	0.1163	**0.0013**	0.0908	0.1014	0.1149	0.0856	0.1089	0.1306	0.1586	0.0914
Yelp	0.1271	0.0206	0.1270	0.1395	0.1567	0.1185	0.137	0.1459	0.0148	**0.0127**

Note: Bold value shows the best mean expert scores on different datasets.

Table 7.11 Rankings of the MLFSs based on the fuzzy score

Dataset	*MLACO*	*Ant-TD*	*BMFS*	*Pareto Cluster*	*LRFS*	*MCLS*	*MDFS*	*MGFS*	*PPT-ReliefF*	*MFS-MCDM*
20NG	5	2	4	6	9	10	7	1	8	3
Arts	5	1	2	9	10	6	4	3	7	8
Bibtex	6	1	3	4	8	10	7	2	9	5
Chemistry	5	1	6	4	9	7	2	3	8	10
Chess	5	1	4	2	8	7	6	3	9	10
Cooking	4	2	6	5	9	7	1	3	8	10
Cs	5	1	4	6	7	8	2	3	9	10
Enron	4	1	2	3	8	9	10	6	7	5
Language log	6	1	8	4	5	9	2	3	7	10
Medical	9	2	1	7	10	8	6	3	5	4
Ohsumed	5	1	3	4	7	9	8	2	6	10
Philosophy	3	1	5	4	8	6	7	2	9	10
Rcv1subset5	8	1	5	4	9	2	6	3	7	10
Recreation	4	1	3	7	10	8	5	2	6	9
Science	5	1	3	8	10	7	4	2	6	9
Slashdot	8	4	2	1	5	10	9	3	6	7
TMC2007	8	1	3	5	7	10	4	2	6	9
Yelp	6	4	5	8	10	1	2	3	7	9

In the final stage of the experiments and results, following the order statistic operation, the final ranking of the algorithms is calculated. This final ranking can be observed in Figure 7.4. The rankings of these algorithms, based on all datasets, are arranged from left to right, with Ant-TD being identified as the best algorithm and MFS-MCDM as the worst.

7.5 DISCUSSION

In this empirical study, we evaluated ten MLFS algorithms using a fuzzy approach, considering six performance experts and ten equal validations. Additionally, we utilized 18 different MTDs to facilitate a comprehensive comparison. Through reading this paper, it becomes apparent that each method must demonstrate good performance across various criteria and datasets, as different methods exhibit varying performances in specific datasets. The evaluation steps' outputs are detailed, with the first step involving the experimentation of methods by six experts based on ten validations. Figures 7.2–7.4 illustrate alternative performance on three datasets, showcasing the advantages of different methods in each scenario. For example, Ant-TD exhibits greater accuracy than BMFS and MGFS in the Bibtex dataset, whereas BMFS and MGFS outperform others based on AP and Coverage experts, respectively, highlighting the importance of multi-criteria evaluation.

Moreover, six tables (Tables 7.5–7.10) provide the mean performance of methods across ten validations. These tables help identify the best method based on experts' perspectives across all datasets. For instance, in Table 7.7, it becomes evident that MGFS requires fewer steps than other methods to cover all relevant labels in most datasets. Additionally, Ant-TD demonstrates significantly fewer losses in RL performance, as seen in Table 7.10.

However, finding the best method isn't straightforward, as each method is evaluated based on 60 scores. The next validation step involves HFG score computation, where all obtained scores serve as input. Table 7.11 ranks all MLFS methods according to their HFG scores in each dataset, with the best method given the first rank. Subsequently, with 18 alternatives ranked, order statistics are used to combine these rankings and determine the best method based on performance across all datasets. Ultimately, Figure 7.4 displays the output of the order statistics, sorting alternatives from best to worst. In this chapter, Ant-TD emerges as the best method for preprocessing before MTC, a conclusion supported by the outputs and experiment results.

7.6 CONCLUSION

Given the significance of selecting MLFS in MTC, it's important to highlight that choosing a method as a preprocessor for MTC is not only crucial but also a hesitant problem, inherently challenging and complex. This arises because finding a suitable MLFS according to a specific criterion is difficult; the method must be assessed across various criteria to make an appropriate choice.

In this empirical study, we adopted a multi-layer evaluation approach for ten MLFS methods using a hesitant fuzzy approach. Initially, all methods underwent evaluation in ten validations. Subsequently, six performance experts – Accuracy, Coverage, AP, HL, OE, and RL – assessed the alternatives based on the previous validations to assign scores. These scores served as input for the hesitant fuzzy method to calculate the HFG score, facilitating the comparison of MLFS algorithms within a particular dataset. Additionally, 18 different MTDs with diverse structures and dimensions were employed to emphasize the significance of the evaluation.

Ranking fusion using order statistics was a crucial step after the computation of HFG scores, wherein the obtained rankings were combined based on their HFG scores. Finally, an overall rank was calculated for each method based on 18 executions. The comparison outputs were detailed step by step in the previous section, revealing that algorithms cannot be selected correctly by considering only one evaluation criterion. As observed, Ant-TD, BMFS, MGFS, and Pareto-clustering emerged as the top performers across all datasets among the ten methods with various structures. Notably, Ant-TD was identified as the top MLFS among all methods and datasets, showcasing its superior performance.

REFERENCES

1. Miri M., Dowlatshahi M. B., Hashemi A., Rafsanjani M. K., Gupta B. B., Alhalabi W. Ensemble feature selection for multi-label text classification: An intelligent order statistics approach. *International Journal of Intelligent Systems* 2022; 37(12): 11319–41. Doi: 10.1002/int.23044
2. Karimi F., Dowlatshahi M. B., Hashemi A. SemiACO: A semi-supervised feature selection based on ant colony optimization. *Expert Systems with Applications* 2023; 214: 119130. Doi: 10.1016/j.eswa.2022.119130
3. Hancer E., Xue B., Zhang M. A survey on feature selection approaches for clustering. *Artif Intell Rev* 2020; 53(6): 4519–45. Doi: 10.1007/s10462-019-09800-w
4. Dowlatshahi M. B.., Hashemi A. Unsupervised feature selection: A fuzzy multi-criteria decision-making approach. *Iranian Journal of Fuzzy Systems* 2023; 20(7): 55–70. Doi: https://doi.org/10.22111/IJFS.2023.7630
5. Hashemi A., Pajoohan M.-R., Dowlatshahi M.B.. NSOFS: a non-dominated sorting-based online feature selection algorithm. *Neural Computing and Applications* 2024; 36(3): 1181–97.

6. Dhal P., Azad C. A comprehensive survey on feature selection in the various fields of machine learning. *Appl Intell* 2022; 52(4): 4543–81. Doi: 10.1007/s10489-021-02550-9
7. Jia W., Sun M., Lian J., Hou S. Feature dimensionality reduction: a review. *Complex Intell Syst* 2022; 8(3): 2663–93. Doi: 10.1007/s40747-021-00637-x
8. Hashemi A., Dowlatshahi M. B., Nezamabadi-Pour H. Minimum redundancy maximum relevance ensemble feature selection: A bi-objective Pareto-based approach. *Journal of Soft Computing and Information Technology* 2023; 12(1): 20–8.
9. Hashemi A., Dowlatshahi M. B., Nezamabadi-Pour H. Ensemble of feature selection algorithms: a multi-criteria decision-making approach. *Int J Mach Learn & Cyber* 2022; 13(1): 49–69. Doi: 10.1007/s13042-021-01347-z
10. Zadeh L. A.. Fuzzy sets. *Information and Control* 1965; 8(3): 338–53. Doi: 10.1016/S0019-9958(65)90241-X
11. Torra V. Hesitant fuzzy sets. *International Journal of Intelligent Systems* 2010; 25(6): 529–39.
12. Hashemi A., Pajoohan M.-R., Dowlatshahi M. B.. A multi-objective optimization approach for online streaming feature selection using fuzzy pareto dominance. *Journal of Mahani Mathematical Research Center* 2024; 13(1).
13. Hashemi A., Pajoohan M.-R., Dowlatshahi M. B.. Online streaming feature selection based on Sugeno fuzzy integral. *2022 9th Iranian Joint Congress on Fuzzy and Intelligent Systems (CFIS)*. 2022. p. 1–6.
14. Miri M., Dowlatshahi M. B., Hashemi A. Evaluation multi label feature selection for text classification using weighted borda count approach. *2022 9th Iranian Joint Congress on Fuzzy and Intelligent Systems (CFIS)*. 2022. p. 1–6.
15. Ali S., Smith K. A. On learning algorithm selection for classification. *Applied Soft Computing* 2006; 6(2): 119–38. Doi: 10.1016/j.asoc.2004.12.002
16. Kou G., Yang P., Peng Y., Xiao F., Chen Y., Alsaadi F. E. Evaluation of feature selection methods for text classification with small datasets using multiple criteria decision-making methods. *Applied Soft Computing* 2020; 86: 105836.
17. Kou G., Lu Y., Peng Y., Shi Y. Evaluation of classification algorithms using MCDM and rank correlation. *International Journal of Information Technology & Decision Making* 2012; 11(01): 197–225.
18. Kashef S., Nezamabadi-Pour H., Nikpour B. Multilabel feature selection: A comprehensive review and guiding experiments. *Wiley Interdisciplinary Reviews: Data Mining and Knowledge Discovery* 2018; 8(2): e1240.
19. Hashemi A., Dowlatshahi M. B., Rafsanjani M. K., Hsu C-H. Ensemble Feature Selection for Multi-label Classification: A Rank Aggregation Method. In: Nedjah N., Martínez Pérez G., and Gupta B. B., editors. *International Conference on Cyber Security, Privacy and Networking (ICSPN 2022)*. Cham: Springer International Publishing; 2023. p. 150–65.
20. Qian W., Huang J., Xu F., Shu W., Ding W. A survey on multi-label feature selection from perspectives of label fusion. *Information Fusion* 2023; 100: 101948. Doi: 10.1016/j.inffus.2023.101948
21. Zou Y., Hu X., Li P. Gradient-based multi-label feature selection considering three-way variable interaction. *Pattern Recognition* 2024; 145: 109900.
22. Bayati H., Dowlatshahi M. B., Hashemi A. MSSL: a memetic-based sparse subspace learning algorithm for multi-label classification. *Int J Mach Learn & Cyber* 2022; 13(11): 3607–24. Doi: 10.1007/s13042-022-01616-5

23. Hashemi A., Bagher Dowlatshahi M., Nezamabadi-Pour H. An efficient Pareto-based feature selection algorithm for multi-label classification. *Information Sciences* 2021; 581: 428–47. Doi: 10.1016/j.ins.2021.09.052
24. Shaikh R., Rafi M., Mahoto N. A., Sulaiman A., Shaikh A. A filter-based feature selection approach in multilabel classification. *Machine Learning: Science and Technology* 2023; 4(4): 045018.
25. Paniri M., Dowlatshahi M. B., Nezamabadi-Pour H. MLACO: A multi-label feature selection algorithm based on ant colony optimization. *Knowledge-Based Systems* 2020; 192: 105285. Doi: 10.1016/j.knosys.2019.105285
26. Paniri M., Dowlatshahi M. B., Nezamabadi-Pour H. Ant-TD: Ant colony optimization plus temporal difference reinforcement learning for multi-label feature selection. *Swarm and Evolutionary Computation* 2021; 64: 100892. Doi: 10.1016/j.swevo.2021.100892
27. Zhang P., Liu G., Gao W.. Distinguishing two types of labels for multi-label feature selection. *Pattern Recognition* 2019; 95: 72–82. Doi: 10.1016/j.patcog.2019.06.004
28. Kashef S., Nezamabadi-Pour H. A label-specific multi-label feature selection algorithm based on the Pareto dominance concept. *Pattern Recognition* 2019; 88: 654–67. Doi: 10.1016/j.patcog.2018.12.020
29. Wajid, M. S., Terashima-Marin, H., Wajid, M. A., Smarandache, F., Verma, S. B., & Wajid, M. K. (2024). Neutrosophic Logic to Navigate Uncertainty of Security Events in Mexico. *Neutrosophic Sets and Systems*, *73*, 120–130.
30. Huang R., Jiang W., Sun G. Manifold-based constraint Laplacian score for multi-label feature selection. *Pattern Recognition Letters* 2018; 112: 346–52.
31. Reyes O., Morell C., Ventura S. Scalable extensions of the ReliefF algorithm for weighting and selecting features on the multi-label learning context. *Neurocomputing* 2015; 161: 168–82.
32. Hashemi A., Dowlatshahi M. B., Nezamabadi-Pour H. A bipartite matching-based feature selection for multi-label learning. *International Journal of Machine Learning and Cybernetics* 2021; 12: 459–75.
33. Hashemi A., Dowlatshahi M. B., Nezamabadi-Pour H. MGFS: A multi-label graph-based feature selection algorithm via PageRank centrality. *Expert Systems with Applications* 2020; 142: 113024. Doi: 10.1016/j.eswa.2019.113024
34. Hashemi A., Dowlatshahi M. B., Nezamabadi-Pour H. MFS-MCDM: Multi-label feature selection using multi-criteria decision making. *Knowledge-Based Systems* 2020; 206: 106365. Doi: 10.1016/j.knosys.2020.106365.
35. Hashemi A., Dowlatshahi M. B. *MLCR: a fast multi-label feature selection method based on K-means and L2-norm*. IEEE; 2020. p. 1–7.
36. Hashemi A., Dowlatshahi M. B. A Fuzzy Integral Approach for Ensembling Unsupervised Feature Selection Algorithms. *2023 28th International Computer Conference, Computer Society of Iran (CSICC)*. 2023. p. 1–6.
37. Torra V., Narukawa Y. *On hesitant fuzzy sets and decision*. IEEE; 2009. p. 1378–82.
38. Wajid, M. A., Camacho-Zuñiga, C., Smarandache, F., Terashima-Marin, H., Wajid, M. S., & Wajid, M. K. (2024). TEC 21 Model and Critical Thinking: An NCM-based Neutrosophic Analysis in Higher Education. *Neutrosophic Sets and Systems*, 73, 140–152.

Chapter 8

Smart home automated face recognition

Abdullah Tahir, Ornel Vojka, Mohammad Umair Rizwan Khan, and Mohd Ovais R Khan

8.1 INTRODUCTION

Nowadays, humans live in the information period. There are various ways in daily life to prepare for the Internet and to make our day-by-day life simpler and more pleasing like Smartphones, computers, tablets, smart cars, and some Smart TVs. The new gadgets can run any project, in a much better and more viable path for doing diverse errands for example power on or power off the device, making cautions through the built-in or outer sensors. These items can be characterized as Smart Objects. Smart Objects can be more helpful to each other and to other objects, for example, actuators or sensors. In like manner, smart systems can be given IoT and can create either large or small systems due to obtaining aggregate intelligence through processing object information. The IoT can be connected in Smart Cities to give various advantages that improve residents.

In other words, brilliant homes can be made using the Internet of Things (IoT), which can control and robotize the correct things in our houses, such as doors, lights, windows, distributed multimedia, fridges, and irrigation systems. Computer Vision gives face detection and recognition for people and is an extremely fascinating application for the IoT. Computer Vision combinations can exhibit greater security systems in an IoT platform for intelligent homes since they have capacities to perceive a human in the incorrect place since this human might be a bad human for the environment [1]. So, How do IoT technologies like Raspberry Pi affect the Smart Home in our present and Future? What benefits does this system offer to homeowners and residents?

To answer all these questions, we have developed a motion detector software system that utilizes advanced algorithms to accurately identify and alert homeowners of potential intruders.

To achieve this project, first, we will present the problem, then we will choose the most suitable technologies and work methodologies. After presenting this, we will define and analyze the requirements of the project while discussing the actors and the different UML diagrams related to the solution. In the end, we will discuss the prototype on the code level, emphasizing

DOI: 10.1201/9781003606055-8

the class and package diagrams, the code architecture, the data model of the different databases, the user interface, and the user experience [2, 3].

8.2 FEASIBILITY AND PROJECT PROPOSITION

8.2.1 Introduction

This chapter is the project proposition entitled "Smart Home Automated Face Recognition Using Raspberry Pi and Camera". In this project, we aim to develop a comprehensive solution utilizing Raspberry Pi and a camera to detect intruders in a smart home environment. We will be addressing the following objectives:

- Design and implement a motion detector software system that integrates the Raspberry Pi and camera for intruder detection.
- Propose a solution that effectively detects and alerts homeowners about potential intrusions in real time.
- Discuss the technologies and tools utilized in the development of the system, including Raspberry Pi, camera integration, and relevant software frameworks.
- Emphasize the added value of our model in enhancing home security and providing peace of mind to homeowners.

In this chapter, we will present the problem of intruder detection in smart homes, describe the technologies employed, and propose our solution, highlighting its benefits and advantages in ensuring a secure and protected living environment.

8.2.2 Problematic

After analyzing the top tools used for home security, it has been observed that homeowners often face challenges in effectively detecting and responding to intrusions in their smart homes. Therefore, the problem we aim to address in this project is: How can we develop a smart home system using Raspberry Pi with a camera that efficiently detects and alerts homeowners about potential intruders, minimizing response time and ensuring optimal performance?

8.2.3 Proposed solution

To address the aforementioned challenge, the system will implement specialized functionality to effectively monitor and detect intrusions within designated areas with a high level of precision and accuracy. Additionally, the solution will integrate a real-time alert mechanism, utilizing the camera,

Raspberry Pi, and other relevant tools, to promptly notify homeowners and authorized entities about any potential security breaches or suspicious activities detected within the smart home environment [4].

Moreover, the system will generate personalized itineraries for individual users, presenting a graphical representation of their movement patterns and activities within the smart home premises on a comprehensive map. This advanced feature enables continuous real-time tracking and monitoring of user behavior, bolstering overall security measures and facilitating swift identification of any irregularities or unauthorized access attempts.

8.2.4 High-level architecture

In this section, we shall present the Global architecture Framework (Figure 8.1), accompanied by succinct elucidations of each constituent component.

8.2.5 Technological choice

In this section, we will delve into an exploration of the employed technologies, providing insights into the specific tools and platforms utilized in the project. Figure 8.2 presents the full system design [5].

8.2.6 Work plan

8.2.6.1 Methodology implementation

The project is structured into three distinct phases, each contributing to the overall objective of utilizing a Raspberry Pi camera in a smart home environment to detect intruders.

The first phase focuses on the installation and configuration of the Raspberry Pi camera. This entails setting up the hardware components, ensuring proper connectivity, and configuring the camera settings to capture high-quality images or video footage.

Moving on to the second phase, the primary objective is to implement a robust face recognition algorithm. This entails developing and fine-tuning an algorithm capable of accurately detecting and recognizing human faces within the captured camera data [6]. The algorithm should be able to distinguish between authorized individuals and potential intruders, enhancing the security measures of the smart home.

In the final phase, the emphasis shifts toward data management and user interface integration. The captured detection results are securely stored in a database, allowing for efficient retrieval and analysis. Additionally, the system is designed to generate real-time alerts, promptly notifying the user interface or relevant stakeholders about potential security breaches [7].

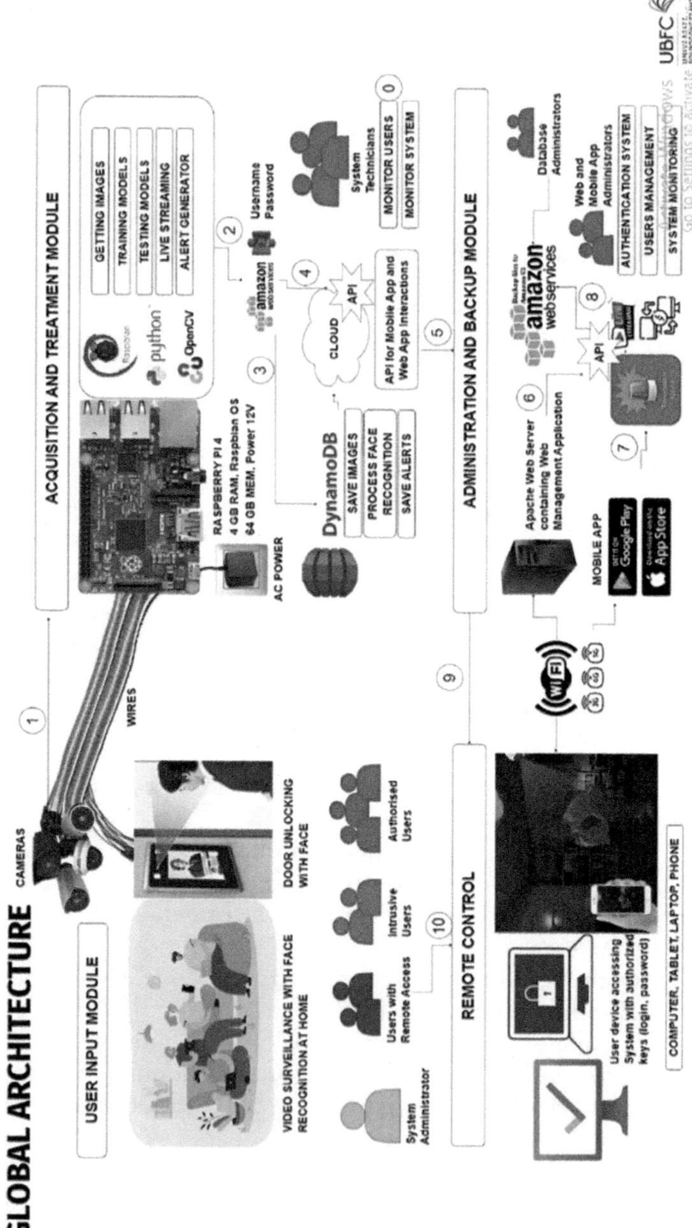

Figure 8.1 Global architecture.

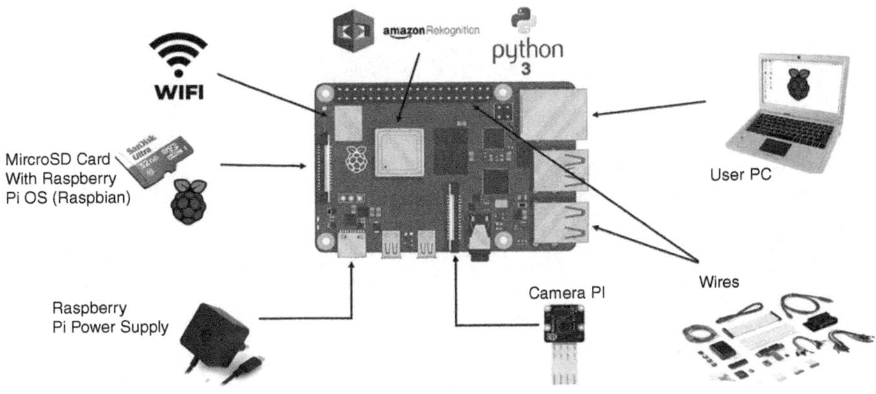

Figure 8.2 System design.

These alerts are actionable notifications, enabling immediate response and appropriate actions to mitigate any detected intrusions.

By dividing the project into these three phases, a systematic approach is taken to ensure the successful deployment and functionality of the Raspberry Pi camera in a smart home environment for intruder detection (Figure 8.3–Figure 8.5).

8.2.7 Conclusion

In this chapter, we introduce the problem statement and provide an overview of the current state-of-the-art solutions. Subsequently, a well-defined problem formulation is presented, accompanied by a selection of appropriate technological tools. Furthermore, a comprehensive work plan has been devised to ensure the systematic execution of the project [8].

```
File Edit Tabs Help
Preview Yes, Full screen Yes
Preview window 0,0,1024,768
Opacity 255
Sharpness 0, Contrast 0, Brightness 50
Saturation 0, ISO 0, Video Stabilisation No, Exposure compensation 0
Exposure Mode 'auto', AWB Mode 'auto', Image Effect 'none'
Flicker Avoid Mode 'off'
Metering Mode 'average', Colour Effect Enabled No with U = 128, V = 128
Rotation 0, hflip No, vflip No
ROI x 0.000000, y 0.000000, w 1.000000 h 1.000000
Camera component done
Encoder component done
Starting component connection stage
Connecting camera preview port to video render.
Connecting camera stills port to encoder input port
Closing down
Close down completed, all components disconnected, disabled and destroyed

pi@raspberrypi:~ $
```

Figure 8.3 Raspberry Pi camera configuration setup.

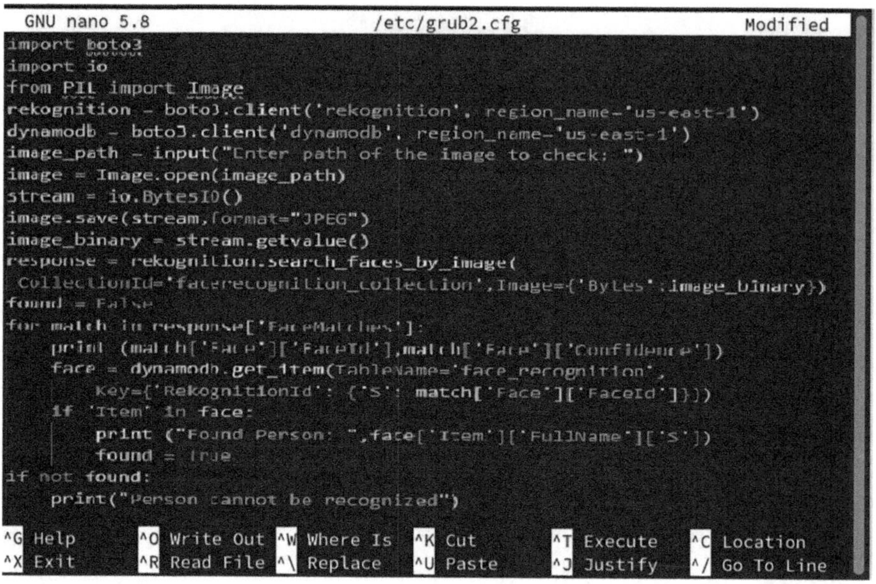

```
GNU nano 5.8                    /etc/grub2.cfg                    Modified
import boto3
import io
from PIL import Image
rekognition = boto3.client('rekognition', region_name='us-east-1')
dynamodb = boto3.client('dynamodb', region_name='us-east-1')
image_path = input("Enter path of the image to check: ")
image = Image.open(image_path)
stream = io.BytesIO()
image.save(stream,format="JPEG")
image_binary = stream.getvalue()
response = rekognition.search_faces_by_image(
 CollectionId='facerecognition_collection',Image={'Bytes':image_binary})
found = False
for match in response['FaceMatches']:
    print (match['Face']['FaceId'],match['Face']['Confidence'])
    face = dynamodb.get_item(TableName='face_recognition',
        Key={'RekognitionId': {'S': match['Face']['FaceId']}})
    if 'Item' in face:
        print ("Found Person: ",face['Item']['FullName']['S'])
        found = True
if not found:
    print("Person cannot be recognized")

^G Help      ^O Write Out  ^W Where Is  ^K Cut    ^T Execute  ^C Location
^X Exit      ^R Read File  ^\ Replace   ^U Paste  ^J Justify  ^/ Go To Line
```

Figure 8.4 Facial recognition using AWS Rekognition.

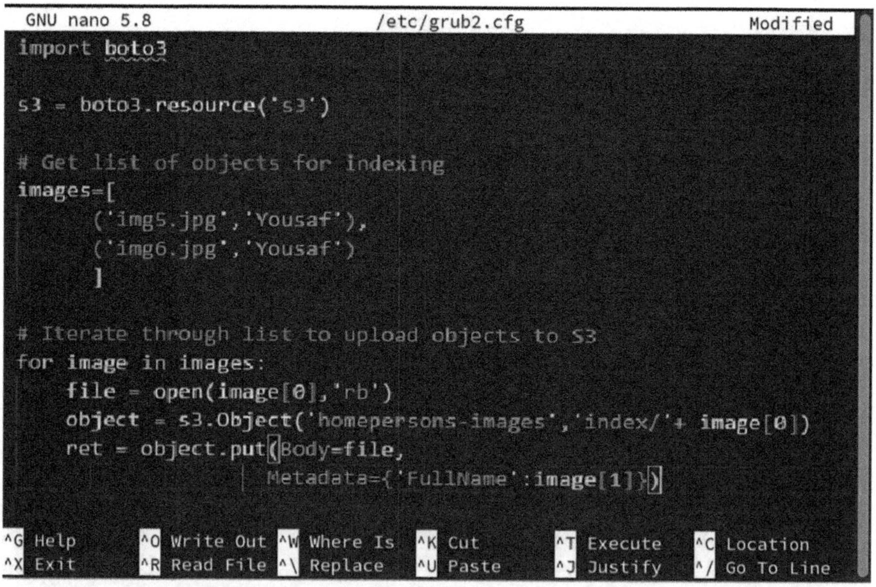

```
GNU nano 5.8                    /etc/grub2.cfg                    Modified
import boto3

s3 = boto3.resource('s3')

# Get list of objects for indexing
images=[
        ('img5.jpg','Yousaf'),
        ('img6.jpg','Yousaf')
        ]

# Iterate through list to upload objects to S3
for image in images:
    file = open(image[0],'rb')
    object = s3.Object('homepersons-images','index/'+ image[0])
    ret = object.put(Body=file,
                    Metadata={'FullName':image[1]})

^G Help      ^O Write Out  ^W Where Is  ^K Cut    ^T Execute  ^C Location
^X Exit      ^R Read File  ^\ Replace   ^U Paste  ^J Justify  ^/ Go To Line
```

Figure 8.5 Uploading images to AWS S3 with metadata.

8.3 REQUIREMENTS DEFINITION AND ANALYSIS

8.3.1 Introduction

This chapter focuses on the comprehensive definition and analysis of the project requirements. The specific functionalities of the system are carefully described and followed by the actor's presentation, and a detailed Nominal Scenario is presented through the use case diagram.

8.3.2 Actors description

The system is designed to distinguish authorized users from potential intruders, allowing seamless and secure interactions within the smart home environment (Figure 8.6).

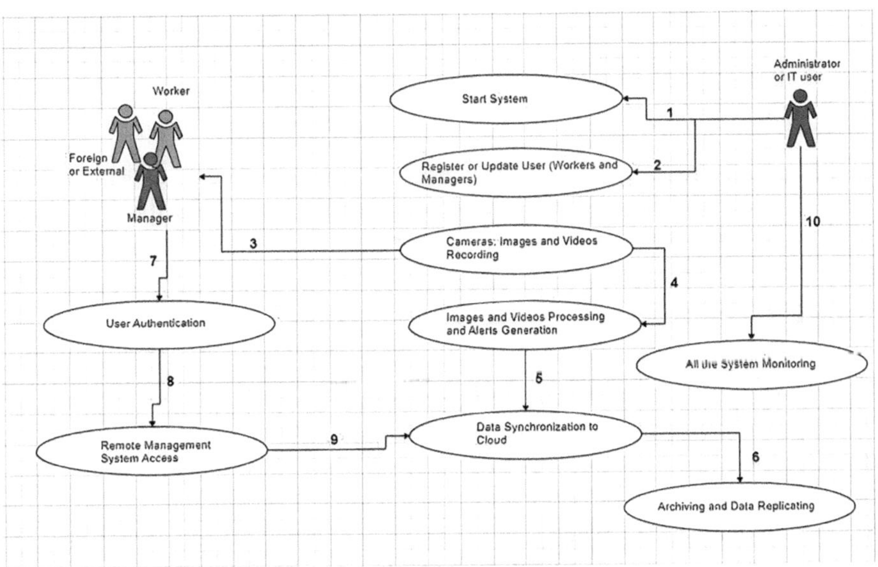

Figure 8.6 Use case diagram.

8.3.3 Conclusion

This chapter encompasses a comprehensive analysis and definition of the project requirements. It begins with a detailed description of the functionalities, followed by the presentation of the "Actors" involved in the system. Subsequently, the use case diagram is meticulously formulated, illustrating the interactions and relationships among the various components of the system [9]. Through this systematic approach, a thorough understanding of the project's scope and functional aspects is established, laying the foundation for successful implementation [10].

8.4 PROTOTYPE

8.4.1 Introduction

As an integral part of the face recognition project, the database plays a crucial role in storing and retrieving images. In this section, provide an overview of the purpose and significance of the database in your project. Emphasize that Amazon Cloud has been selected as the cloud service provider, offering a reliable and scalable infrastructure.

DynamoDB, a NoSQL database, has been chosen for its efficient storage and retrieval capabilities specifically tailored to handle image data.

8.4.2 Image storage process

8.4.2.1 Introduction

In the image storage process, we ensure efficient storage and retrieval of captured face images for our face recognition project. We will see the steps outline how we store the images in our database.

8.4.2.2 Image capture

We capture an image using a camera connected to our system: To acquire face images, we utilize a Raspberry Pi camera module integrated with our system. We configure the camera settings, including resolution and frame rate, to ensure optimal image capture quality [11].

8.4.2.3 Encoding the image

We encode the image in an appropriate format: After capturing the image, we encode it to transform it into a suitable format for storage and transmission. Common formats used for image encoding include JPEG and PNG. This encoding process optimizes the image size while preserving essential visual details [12].

8.4.2.4 Upload to Amazon cloud

We upload the encoded image to our Amazon Cloud infrastructure: Leveraging the power and scalability of Amazon Web Services (AWS), we utilize Amazon S3 (Simple Storage Service) to securely store and manage the captured face images. We upload the encoded image to an S3 bucket, which serves as a reliable and scalable storage container.

8.4.2.5 *Data integrity and access controls*

We ensure data integrity and access controls: As part of the image storage process, we prioritize data integrity and security. Amazon S3 provides various features to enforce access controls and permissions. We configure appropriate access policies to ensure that only authorized entities can retrieve or modify the stored images.

8.4.2.6 *Amazon DynamoDB*

We utilize Amazon DynamoDB for metadata storage: In addition to storing the images in Amazon S3, we utilize Amazon DynamoDB, a NoSQL database, to store associated metadata for each captured image. DynamoDB's fast and scalable storage and retrieval capabilities allow for efficient querying based on attributes such as timestamp, user ID, or image quality scores [13].

8.4.3 Data Model

To effectively organize and manage the data for our face recognition project, we have designed a comprehensive data model. This data model serves as a blueprint for structuring and connecting the various components of our database (Figure 8.7). It defines the tables, attributes, and relationships between entities, providing a solid foundation for efficient data storage and retrieval. The data model incorporates two tables: "face recognition" and S3, each playing a specific role in capturing and storing essential information

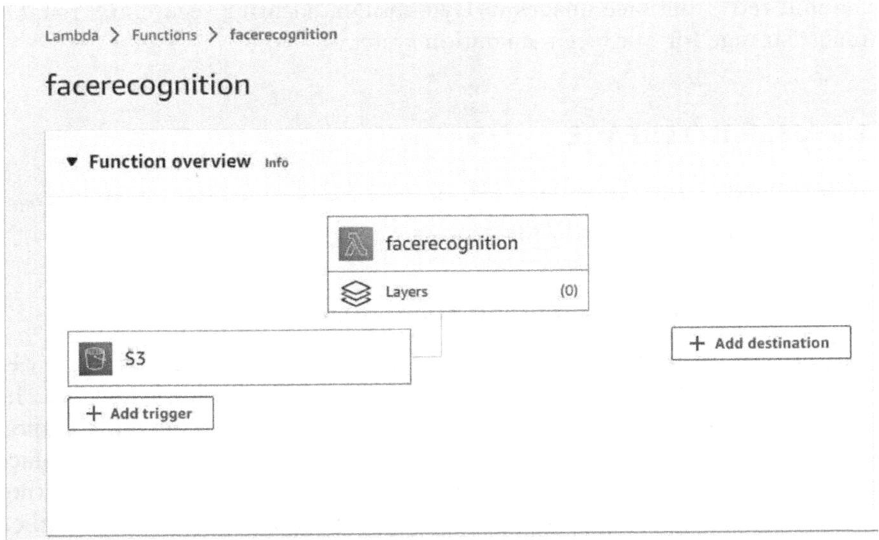

Figure 8.7 Data model: structure and relationships.

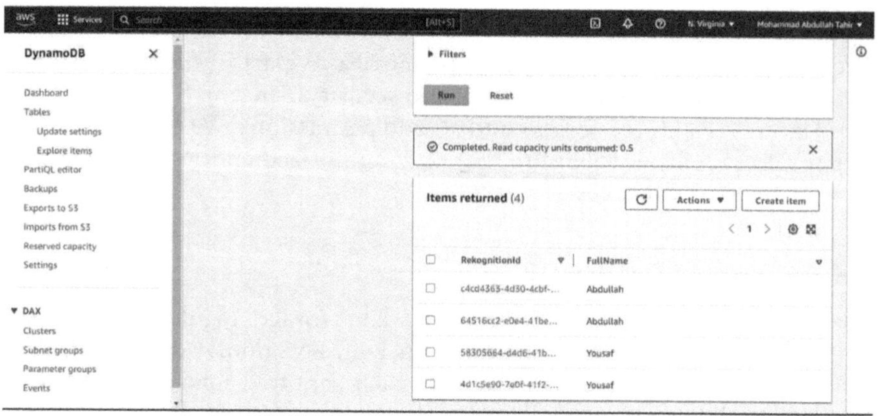

Figure 8.8 Image storage process in DynamoDB.

related to our face recognition system [14]. In the following section, we will present a visual representation of this data model, highlighting the connections between the relevant tables.

8.4.3.1 *DynamoDB*

In our face recognition project, the storage and management of captured face images are integral components. To efficiently handle the image data, we have implemented a robust storage solution using DynamoDB, our chosen NoSQL database (Figure 8.8). DynamoDB offers a scalable and highly available infrastructure that allows us to store and retrieve images seamlessly [15]. In the following section, we will explore the process of storing and retrieving face images in DynamoDB, ensuring secure and reliable image storage for our face recognition system.

8.5 USER INTERFACE

In this section, we will go through the system architecture, which libraries were used, and how they fit the process.

8.5.1 The Raspberry Pi face recognition system

The face recognition system in the smart home automation project utilizes the Raspberry Pi camera module and Python programming language. The Amazon Rekognition service, along with the boto3, io, and PIL (Python Imaging Library) modules, are employed for efficient and accurate face detection [16]. The system verifies whether the detected individual is authorized or not, and the result is stored in a database. These alerts are then displayed in a web application interface.

8.5.2 Libraries and technologies used

8.5.2.1 Amazon Rekognition (boto3, io, and PIL)

Amazon Rekognition is an integral part of the face recognition system, allowing for advanced image analysis and identification capabilities. The boto3 library is utilized to interact with the Amazon Rekognition service, while the io and PIL modules assist in processing and handling images [17].

8.5.2.2 OpenCV

OpenCV (Open Source Computer Vision Library) is a crucial component of our face recognition system, offering a wide range of computer vision algorithms and tools. It provides extensive support for image and video analysis, including face detection, recognition, and tracking. With OpenCV, we can leverage pre-trained models and techniques to identify and analyze faces within captured images or video streams [18]. This library plays a fundamental role in our system by enabling advanced image processing, feature extraction, and facial attribute recognition. It offers a comprehensive set of functions for image manipulation, transformation, and enhancement. Furthermore, OpenCV's integration with various programming languages, including Python, allows for seamless integration with our Raspberry Pi-based solution [19].

8.5.2.3 Python

Python, a widely used programming language, serves as a robust foundation for our face recognition project. Its rich ecosystem of libraries and frameworks, combined with its simplicity and readability, makes it an excellent choice for implementing face recognition algorithms [20]. In particular, the availability of libraries such as OpenCV, dlib, and face_recognition provides powerful tools and pre-trained models for face detection, landmark localization, and facial feature extraction. Python's versatility enables us to leverage these libraries to develop sophisticated face recognition pipelines. Additionally, Python's extensive community support and active developer community ensure that we have access to resources, tutorials, and updates, facilitating the development and refinement of our face recognition system.

8.5.2.4 Raspberry Pi with camera

The combination of Raspberry Pi and the camera module forms a powerful duo for our face recognition project. Raspberry Pi, a compact and versatile single-board computer, serves as the brains behind our system. With its GPIO (General Purpose Input/Output) pins, we can easily connect and control external components, including the camera module. The camera module, specifically designed for Raspberry Pi, enables us to capture high-resolution images for face recognition purposes. Its small form factor and flexible mounting options make it suitable for various deployment scenarios. The Raspberry Pi's processing power, combined with the camera module's image quality, allows us to implement real-time face detection, recognition, and tracking algorithms efficiently. Additionally, the camera module's adjustable focus and other settings provide us with the flexibility to optimize image capture for different lighting conditions and distances. The seamless integration of Raspberry Pi and the camera module empowers our face recognition system with a cost-effective, compact, and customizable solution.

8.5.2.5 Database Storage

The system employs a database to store the detected face recognition results. The exact database technology used will depend on the specific requirements of the project. In our Project we used DaynamoDB since it is a vital component of our face recognition system, providing a scalable and reliable solution for storing and managing face images. It offers efficient storage and retrieval capabilities, ensuring that our face recognition algorithms can access images quickly and without delay. With automatic scaling,

DynamoDB can handle varying workloads and accommodate our growing storage needs. The database's security features, including encryption and access controls, safeguard the privacy and integrity of our stored face images. Furthermore, its seamless integration with other AWS services allows us to leverage additional functionalities like data backup and synchronization, enhancing the overall robustness of our image storage solution.

8.5.2.6 Web application interface

A web application interface is developed to display the alerts and provide a user-friendly experience. The specific technologies and frameworks used in creating the interface will depend on the project requirements and development choices.

8.5.3 The system features

The smart home automated face recognition system aims to provide enhanced home security by leveraging advanced technologies such as the Raspberry Pi camera and Amazon Rekognition. By incorporating face recognition capabilities, the system offers several benefits:

- Enhanced Home Security: The system acts as an intelligent security measure, capable of accurately identifying authorized individuals and detecting any unauthorized access attempts. This advanced level of security minimizes the risk of potential intrusions and enhances the overall safety of the smart home.

- Real-time Alerts to Homeowners and Authorities: In the event of any suspicious activity or unauthorized access, the system triggers real-time alerts. Homeowners receive immediate notifications on their preferred devices, such as smartphones or computers, ensuring they are promptly informed of potential security breaches. Additionally, relevant authorities, such as local law enforcement, can also be notified, enabling a rapid response to any security incidents.
- Improved Evidence of Unauthorized Access Attempts: The automated face recognition system provides robust evidence in the form of captured images or videos, enabling homeowners to have a comprehensive record of any unauthorized access attempts. This evidence can serve as valuable proof in the event of legal proceedings or insurance claims, increasing the effectiveness of security measures and facilitating a higher level of accountability [2].

By combining the power of smart home automation, face recognition technology, and real-time alerts, the system offers a comprehensive solution for enhancing home security and providing homeowners with a sense of safety and peace of mind.

8.5.4 Conclusion

This chapter provides an overview of the system architecture, libraries used, and how they fit into the overall process.

8.6 CONCLUSION AND PERSPECTIVE

This report marks the culmination of an extensive software engineering process aimed at developing a robust security solution for homeowners. The project has successfully achieved its primary objective of enhancing security, providing real-time alerts, and improving evidence of unauthorized access attempts. The conception of this project has gone through different engineering stages: at first, we have chosen the most suitable technologies and work methodologies. After presenting this, we defined and analyzed the requirements of the project while discussing the actors and the UML diagram related to the solution. In the end, we talked about the prototype on the code level, emphasizing the class and package diagrams, the code architecture, and the data model of the different databases.

Throughout the project's lifecycle, various engineering stages were meticulously executed to ensure the seamless implementation of the solution. Careful consideration was given to the selection of appropriate technologies and work methodologies, which laid a solid foundation for the project's development.

Looking ahead, there are several promising avenues for future work and enhancement within this security solution. These areas of focus include:

1. Integration of Additional Security Features: The integration of advanced security features, such as motion sensors, smart locks, or biometric authentication, can significantly fortify the overall security system, reinforcing protection against potential intrusions.
2. Enhancing Machine Learning Algorithms: Continuous refinement and improvement of the face recognition algorithms using machine learning techniques hold the potential to enhance accuracy and reliability in identifying authorized individuals, ensuring a more robust security mechanism.
3. Integration with Home Automation Systems: Seamless integration of the security solution with existing home automation systems would allow for comprehensive control and monitoring of various smart home devices. This integration would elevate the overall user experience and convenience in managing security settings.
4. Expansion to Multi-Platform Support: Extending the solution's support to multiple platforms, such as Android and web browsers, would broaden its user base and improve accessibility, enabling homeowners to monitor and control their smart home security from a range of devices.
5. User Interface Enhancements: Continual improvement of the user interface in the web application is crucial. By prioritizing an intuitive and user-friendly design, homeowners can easily access real-time alerts and efficiently manage their security settings, enhancing the overall usability and effectiveness of the solution.

By focusing on these areas of future work, the security solution can be further refined and expanded, providing homeowners with enhanced peace of mind and advanced security measures for their smart homes. Continuous innovation and development will ensure the solution remains at the forefront of home security technology, addressing evolving threats and meeting the ever-changing needs of homeowners.

REFERENCES

[1] Jain, Arihant Kumar et al. A Review of Face Recognition System Using Raspberry Pi in the Field of IoT. (2018). https://easychair.org/publications/paper/pmvP DOI:https://doi.org/10.29007/1xq9

[2] Smart Home Security System using IOT, Face Recognition and Raspberry Pi. https://www.ijcaonline.org/archives/volume176/number13/dhobale-2020-ijca-920105.pdf DOI: 10.5120/ijca2020920105

[3] Python Software Foundation. (2021). Python 3 Documentation. Retrieved https://docs.python.org/3/
[4] Zhang, J. L., Zhu, Q. H. and Yang, X. Q., "Design of Intelligent Home Control System Based on Machine Learning," *2019 International Conference on Intelligent Transportation, Big Data & Smart City (ICITBS)*, Changsha, China, 2019, pp. 498–503, DOI: 10.1109/ICITBS.2019.00126
[5] Bhattacharya, Shubhobrata et al. "Smart Attendance Monitoring System (SAMS): A Face Recognition Based Attendance System for Classroom Environment." *2018 IEEE 18th International Conference on Advanced Learning Technologies (ICALT)* (2018): 358–360.https://www.semanticscholar.org/paper/Smart-Attendance-Monitoring-System-%28SAMS%29%3A-A-Face-Bhattacharya-Nainala/eceb5596536a65dbc1977b5d9191180a346f609a
[6] Wajid, M. A., & Zafar, A. (2022). Neutrosophic Image Segmentation: An Approach for the Treatment of Uncertainty in Multimodal Information Systems. *International Journal of Neutrosophic Science (IJNS)*, 19(1).
[7] Wajid, M. S., & Wajid, M. A. (2021). The Importance of Indeterminate and Unknown Factors in Nourishing Crime: A Case Study of South Africa using Neutrosophy. *Neutrosophic Sets and Systems*, 41(2021), 15.
[8] McGuinness, K., et al. (2020). YOLOv4: Optimal Speed and Accuracy of Object Detection. arXiv preprint arXiv:2004.10934. DOI: https://doi.org/10.48550/arXiv.2004.10934
[9] Wajid, M. A., & Zafar, A. (2022). A Multimodal Approach of Information Access and Retrieval Using Neutrosophic Sets. In *Emerging Trends in IoT and Computing Technologies* (pp. 382–388). Routledge.
[10] L. Li, X. Mu, S. Li and H. Peng, "A Review of Face Recognition Technology," *in IEEE Access*, vol. 8, pp. 139110–139120, 2020, doi: 10.1109/ACCESS.2020.3011028. https://ieeexplore.ieee.org/abstract/document/9145558#citations
[11] Venkata Kranthi B and Surekha Borra. 2022. Real-Time Face Detection and Recognition on Raspberry Pi using LBP and Deep Learning. In *Proceedings of the International Conference on Data Science, Machine Learning and Artificial Intelligence (DSMLAI '21'). Association for Computing Machinery*, New York, NY, USA, 124–129. https://doi.org/10.1145/3484824.3484903
[12] Wajid, M. S., Terashima-Marin, H., Najafirad, P., Pablos, S. E. C., & Wajid, M. A. (2024). DTwin-TEC: An AI-based TEC District Digital Twin and Emulating Security Events by Leveraging Knowledge Graph. *Journal of Open Innovation: Technology, Market, and Complexity*, 10(2), 100297.
[13] Szegedy, C. et al., "Going Deeper with Convolutions," *2015 IEEE Conference on Computer Vision and Pattern Recognition (CVPR)*, Boston, MA, USA, 2015, pp. 1–9, doi: 10.1109/CVPR.2015.7298594. https://ieeexplore.ieee.org/document/7298594
[14] Sruthy, S. and George, S. N., "WiFi Enabled Home Security Surveillance System Using Raspberry Pi and IoT Module," *2017 IEEE International Conference on Signal Processing, Informatics, Communication and Energy Systems (SPICES)*, Kollam, India, 2017, pp. 1–6, doi: 10.1109/SPICES.2017.8091320. https://ieeexplore.ieee.org/document/8091320
[15] Kakumanu, P., Makrogiannis, S., & Bourbakis, N. G. (2007). A Survey of Skin-color Modeling and Detection Methods. *Pattern Recognition*, 40(3), 1106–1122.

https://www.sciencedirect.com/science/article/abs/pii/S0031320306002767?via%3Dihub

[16] Ruhitha, V., Prudhvi Raj, V. N. and Geetha, G., "Implementation of IOT based Attendance Management System on Raspberry Pi," *2019 International Conference on Intelligent Sustainable Systems (ICISS)*, Palladam, India, 2019, pp. 584–587,doi: 10.1109/ISS1.2019.8908092. https://ieeexplore.ieee.org/document/8908092

[17] Alhajim, D., Salman Fahama, H., & Ta, M. (2024). Enhancing and Securing a Real-Time Embedded Face Recognition System using Raspberry Pi. *Journal of Al-Qadisiyah for Computer Science and Mathematics*, 16(1), Comp. 92–104. https://doi.org/10.29304/jqcsm.2024.16.11438 https://jqcsm.qu.edu.iq/index.php/journalcm/article/view/1438

[18] Chauhan, R., Negi, S., Kumar, A. and Devliyal, S., "A Raspberry pi Based Smart Attendance System," *2024 International Conference on Automation and Computation (AUTOCOM)*, Dehradun, India, 2024, pp. 612–617, DOI: 10.1109/AUTOCOM60220.2024.10486161. https://ieeexplore.ieee.org/document/10486161

[19] Dass, Sanchit et al. "Real Time Face Recognition using Raspberry Pi." *International Journal of Computer Applications* (2020): n. pag. https://www.semanticscholar.org/paper/Real-Time-Face-Recognition-using-Raspberry-Pi-Dass-Sadrulhuda/3b7132419a8926e24fbdb535cd3c13bfd0fed091

[20] Faisal, F. and Hossain, S. A., "Smart Security System Using Face Recognition on Raspberry Pi," *2019 13th International Conference on Software, Knowledge, Information Management and Applications (SKIMA)*, Island of Ulkulhas, Maldives, 2019, pp. 1–8, doi: 10.1109/SKIMA47702.2019.8982466. https://ieeexplore.ieee.org/document/8982466

Chapter 9

Synthesis of machine learning applications cutting-edge modern agriculture

Projecting fuzzy logic crop health monitoring via soil and plant sensors technologies

Bhupinder Singh and Christian Kaunert

9.1 INTRODUCTION

Fuzzy logic models identify and quantify essential indicators of crop health, translating sensor data into actionable insights. Parameters such as soil moisture, nutrient levels, and plant physiological conditions become quantifiable metrics. Fuzzy rules define the relationships between these variables, allowing for nuanced and context-aware assessments of crop health. Armed with fuzzy logic-based insights, farmers can embrace precision agriculture practices [1]. Resource optimization becomes a reality as decisions on irrigation, fertilization, and pest control are guided by real-time, data-driven recommendations. The result is a more sustainable and environmentally conscious approach to farming, minimizing waste and maximizing yield. This chapter explores the application of machine learning (ML), specifically fuzzy logic, in crop health monitoring using soil and plant sensors. By leveraging data-driven insights, fuzzy logic models enhance the precision of agricultural practices, optimizing resource utilization and contributing to sustainable farming [2].

Farmers can now remotely monitor and manage their fields using smartphones and tablets, thanks to specific mobile apps built for agricultural real-time monitoring. These applications combine data from a variety of sources, like as satellite imagery, weather stations, and Internet of Things (IoT) sensors, to provide up-to-date information on crop health, soil moisture, and other relevant aspects [3]. Drones, or unmanned aircraft systems, are used for airborne data collecting and high-resolution imagery. Drones play an important role in agriculture by analyzing field conditions, monitoring crop health, and spotting possible difficulties such as insect infestations or nutrient deficits. Drone data collected in real-time and in great detail allows farmers to apply targeted interventions, make educated decisions, and improve agricultural practices for greater crop yields and resource efficiency [4].

DOI: 10.1201/9781003606055-9

The development of digitization has resulted in an avalanche of data across several industries inside data-driven organizations [5]. The management of digital data in man-to-machine (M2M) interactions has substantially accelerated the information flood. Notably, there has been significant advancement in digital agricultural management applications, affecting the field of information and communication technology (ICT) to give benefits to both farmers and consumers and this progress has also brought technical solutions to rural regions. There are several issues associated with the collecting, processing, and application of data to improve agricultural output. The assurance of data security and privacy is a major impediment for farmers operating in the digital age. Agricultural information systems frequently face data availability and quality concerns. When dealing with real-time data influx, the complexity increases. There are additional issues with data efficiency and geographical semantic integration, which make efficient data mining challenging [6].

9.1.1 Major contributions of this study

This chapter tackles essential strategies for automating agricultural pre-harvesting processes, including soil and seed management, crop disease detection, irrigation, weed control, insecticides, and yield management [7]. It investigated the roles of ML in general, and deep learning (DL) in particular, in the detection of plant fruit and leaf diseases. It also addresses the significance of ML, Fuzzy Logics, IoT, smart agriculture, and drones in improving duties connected to increased agricultural productivity [8]. This encompasses the many methodologies used in these operations, such as image processing. The chapter looks at many illnesses and infections that damage crops including apples, grapes, bananas, and other fruits and vegetables [9]. It examines symptoms for classification, defines the phases required in automated detection and classification of plant diseases, and investigates various techniques and algorithms applicable at each stage [10]. The thorough study not only reveals future promises and conversations, but it also analyzes obstacles and issues associated with the implementation of ML models for smart agricultural operations.

9.1.2 Role of technology in modern agriculture

As technology advances, ML is finding more uses in agriculture. The IoT and the Fourth Industrial Revolution (Industry 4.0) are both facilitating the development of new technologies and breakthroughs [11]. These contemporary agricultural technologies and developments, known as "smart agriculture" or "Agriculture 4.0" can increase crop yield while simultaneously lowering water and energy use. Agriculture has experienced a number of obstacles, including inadequate irrigation infrastructure, climate change, groundwater depletion, food shortages, and waste management issues [12].

The adoption of various cognitive methods has a substantial impact on cultivation outcomes [13]. Despite continuous intensive research and the availability of several apps, the agriculture industry remains underserved in terms of addressing real-world challenges and adopting autonomous solutions. Technological improvements have a significant influence on many aspects of agriculture, including fertilizers, insecticides, seed technology, and others [14]. Pest-resistant crops and increased yields have resulted from biotechnology and genetic engineering, for example. Mechanization has resulted in more efficient tilling and harvesting, lessening reliance on manual labor. Improvements in irrigation technologies, transportation networks, and processing gear have reduced waste, making a noticeable difference across all agricultural disciplines [15]. Emerging technologies such as robots, precision agriculture, artificial intelligence, and blockchain technology are at the forefront of agricultural innovation in the modern period [16].

The adoption of appropriate crop management strategies increases agricultural production and can result in higher yields with higher quality [17]. Crop management refers to a set of agricultural practices that aim to improve crop growth, development, and production. The process begins with seedbed preparation, continues with seed sowing, and concludes with crop harvest, storage, and commercialization [18]. Various factors influence the timing and sequencing of these agricultural practices, such as whether the crops are winter or spring crops, the type of harvested products (such as grain, hay, and silage), sowing methods (broadcast or row crops), and considerations such as plant age, soil conditions, climate, and weather patterns [19]. The farmers often follow the following sequence of stages during agricultural activities:

Crop Evaluation.
Preparation of Land.
Sowing Seeds.
Fertilization and Irrigation.
Crop Upkeep (including pesticide usage, crop trimming, and so on).
Harvesting.
Activities Following Harvest.

9.1.3 Need for precise crop health monitoring for sustainable farming

Crop monitoring is critical in agriculture because it enables the early detection of pests and diseases, improves resource consumption, and promotes sustainable practices [20]. This monitoring assists farmers in making educated decisions, increasing productivity, and lowering environmental impact, all of which contribute to improved economic outcomes and long-term agricultural sustainability. It provides critical information on crop health, growth, and environmental conditions [21]. Crop monitoring has entered a new phase in the changing agricultural landscape, driven by data, accuracy, and technology. The capacity to methodically examine and assess crop growth has become an invaluable resource for farmers and agronomists from the first phases of planting to the final harvest [22].

Crop monitoring in the modern day is heavily reliant on technology, with advanced instruments and methodologies used for data collecting, analysis, and decision-making [23]. Satellite photography, drones, IoT sensors, and artificial intelligence (AI)-powered analytics allow for real-time monitoring of crop health, soil conditions, and weather trends. This technology enables farmers to conduct precision agriculture, maximize resource use, and respond quickly to emergencies, resulting in improved output, lower costs, and the adoption of more environmentally friendly agricultural methods [24].

9.1.4 Objectives of the chapter

This chapter has the following the following objectives:

- Investigates the possibilities of ICT technologies in traditional agriculture, providing light on both the benefits and drawbacks of incorporating them into farming techniques.
- Goes into robots, IoT devices, and ML difficulties and the roles of artificial intelligence and sensors in agriculture.
- Explores the possible use of drones for crop surveillance and production improvement.
- Showcases worldwide and cutting-edge IoT-based farming systems and platforms when applicable.

9.1.5 Organization of the chapter

This chapter comprehensively explores the various dimensions of Modern Agriculture via ML Applications and integration of Fuzzy Logic Crop Health Monitoring via Soil and Plant Sensors Technologies. Section 9.2 elaborates the ML in Agriculture. Section 9.3 explains Fuzzy Logic in Crop Health Monitoring. Section 9.4 lays down Soil and Plant Sensors: Overview of soil sensors for data collection. Section 9.5 specifies the Development of Fuzzy Logic Models. Section 9.6 travels the Crop Health Indicators: Identification of key indicators for crop health assessment. Section 9.7 expands the Precision Agriculture and Resource Optimization. Section 9.8 discusses the Challenges in the Deployment of Fuzzy Logic Systems in Agriculture. And, finally, Section 9.9 concludes the chapter with Future Scope.

9.2 MACHINE LEARNING IN AGRICULTURE

Farmers strive to produce the same amount of food with fewer resources, resulting in less water usage and less reliance on pesticides. Farmers attempt to maximize both quantity and profitability, aligning with the public's desire for healthful food, while seeking for increasing output and reduced costs. To fulfill these different demands, the agriculture business is continually seeking innovative goods, techniques, and technology. Precision agriculture is becoming increasingly important for farmers to handle these difficulties. Agriculture must use a variety of technologies to gather and analyze data effectively in order to remain creative [25]. Through the integration of current technology such as sensor and actuator networks, unmanned aircraft systems, satellite images, IoT, drones, and others, smart agriculture intends to reduce farmers' workloads while increasing agricultural output. It is a crop management concept that enables farmers to address geographical and temporal changes in agriculture, including topics such as water management, production management, fertilizer management, intrusion assaults, and real-time data monitoring [26]. A range of sensors, including those for chemical levels, pH, wind, rain, temperature, moisture, and sound, are extensively used in modern agricultural techniques. Each sensor is outfitted with capabilities and hardware that are customized to the exact conditions of its location [27]. IoT peripherals, image processing systems, big data capabilities, data analytics, and other technologies are also helping to advance smart farm management and wireless technologies. Several developing technologies, particularly web-based ones, create large amounts of data that may be widely accessed and shared [28]. The following are the important characteristics of data mining in the context of smart agriculture: (i) The need to handle huge volumes of data efficiently and quickly for research. (ii) Data source diversity is critical for smart agriculture; this includes dealing with sensors in a variety of formats, and various forms of data such as pictures, strings, and numbers. (iii) The capacity to interpret and interact with various device types, websites, and web-based information extraction [29].

The soil is important in forecasting climatic illnesses, and weather data can alert farmers to impending infection outbreaks [30]. Using acquired data for crop protection can result in greater production and less environmental impact. Given human's limited ability to digest this data, tools and approaches that ease analysis are critical for making educated decisions. The use of data mining tools is essential for analyzing this data. Big data has patterns, and finding these patterns requires using a variety of approaches to study the data [31].

9.2.1 Machine learning algorithms and their applications in precision farming

Smart and precision agriculture, which represent the confluence of the IoT and information technology (IT), are critical in many aspects of farming [32]. The major goal is to collect data from many sources, evaluate it, forecast it, and arrange agricultural activities depending on environmental circumstances. Sensors are now widely used to capture a wide range of data, including soil temperature, moisture, foliage, sunlight levels, and direction

[33]. The combination of digital automation, data gathering, data transmission, decision-making, data processing, and data analysis underpins smart agriculture (SA). Sensor networks are a popular data-gathering and communication tool in this industry and environmental data is critical in smart agriculture [34].

Crop development is influenced by pre-harvesting circumstances [35]. ML is used in pre-harvesting operations to collect data on numerous elements like as soil quality, seed quality, fertilizer treatment, pruning, genetic and environmental variables, and irrigation conditions. Concentrating on each of these aspects is critical for reducing total production losses [36]. This lecture focuses on major pre-harvesting components and how neural networks and ML are used to record and evaluate the characteristics related with each of these components [37].

9.2.2 Data-driven insights in optimizing agricultural practices

Data mining techniques have been used in a variety of agricultural applications, including pest identification, disease detection, yield prediction, and fertilizer and pesticide usage planning [38]. They also contribute to crop management and can investigate alternate approaches to improve farm organization. As a result, gathering this data constitutes a new input that has a direct impact on agricultural efficiency, providing better outcomes [39]. This schematic depiction of the smart agricultural system includes a perceptual device, network functions, and capabilities available at the edge, all of which are integrated into cloud-based services. Data is gathered and used on the perception layer of these systems, where sensors, global positioning system (GPS), RFID tags, and cameras may be put on the farm [40]. Because of this setup, these devices cannot process or store data close to the user or work in the cloud, limiting their value. Because this component is linked to internal sensor networks via networking technologies, it is compatible with wireless sensor networks. Security features, data filters, intelligence, diversified processing, numerous in/outbound interfaces, and the gateway are all part of the creative edge [41].

The advancement of computer vision, ML, fuzzy logics technologies allows for the accurate and rapid identification of many crop diseases in modern farming conditions [42]. This tool tackles contemporary agricultural difficulties by giving quick and exact findings via automated evaluations utilizing image processing techniques. Implementing this technology provides cost and time savings by removing the need for substantial manual effort. It also helps to improve crop quality and overall productivity. Early insights into crop conditions, as well as the identification of disease-affected regions, can aid in disease control through prompt intervention strategies [43].

9.3 FUZZY LOGIC IN CROP HEALTH MONITORING

Soil, geographical, and climatic conditions all have a substantial impact on crop output [44]. Many rural farmers are unaware of the influence of these characteristics on crop productivity and frequently rely on traditional knowledge for crop selection, resulting in significant economic losses [45]. To solve this, a scientific approach that incorporates site-specific elements as well as traditional knowledge may provide an efficient answer. The deployment of a fuzzy logic-based crop recommendation system to aid rural farmers in making informed decisions. In the field, monitoring devices collect data regarding farming variables such as light intensity, humidity, and temperature with the purpose of boosting crop yield. The IoT technology has recently grown popular in a variety of industries, including agricultural monitoring systems. Manual work is frequently necessary in conventional agricultural systems for handling crops and cattle, resulting in poor resource usage [46]. This disadvantage may be reduced by implementing smart farming, in which farmers are taught the use of IoT, have access to the GPS, and have data management skills to improve the quantity and quality of their agricultural goods [47].

The traditional agricultural practices are labor-intensive, with farmers bearing a large weight of continual crop monitoring. Implementing smart farming with IoT technology provides a solution, allowing farmers to use a mobile application for real-time observation and monitoring of parameters such as air humidity, temperature, and soil moisture, all of which affect plant development [48]. Also, depending on timers to regulate pumps in traditional watering systems is unworkable in real-world circumstances. It describes a framework for adjusting a pump's switching time based on user-defined variables, with sensors acting as the key aspect and contributor to the system [49].

9.3.1 Adaptability of fuzzy logic for handling uncertainty in agriculture

The successful management of agricultural supply chains has been a major topic for both scholars and practitioners as the agricultural industry transforms and new difficulties emerge [50]. Given this background, incorporating uncertain components into management decision-making has become increasingly important, as it may improve efficiency, responsiveness, corporate integration, and overall market competitiveness. An increasing research effort is given to dealing with uncertainty in order to appropriately meet the unexpected character of several real-world agricultural applications [51].

Fuzzy logic, which is well-known for its capacity to handle uncertainty in a variety of scientific and technical disciplines, may provide a solution for handling highly variable aspects in the agricultural supply chain, such as soil content, rainfall, humidity, and production and yield estimates [52]. This study investigates the use of fuzzy logic in the agricultural supply chain, taking into account factors such as land suitability, production techniques, irrigation, cold storage deficiencies, transportation, waste management, environmental and sustainability concerns, and drought management [53]. The chapter develops an integrated framework with the goal of analyzing the overall performance of the supply chain. Finally, the framework finds that there is a lack of efficient knowledge-based models in this sector. To solve this, the chapter advises adding a larger degree of heuristic and meta-heuristic simulations in the agricultural supply chain to construct more robust models using collaborative fuzzy applications with big data and Geographic Information System [54].

9.3.2 Advantages of fuzzy logic in modeling complex agricultural systems

Fuzzy logic modeling is comparable to classical modeling, where variables are important in determining how a system behaves and have relationships between them to characterize the system [55]. In traditional models, connections are expressed using mathematical functions, variables have real number values, and the outputs are exact numerical values. Alternatively, fuzzy logic creates fuzzy subsets as outputs and uses language phrases like "large, medium and small" to express variable values. It also forms linkages using if-then rules [56]. Defuzzification methods can then be used to make these fuzzy subsets exact. Fuzzification is the process of converting system variable values into language phrases by figuring out how much a value belongs to a certain fuzzy set. The application of modern technology and data analytics results in improved yield forecast and planning in agriculture [57, 58].

A. Prospective Crop Yields: Using historical data, weather trends, and crop monitoring information, farmers may generate more accurate predictions about prospective crop yields in the next season [59]. This skill allows businesses to more efficiently organize resources and operations, optimize planting dates, and adapt cultivation procedures to enhance production and profitability while avoiding risks associated with unknown crop outcomes [60].
B. Early Pest and Disease Identification: It is critical in agriculture to avoid significant crop damage and the need for excessive pesticide usage [61]. Using technology like as remote sensing, IoT sensors, and other monitoring tools, farmers may detect early signs of pests and illnesses. Not only does early identification protect crop health, but it also maximizes output potential, supporting more ecologically friendly and sustainable farming methods [62].
C. Enhancing Agricultural Resource Management: Precision agriculture technologies and data-driven decision-making enable successful

agricultural resource management, including the focused use of fertilizers and water [63, 64]. Farmers may continually monitor real-time data on soil moisture levels, crop nutritional requirements, and meteorological conditions by using IoT sensors, satellite imagery, and AI analytics. This vital data enables the exact and strategic application of resources, improving water efficiency and reducing fertilizer waste [65].

9.4 SOIL AND PLANT SENSORS

During the "Green Revolution" the use of synthetic fertilizers was critical in increasing agricultural output. However, because the guidelines were established for larger agro-ecological zones, the use of these fertilizers did not take into consideration local soil and water circumstances [66, 67]. The major goal was to increase output, with little regard for any negative environmental implications. This legacy continues, with recommendations still focused on "top-down" techniques based on limited data and general empirical connections between soil nutrient concentrations, fertilization rates, and yields [68, 69].

The use of soil sensors in agriculture has the potential to transform this strategy [70]. It provides novel "bottom-up" methodologies for characterizing local soil and environmental conditions in space and time. This adjustment in strategy improves production efficiency with the goal of increasing farm revenue while avoiding negative environmental consequences. The proximal or ground-based soil sensors, whether invasive or noninvasive, may collect high-resolution data quickly [71, 72]. In certain cases, these sensors can even allow for real-time analysis and processing, with readings occurring as often as once per second. When compared to standard laboratory procedures, sensor-based soil analysis has numerous potential advantages, including cheaper costs, increased efficiency, faster findings, and the ability to collect rich datasets while crossing a field [73].

9.4.1 Types of plant sensors and their applications

The sensors may be portable, field-fixed, or incorporated into agricultural gear such as tractors, combines, and sprayers, and perform a variety of functions such as obstacle avoidance, crop yield quantification, and pesticide volume measurement [74]. The incorporation of IoT-based agricultural sensors, such as temperature, moisture, depth, and humidity sensors, among others, improves environmental monitoring applications. These sensors, known collectively as agricultural sensors, provide crucial data that allows farmers to monitor and enhance crops by responding to changes in environmental circumstances [75, 76]. They are used in the agriculture business on weather stations, drones, and robotics. Sensors that measure air temperature and humidity, soil moisture, soil pH, light intensity, and carbon dioxide are used in all stages of crop growth, from nursery to growth and harvest [77].

Water and fertilizer levels are monitored using agricultural conductivity and pH sensors [78]. Temperature sensors are useful for growing crops that require certain moist or dry conditions [79]. These sensors detect the presence of thermal energy in the soil, assisting in the identification of ideal locations for plant growth as an important component in assuring product quality [80, 81]. GPS-based sensors, GIS-based sensors, electrochemical sensors, mechanical sensors, soil moisture sensors, airflow sensors, yield monitoring sensors, VRT sensors, and other types of sensors are also extensively used in precision farming and smart agriculture. Agriculture sensors, which include air temperature and humidity, soil moisture, soil pH, light intensity, and carbon dioxide, are essential for gathering data at various phases of crop growth, including nursery, growth, and harvest [82].

Electrochemical sensors are essential for collecting, analyzing, and mapping soil chemical data. These sensors, which are often installed on specially built sleds, give accurate information that is critical in agriculture [83]. Smart agriculture also known as precision agriculture, allows farmers to maximize harvests while using minimal resources such as water, fertilizer, and seeds. Farmers obtain a micro-scale awareness of their crops through the use of sensors and field mapping, allowing for resource-saving and decreased environmental impact [84, 85]. Sensors in digital agriculture assess crop growth and production characteristics. These devices can be integrated with machines for input applications, plant growth monitoring, and harvesters. IoT sensors help with real-time crop monitoring by allowing farmers to calculate the precise water and fertilizer needs of each plant at any given time [86]. The list of types of sensors is as follows:

Airflow Sensors
Sensors Utilizing Optical Technology in Agriculture
Soil Nutrient Detection through Electrochemical Sensors
Mechanical Sensors for Soil Applications in Agriculture
Dielectric Sensors for Measuring Soil Moisture
Location-Based Sensors in Agriculture
Electronic Sensors in Agricultural Applications

9.5 DEVELOPMENT OF FUZZY LOGIC MODELS: INCORPORATING SOIL AND PLANT SENSOR DATA INTO FUZZY LOGIC MODELS

Remote sensing and satellite photography are advanced methods used to acquire critical data from the Earth's surface from a distance. Satellites outfitted with various sensors collect photographs and data on the state of the land, sea, and atmosphere [87, 88]. These technologies are used in agriculture to monitor soil moisture levels, diagnose soil pests and diseases, and assess crop health. Field IoT devices are a network of networked items that include internet connectivity and sensors for real-time data collection and sharing [89]. These IoT devices are used in agriculture to monitor a variety of elements such as crop health, soil moisture, temperature, and humidity [90]. These solutions give farmers remote access to critical information, allowing them to make data-driven decisions, optimize resource consumption, and deploy precision agricultural approaches. In the long run, these approaches lead to improved productivity, lower prices, and the adoption of more ecologically friendly farming methods [91].

9.6 CROP HEALTH INDICATORS: IDENTIFICATION OF KEY INDICATORS FOR CROP HEALTH ASSESSMENT

Agriculture is vital to the economy of many countries, acting as a main source of food as well as a crucial contributor to economic growth. However, the global agricultural community is today facing a major challenge in the shape of diminishing output [92]. Disease management tactics are often decided by criteria such as disease kind, crop economic value, and market quality or demand. As a result, developing plant disease resistance and successfully controlling them is critical for assuring a stable food supply. Identifying crop diseases and quantifying the amount of infection are critical steps for mitigating possible dangers to plant life and resolving economic, biological, socioeconomic, and ecological losses. The accurate diagnosis of illnesses and the severity of symptoms is critical [93].

Drought is a major threat to agricultural productivity, affecting all phases of plant growth. Early seedling development factors can have a significant impact on eventual production. Despite the fact that low rainfall occurs often during growing seasons, the reactions of plants to different amounts of early-season soil moisture content have not been adequately investigated [94]. Two plant hybrids were grown in a pot-culture facility under five different soil moisture treatments to see how growth and development respond to varied soil moisture levels throughout the early phases of the growing season. Inadequate soil moisture content hampered plant growth and development, resulting in decreased physiological and phenotypic expression [95].

9.7 PRECISION AGRICULTURE AND RESOURCE OPTIMIZATION

Precision agriculture became a key component in the 1980s during the third wave of the modern agricultural revolution. PA was first used to focus fertilizer delivery depending on specific soil conditions, but it has since grown to include the creation of automated guidance systems for agricultural trucks and tools, as well as autonomous machinery and processes. It is also used in farm research and agricultural production system management [96]. The key idea of ML in agriculture is its adaptability, speed, precision, and cost-effectiveness. Artificial intelligence in agriculture not only assists farmers in improving their farming operations, but it also moves to targeted farming to obtain higher yields and better quality while using fewer resources. Climate variations (sunlight, rain, hail), agricultural machinery engines, animals, human traffic, and other exterior variables subject smart devices on farms to a variety of environmental influences [97]. These components highlight weaknesses in smart farm technologies that may not be found in other scenarios [98]. Deploying sensors and actuators in open regions where external agents, such as humans or machines, may interact with the surrounding environment is an example of novel application of artificial intelligence in agriculture. If an agent accidentally displaces or destroys the sensor, the corresponding controls will stop working. External agents may easily circumvent these devices, especially when they are not physically attached, as is frequently the case with smart cities. This absence of safeguards creates a significant risk in agricultural systems. Larger-scale ML applications in farming are now being developed, and amazing results are projected to be achieved throughout all stages of agricultural research and development over time [99].

9.8 CHALLENGES IN DEPLOYMENT OF FUZZY LOGIC SYSTEMS IN AGRICULTURE

In the digital era, protecting sensitive information from unwanted access, loss, or abuse is critical, making data privacy and security top priorities [100]. This is true across industries, including agriculture, where strong data protection procedures are essential for preserving trust, regulatory compliance, and operational consistency [101]. It is critical to use robust encryption, set access restrictions, perform frequent audits, and follow data protection legislation to limit risks and preserve the privacy and security of personal and private data. The integration of many tools and systems enables the smooth use of various technologies to improve performance and create advantageous synergies. In agriculture, the use of IoT sensors, drones, GPS, GIS, satellite imagery, and AI analytics allows for complete data collecting, analysis, and informed decision-making [102, 103].

9.9 CONCLUSION AND FUTURE SCOPE

The importance of ML is generally recognized in many industries, including agriculture. A subset of agricultural technology is becoming increasingly important for survival. Data is at the heart of farming decision-making, and it has immense promise. This illustrates the benefit of creating rules using mathematical equations and language factors in order to reduce human interference in farming. The goal is to educate farmers on the use of an integrated technology system to monitor and control activities. This method not only reduces manual input but also produces a competent group of decision-makers. Furthermore, the results provide light on the importance of each component in nurturing healthy plants, eventually leading to wise water management. Notably, previous research has not looked at chili plants, particularly those planted in containers, which have unique requirements. Chili plants may survive in ideal conditions, such as higher temperatures for seed germination and periodic soil drying between waterings. Improving small-holder farmers' access and affordability, adopting AI and ML applications, utilizing blockchain for traceability, and monitoring climate-resilient crops are all critical steps toward supporting sustainable agriculture and a resilient, transparent food system. These technologies and techniques have the potential to increase agricultural productivity, give farmers more control, create trust and responsibility, and pave the way for a more positive and secure future in farming and food production.

ML is expected to play a critical role in increasing resource efficiency, supporting sustainable resource consumption, and enabling major environmental advantages. Farmers will strive to produce the same amount of food with fewer resources in the future, which will result in less water consumption and less chemical use. Farmers seek improved production efficiency and lower costs in order to maximize both quantity and profitability while meeting the public's need for healthful food. To meet these changing demands, the agricultural business is continually seeking new goods, techniques, and technology. Precision agriculture is emerging as a response to farmers' different demands. There is a need for crop size and farm structure growth, as well as a modern approach to technology. While this is possible, it is unclear if it can be done in a sustainable and inclusive manner. Nonetheless, a significant change is necessary to reengineer agricultural operations on a large scale and at a quick speed. Agricultural sectors are eager to adopt new technologies in order to improve yields, owing to variables such as unpredictable weather, rising food demand, and rapid population expansion.

REFERENCES

1. Shaikh, T. A., Rasool, T., & Lone, F. R. (2022). Towards leveraging the role of machine learning and artificial intelligence in precision agriculture and smart farming. *Computers and Electronics in Agriculture*, *198*, 107119.
2. Shaikh, T. A., Mir, W. A., Rasool, T., & Sofi, S. (2022). Machine learning for smart agriculture and precision farming: towards making the fields talk. *Archives of Computational Methods in Engineering*, *29*(7), 4557–4597.
3. Rani, S., Das, K., Aminuzzaman, F. M., Ayim, B. Y., & Borodynko-Filas, N. (2023). Harnessing the future: cutting-edge technologies for plant disease control. *Journal of Plant Protection Research*, 387–398.

4. Kowalska, A., & Ashraf, H. (2023). Advances in deep learning algorithms for agricultural monitoring and management. *Applied Research in Artificial Intelligence and Cloud Computing*, 6(1), 68–88.
5. Kose, U., Prasath, V. S., Mondal, M. R. H., Podder, P., & Bharati, S. (Eds.). (2022). *Artificial Intelligence and Smart Agriculture Technology*. CRC Press.
6. AlZubi, A. A., & Galyna, K. (2023). Artificial intelligence and internet of things for sustainable farming and smart agriculture. *IEEE Access*.
7. Wanyama, J., Kiraga, S., Bwambale, E., & Katimbo, A. (2023). Improving nutrient use efficiency through fertigation supported by machine learning and internet of things in a context of developing countries: Lessons for Sub-Saharan Africa. *Journal of Biosystems Engineering*, 48(4), 375–391.
8. Dinesh, R. A., Shanmugam, J., & Biswas, K. (2023). Integration of Technology and Nanoscience in Precision Agriculture and Farming. In *Contemporary Developments in Agricultural Cyber-Physical Systems* (pp. 149–171). IGI Global.
9. Adli, H. K., Remli, M. A., Wan Salihin Wong K. N. S., Ismail, N. A., González-Briones, A., Corchado, J. M., & Mohamad, M. S. (2023). Recent advancements and challenges of AIoT application in smart agriculture: A review. *Sensors, 23*(7), 3752.
10. Balkrishna, A., Pathak, R., Kumar, S., Arya, V., & Singh, S. K. (2023). A comprehensive analysis of the advances in Indian digital agricultural architecture. *Smart Agricultural Technology*, *5*, 100318.
11. Ramamurthy, S., Chitraputhirapillai, S., Sampathkumar, T., & Jayaraj, J. (2022). Chapter-4 Artificial Intelligence (AI): Future of Indian Agriculture and Its Challenges. *Chief Editor Dr. RK Naresh*, 49.
12. Pathmudi, V. R., Khatri, N., Kumar, S., Abdul-Qawy, A. S. H., & Vyas, A. K. (2023). A systematic review of IoT technologies and their constituents for smart and sustainable agriculture applications. *Scientific African*, 19, e01577.
13. Mohan, D. S., Dhote, V., Mishra, P., Singh, P., & Srivastav, A. (2023). IoT Framework for Precision Agriculture: Machine Learning Crop Prediction. *International Journal of Intelligent Systems and Applications in Engineering, 11*(5s), 300–313.
14. Mondejar, M. E., Avtar, R., Diaz, H. L. B., Dubey, R. K., Esteban, J., Gómez-Morales, A., ... & Garcia-Segura, S. (2021). Digitalization to achieve sustainable development goals: Steps towards a Smart Green Planet. *Science of The Total Environment*, *794*, 148539.
15. Mohanty, M. N. (2021). Machine intelligence techniques for agricultural production. *Smart Agriculture: Emerging Pedagogies of Deep Learning, Machine Learning and Internet of Things*, 175.
16. Kamarudin, M. H., Ismail, Z. H., & Saidi, N. B. (2021). Deep learning sensor fusion in plant water stress assessment: A comprehensive review. *Applied Sciences*, *11*(4), 1403.
17. Mitra, A., Vangipuram, S. L., Bapatla, A. K., Bathalapalli, V. K., Mohanty, S. P., Kougianos, E., & Ray, C. (2022). Everything you wanted to know about smart agriculture. arXiv preprint arXiv:2201.04754.
18. Panwar, E., Kukunuri, A. N. J., Singh, D., Sharma, A. K., & Kumar, H. (2022). An Efficient Machine Learning Enabled Non-Destructive Technique for Remote Monitoring of Sugarcane Crop Health. *IEEE Access*, *10*, 75956–75970.

19. Kaswan, K. S., Dhatterwal, J. S., Baliyan, A., & Jain, V. (2022). Special sensors for autonomous navigation systems in crops investigation system. In *Virtual and Augmented Reality for Automobile Industry: Innovation Vision and Applications* (pp. 65–86). Cham: Springer International Publishing.
20. Kuppusamy, P., Suresh, J. K., & Shanmugananthan, S. (2023). Machine Learning-Enabled Internet of Things Solution for Smart Agriculture Operations. In *Handbook of Research on Machine Learning-Enabled IoT for Smart Applications Across Industries* (pp. 84–115). IGI Global.
21. Apat, S. K., Mishra, J., Raju, K. S., & Padhy, N. (2022). The robust and efficient Machine learning model for smart farming decisions and allied intelligent agriculture decisions. *Journal of Integrated Science and Technology*, *10*(2), 139–155.
22. Bhardwaj, A., Kishore, S., & Pandey, D. K. (2022). Artificial intelligence in biological sciences. *Life*, *12*(9), 1430.
23. Shamia, D., Suganyadevi, S., Satheeswaran, V., & Balasamy, K. (2023). Digital twins in precision agriculture monitoring using artificial intelligence. In *Digital Twin for Smart Manufacturing* (pp. 243–265). Academic Press.
24. Salunkhe, S. S., Agarkar, A. A., Karyakarte, M., Mulerikkal, J., Jonnala, P., & Mishra, D. K. (2023). Integrated agriculture IoT based farm monitoring and management systems. *International Journal of Nanotechnology*, *20*(5–10), 556–568.
25. Hossen, M. I., Fahad, N., Sarkar, M. R., & Rabbi, M. R. (2023). Artificial Intelligence in Agriculture: A Systematic Literature Review. *Turkish Journal of Computer and Mathematics Education (TURCOMAT)*, *14*(1), 137–146.
26. Bharti, V., Biswas, B., & Shukla, K. K. (2021). Computational intelligence in Internet of things for future healthcare applications. In *IoT-Based Data Analytics for the Healthcare Industry* (pp. 57–78). Academic Press.
27. Haridasan, A., Thomas, J., & Raj, E. D. (2023). Deep learning system for paddy plant disease detection and classification. *Environmental Monitoring and Assessment*, *195*(1), 120.
28. Bhavani, Y. V. K., Hatture, D. S. M., Pagi, D. V., & Saboji, D. S. (2023). An analytical review on traditional farming and smart farming: Various technologies around smart farming. *International Journal for Innovative Engineering & Management Research, Forthcoming.*
29. Xie, D., Chen, L., Liu, L., Chen, L., & Wang, H. (2022). Actuators and sensors for application in agricultural robots: A review. *Machines*, *10*(10), 913.
30. Abuzanouneh, K. I. M., Al-Wesabi, F. N., Albraikan, A. A., Al Duhayyim, M., Al-Shabi, M., Hilal, A. M., & Muthulakshmi, K. (2022). Design of Machine Learning Based Smart Irrigation System for Precision Agriculture. *Comput. Mater. Contin.*, *72*(1), 109–124.
31. Vadivelu, V., Nachimuthu, P., Ravi, A., & Thatchanamurthy, A. K. (2023, August). An IoT adoption in agriculture: Challenges and futuristic directions. In *AIP Conference Proceedings* (Vol. 2857, No. 1). AIP Publishing.
32. Naqvi, R. Z., Farooq, M., Naqvi, S. A. A., Siddiqui, H. A., Amin, I., Asif, M., & Mansoor, S. (2020). Big data analytics and advanced technologies for sustainable agriculture. *Handbook of Smart Materials, Technologies, and Devices: Applications of Industry 4.0*, 1–27.
33. Nabwire, S., Suh, H. K., Kim, M. S., Baek, I., & Cho, B. K. (2021). Application of artificial intelligence in phenomics. *Sensors*, *21*(13), 4363.

34. Peladarinos, N., Piromalis, D., Cheimaras, V., Tserepas, E., Munteanu, R. A., & Papageorgas, P. (2023). Enhancing smart agriculture by implementing digital twins: A comprehensive review. *Sensors*, *23*(16), 7128.
35. Shamshiri, R., Kalantari, F., Ting, K. C., Thorp, K. R., Hameed, I. A., Weltzien, C., & Shad, Z. M. (2018). Advances in greenhouse automation and controlled environment agriculture: A transition to plant factories and urban agriculture, *International Journal of Agricultural and Biological Engineering*, *11*(1), 1–22.
36. Omia, E., Bae, H., Park, E., Kim, M. S., Baek, I., Kabenge, I., & Cho, B. K. (2023). Remote Sensing in Field Crop Monitoring: A Comprehensive Review of Sensor Systems, Data Analyses and Recent Advances. *Remote Sensing*, *15*(2), 354.
37. Singh, S. (2018). Internet of Things Enabled Intelligent Mote with LoRa Architecture for Agricultural Use. *Turkish Journal of Computer and Mathematics Education (TURCOMAT)*, *9*(3), 1136–1144.
38. Saad, M. H. M., Hamdan, N. M., & Sarker, M. R. (2021). State of the art of urban smart vertical farming automation system: Advanced topologies, issues and recommendations. *Electronics*, *10*(12), 1422.
39. Masood, F., Khan, W. U., Jan, S. U., & Ahmad, J. (2023). AI-enabled traffic control prioritization in software-defined IoT networks for smart agriculture. *Sensors*, *23*(19), 8218.
40. Pandey, K., & Chudasama, D. (2022). Hybrid Au-thentication scheme based on cyber-physical system use in agriculture. *Journal of Artificial Intelligence Research & Advances*, *9*(2), 24–35p.
41. Shingade, S. D., & Mudhalwadkar, R. P. (2024). Analysis of crop prediction models using data analytics and ML techniques: a review. *Multimedia Tools and Applications*, *83*(13), 37813–37838.
42. Kabir, M. S., Islam, S., Ali, M., Chowdhury, M., Chung, S. O., & Noh, D. H. (2022). Environmental sensing and remote communication for smart farming: A review. *Precis Agric*, *4*, 82.
43. Jayasinghe, S. L., Thomas, D. T., Anderson, J. P., Chen, C., & Macdonald, B. C. (2023). Global Application of Regenerative Agriculture: A Review of Definitions and Assessment Approaches. *Sustainability*, *15*(22), 15941.
44. Wyawahare, M., Madake, J., Sarkar, A., Parkhe, A., Khuspe, A., & Gaikwad, T. (2023, April). Crop-weed detection, depth estimation and disease diagnosis using YOLO and Darknet for Agribot: A precision farming Robot. In *International Conference on Paradigms of Communication, Computing and Data Analytics* (pp. 57–69). Singapore: Springer Nature Singapore.
45. Corceiro, A., Alibabaei, K., Assunção, E., Gaspar, P. D., & Pereira, N. (2023). Methods for detecting and classifying weeds, diseases and fruits using AI to Improve the sustainability of agricultural crops: A review. *Processes*, *11*(4), 1263.
46. Akhtar, M. N., Ansari, E., Alhady, S. S. N., & Abu Bakar, E. (2023). Leveraging on advanced remote sensing-and artificial intelligence-based technologies to manage palm oil plantation for current global scenario: A review. *Agriculture*, *13*(2), 504.
47. Rehman, S. U. (2022). A Comparative Analysis of Fruits and Vegetables Quality Using AI-Assisted Technologies: A review. *Foundation University Journal of Engineering and Applied Sciences (HEC Recognized Y Category, ISSN 2706-7351)*, *3*(2).

48. Akensous, F. Z., Sbbar, N., Ech-Chatir, L., & Meddich, A. (2023). Artificial Intelligence, Internet of Things, and Machine-Learning: To Smart Irrigation and Precision Agriculture. In *Artificial Intelligence Applications in Water Treatment and Water Resource Management* (pp. 113–145). IGI Global.
49. Bhavsar, D., Limbasia, B., Mori, Y., Aglodiya, M. I., & Shah, M. (2023). A comprehensive and systematic study in smart drip and sprinkler irrigation systems. *Smart Agricultural Technology*, *5*, 100303.
50. Bhattacharyya, D., Joshua, E. S. N., Rao, N. T., & Kim, T. H. (2023). Hybrid CNN-SVM Classifier Approaches to Process Semi-Structured Data in Sugarcane Yield Forecasting Production. *Agronomy*, *13*(4), 1169.
51. Sidhu, R. K., Kumar, R., Rana, P. S., & Jat, M. L. (2021). Automation in drip irrigation for enhancing water use efficiency in cereal systems of South Asia: Status and prospects. *Advances in agronomy*, *167*, 247–300.
52. Sidhu, R. K., Kumar, R., Rana, P. S., & Jat, M. L. (2021). Automation in drip irrigation for enhancing water use efficiency in cereal systems of South Asia: Status and prospects. *Advances in Agronomy*, *167*, 247–300.
53. Sidhu, R. K., Kumar, R., Rana, P. S., & Jat, M. L. (2021). Automation in drip irrigation for enhancing water use efficiency in cereal systems of South Asia: Status and prospects. *Advances in Agronomy*, *167*, 247–300.
54. Polymeni, S., Plastras, S., Skoutas, D. N., Kormentzas, G., & Skianis, C. (2023). The Impact of 6G-IoT Technologies on the Development of Agriculture 5.0: A Review. *Electronics*, *12*(12), 2651.
55. Tripathy, A. S., & Sharma, D. K. (2020). Image processing techniques aiding smart agriculture. In *Modern Techniques for Agricultural Disease Management and Crop Yield Prediction* (pp. 23–48). IGI global.
56. Padhy, S., Alowaidi, M., Dash, S., Alshehri, M., Malla, P. P., Routray, S., & Alhumyani, H. (2023). AgriSecure: A Fog Computing-Based Security Framework for Agriculture 4.0 via Blockchain. *Processes*, *11*(3), 757.
57. Sharma, A., Sharma, V., Jaiswal, M., Wang, H. C., Jayakody, D. N. K., Basnayaka, C. M. W., & Muthanna, A. (2022). Recent trends in AI-based intelligent sensing. *Electronics*, *11*(10), 1661.
58. Sharma, A., Sharma, V., Jaiswal, M., Wang, H. C., Jayakody, D. N. K., Basnayaka, C. M. W., & Muthanna, A. (2022). Recent trends in AI-based intelligent sensing. *Electronics*, *11*(10), 1661.
59. Krishna, I. (2023). Examining the applications of artificial intelligence in agriculture. *Journal of Informatics Education and Research*, *3*(2).
60. Abdalla, A., Cen, H., Abdel-Rahman, E., Wan, L., & He, Y. (2019). Color calibration of proximal sensing RGB images of oilseed rape canopy via deep learning combined with K-means algorithm. *Remote Sensing*, *11*(24), 3001.
61. Sahil, K., Mehta, P., Bhardwaj, S. K., & Dhaliwal, L. K. (2023). Development of mitigation strategies for the climate change using artificial intelligence to attain sustainability. In *Visualization Techniques for Climate Change with Machine Learning and Artificial Intelligence* (pp. 421–448). Elsevier.
62. Khang, A. W. Y., Alsayaydeh, J. A. J., Gani, J. A. B. M., Bin Pusppanathan, J., Teh, A. A., Ismail, A. F. M. F., & Geok, T. K. (2023). Reliable multi-path communication for IoT based solar automated monitoring as motivation towards multi-farming hydroponic. *International Journal of Interactive Mobile Technologies*, *17*(21).

63. Albahar, M. (2023). A Survey on Deep Learning and Its Impact on Agriculture: Challenges and Opportunities. *Agriculture*, *13*(3), 540.
64. Kavithamani, V., & UmaMaheswari, S. (2023). Investigation of deep learning for whitefly identification in coconut tree leaves. *Intelligent Systems with Applications*, *20*, 200290.
65. Bandi, R., Swamy, S., & Arvind, C. S. (2023). Leaf disease severity classification with explainable artificial intelligence using transformer networks. *International Journal of Advanced Technology and Engineering Exploration*, *10*(100), 278.
66. Chen, L., Chen, Z., Zhang, Y., Liu, Y., Osman, A. I., Farghali, M., & Yap, P. S. (2023). Artificial intelligence-based solutions for climate change: a review. *Environmental Chemistry Letters*, 1–33.
67. Ji, W., Huang, X., Wang, S., & He, X. (2023). A comprehensive review of the research of the "Eye–Brain–Hand" harvesting system in smart agriculture. *Agronomy*, *13*(9), 2237.
68. Rehman, Z., Tariq, N., Moqurrab, S. A., Yoo, J., & Srivastava, G. (2023). Machine learning and internet of things applications in enterprise architectures: Solutions, challenges, and open issues. *Expert Systems*, e13467.
69. Raut, R., Kautish, S., Polkowski, Z., Kumar, A., & Liu, C. M. (Eds.). (2022). *Green Internet of Things and Machine Learning: Towards a Smart Sustainable World*. John Wiley & Sons.
70. Shelare, S. D., Belkhode, P. N., Nikam, K. C., Jathar, L. D., Shahapurkar, K., Soudagar, M. E. M., ... & Rehan, M. (2023). Biofuels for a sustainable future: Examining the role of nano-additives, economics, policy, internet of things, artificial intelligence and machine learning technology in biodiesel production. *Energy*, 128874.
71. Nayak, A. K. (2022). Advanced Techniques for Precision Farming in Rice. *Climate Resilient Technologies for Rice based Production Systems in Eastern India*.
72. Cheah, C. G., Chia, W. Y., Lai, S. F., Chew, K. W., Chia, S. R., & Show, P. L. (2022). Innovation designs of industry 4.0 based solid waste management: Machinery and digital circular economy. *Environmental Research*, *213*, 113619.
73. Mohammadi, M., Al-Fuqaha, A., Sorour, S., & Guizani, M. (2018). Deep learning for IoT big data and streaming analytics: A survey. *IEEE Communications Surveys & Tutorials*, *20*(4), 2923–2960.
74. Choudhary, H., Rajput, U., Kumar, A., & Dhiman, P. (2023). Crop recommendation system using WSN and ML algorithms.
75. Wójcik-Czerniawska, A. (2022). The role of Artificial Intelligence (AI) in agriculture and its impact on economy. *Agri-Tech Economics for Sustainable Futures*, *19*, 147.
76. Saheb, T., Dehghani, M., & Saheb, T. (2022). Artificial intelligence for sustainable energy: A contextual topic modeling and content analysis. *Sustainable Computing: Informatics and Systems*, *35*, 100699.
77. Rai, B. (2023). *Development of autonomous mobile robot for pesticide application and irrigation* (Doctoral dissertation, Mizoram University).
78. Ten, S. T., Shafie, K. A., Rani, R. A., & Hashim, M. H. (2023). Agriculture modernization and enhancement using advanced technologies. *Advances in Agricultural and Food Research Journal*, *4*(1).

79. Mohapatra, S., & Chaudhary, N. (2024). A fuzzy machine learning based approach for promoting sustainability in agriculture. *International Journal of Intelligent Systems and Applications in Engineering*, *12*(16s), 730–736.
80. Amertet Finecomess, S., Gebresenbet, G., & Alwan, H. M. (2024). Utilizing an Internet of Things (IoT) device, intelligent control design, and simulation for an agricultural system. *IoT*, *5*(1), 58–78.
81. Schutz, D. R., Mercaldi, H. V., Peñaloza, E. A., Silva, L. J., Oliveira, V. A., & Cruvinel, P. E. (2024). Advanced embedded generalized predictive controller based on fuzzy gain scheduling for agricultural sprayers with dead zone nonlinearities. *Journal of Process Control*, *135*, 103164.
82. Abidi, J. H., Elzain, H. E., Sabarathinam, C., Selmane, T., Slevam, S., Farhat, B., … & Senapathi, V. (2024). Evaluation of groundwater quality indices using multi-criteria decision-making techniques and a fuzzy logic model in an irrigated area. *Groundwater for Sustainable Development*, 101122.
83. Celekli, A., Lekesiz, Ö., Yavuzatmaca, M., & Dügel, M. (2024). Fuzzy logic as a novel approach to predict biological condition gradient of various streams in Ceyhan River Basin (Turkey). *Science of The Total Environment*, *916*, 170069.
84. Mana, A. A., Allouhi, A., Hamrani, A., Rahman, S., el Jamaoui, I., & Jayachandran, K. (2024). Sustainable AI-based production agriculture: Exploring AI applications and implications in agricultural practices. *Smart Agricultural Technology*, 100416.
85. Abilda, S., Kaliyeva, A., Ilyashova, G., & Yerezhepova, A. (2024). Corporate strategies in agricultural enterprises: Adaptation and development in the COVID-crisis environment. *Heliyon*, *10*(2).
86. Sarkar, R. (2024). Classification of Finding Degree of Truth Using Fuzzy Logic System for Generating a Systematic Flow of Heat Moisture System. In *Agriculture and Aquaculture Applications of Biosensors and Bioelectronics* (pp. 416–427). IGI Global.
87. Bhoyar, C., Kanojia, K. P., & Chourasia, B. (2024). Fuzzy rules based smart irrigation system using adaptive bacterial foraging optimization. In *IOP Conference Series: Earth and Environmental Science* (Vol. 1285, No. 1, p. 012019). IOP Publishing.
88. Hamouda, Y. E. (2024). Optimally sensors nodes selection for adaptive heterogeneous precision agriculture using wireless sensor networks based on genetic algorithm and extended kalman filter. *Physical Communication*, *63*, 102290.
89. Kasera, R. K., Gour, S., & Acharjee, T. (2024). A comprehensive survey on IoT and AI based applications in different pre-harvest, during-harvest and post-harvest activities of smart agriculture. *Computers and Electronics in Agriculture*, *216*, 108522.
90. Felizardo, K. B., Paredes, A. M. C., & Arboleda, E. R. (2024). Advancements in Artificial Intelligence (AI) for enhanced insights and automation in rice agriculture: A systematic review. *International Journal of Science and Research Archive*, *11*(1), 444–463.
91. Singh, R. R., & Hati, A. J. (2024). Soilless Smart Agriculture Systems for Future Climate. In *Digital Agriculture: A Solution for Sustainable Food and Nutritional Security* (pp. 61–111). Cham: Springer International Publishing.
92. Krishnan, V. G., Rao, B. S., Prasad, J. R., Pushpa, P., & Kumari, S. (2024). Sugarcane yield prediction using NOA-based swin transformer model in

IoT smart agriculture. *Journal of Applied Biology and Biotechnology*, *12*(2), 239–247.

93. Cartolano, A., Cuzzocrea, A., & Pilato, G. (2024). Analyzing and assessing explainable AI models for smart agriculture environments. *Multimedia Tools and Applications*, 1–22.
94. Budak, M., Kılıç, M., Günal, H., Çelik, İ., & Sırrı, M. (2024). Land suitability assessment for rapeseed potential cultivation in upper Tigris basin of Turkiye comparing fuzzy and boolean logic. *Industrial Crops and Products*, *208*, 117806.
95. Polyakov, V., & Aleksandrovskaya, L. (2024). Smart agriculture as the embodiment of the symbiosis of the technological and intellectual potential of the agricultural sector. In *E3S Web of Conferences* (Vol. 480, p. 03017). EDP Sciences.
96. Benaiche, M., Mokhtari, E., Berghout, A., Abdelkebir, B., & Engel, B. (2024). Identification of soil erosion-susceptible areas using revised universal soil loss equation, analytical hierarchy process and the fuzzy logic approach in sub-watersheds Boussellam and K'sob Algeria. *Environmental Earth Sciences*, *83*(1), 34.
97. Kumar, G. K., Bangare, M. L., Bangare, P. M., Kumar, C. R., Raj, R., Arias-Gonzáles, J. L., ... & Mia, M. S. (2024). Internet of things sensors and support vector machine integrated intelligent irrigation system for agriculture industry. *Discover Sustainability*, *5*(1), 6.
98. Mohamed-Amine, N., Abdellatif, M., & Belaid, B. (2024). Artificial intelligence for forecasting sales of agricultural products: A case study of a moroccan agricultural company. *Journal of Open Innovation: Technology, Market, and Complexity*, *10*(1), 100189.
99. Haghiri, M., Raeisi, N., Azizi, R., Shabani, K., & Ghadiri, M. (2024). Evaluation of karst aquifer development and karst water resource potential using fuzzy logic model (FAHP) and analysis hierarchy process (AHP): a case study, North of Iran. *Carbonates and Evaporites*, *39*(2), 11.
100. Hussain, S., Nasim, W., Mubeen, M., Fahad, S., Tariq, A., Karuppannan, S., & Ghassan Abdo, H. (2024). Agricultural land suitability analysis of Southern Punjab, Pakistan using analytical hierarchy process (AHP) and multi-criteria decision analysis (MCDA) techniques. *Cogent Food & Agriculture*, *10*(1), 2294540.
101. Wajid, M. S., Terashima-Marin, H., Wajid, M. A., Smarandache, F., Verma, S. B., & Wajid, M. K. (2024). Neutrosophic Logic to Navigate Uncertainty of Security Events in Mexico. Neutrosophic Sets and Systems, *73*, 120–130.
102. Raihan, A. (2024). A Systematic Review of Geographic Information Systems (GIS) in Agriculture for Evidence-Based Decision Making and Sustainability. *Global Sustainability Research*, *3*(1), 1–24.
103. Wajid, M. A., Camacho-Zuñiga, C., Smarandache, F., Terashima-Marin, H., Wajid, M. S., & Wajid, M. K. (2024). TEC 21 model and critical thinking: an NCM-based neutrosophic analysis in higher education. *Neutrosophic Sets and Systems*, 73, 140–152.

Chapter 10

Revolutionizing data analytics

The cutting-edge role of soft computing techniques

Wasim Khan, Mohammad Ishrat, and Monia Mohammed Al Farsi

10.1 INTRODUCTION

In contrast to conventional algorithms, the set of computational methods known as "soft computing" is specifically designed to deal with data that is vague, unclear, or otherwise difficult to interpret [1]. Because approximation is fundamental to soft computing, it is very useful in situations where exact solutions are either not practicable or not required. This is in sharp contrast to the exacting standards of conventional computers, which depend on predetermined inputs and outcomes. To simulate the way the human brain works when faced with ambiguity or uncertainty, soft computing relies on its incredibly adaptable nature [2]. The incorporation of essential components like neural networks, fuzzy logic, and evolutionary algorithms allows the system to adapt and is good at handling uncertainty and imprecision. One alternative to classical logic's black-and-white true/false dichotomy is fuzzy logic, which provides degrees of truth or probability. When it comes to pattern recognition and predictive modeling, neural networks shine because they mimic the brain's learning capacities. Biological evolution serves as an inspiration for evolutionary algorithms, which are highly effective in searching complicated, multidimensional areas for resilient solutions.

As we go into the world of data analytics, where big data analysis is the name of the task, soft computing becomes even more important. In data analytics, a variety of methods from computer science, statistics, and machine learning are used to find trends, patterns, and correlations in the data. This allows for better decision-making and the ability to spot hidden insights [3]. Given the importance of big data, the merging of soft computing with data analytics is an inevitable and vital development. Common problems with big data include data that is not organized, is noisy, or is incomplete, all of which traditional data analytics approaches struggle with. Soft computing, which has an inherent ambiguity tolerance and may produce approximations, significantly improves data analytics procedures. It allows for a more sophisticated analysis of data, which can provide insights that stricter computational methods could overlook. The authors of Ref. [4]proved that this synergy

DOI: 10.1201/9781003606055-10

opens up new possibilities in many different disciplines, such as scientific research that derives deeper insights from complicated data sets and commercial analytics that forecast customer behavior.

When applied to data analytics, soft computing essentially signifies a paradigm shift in problem-solving in the age of big data. With its foundation in approximation, tolerance of uncertainty, and experience learning, soft computing offers a strong and adaptable framework for data analytics that can go beyond what traditional approaches can achieve. This chapter highlights the importance of soft computing in the ever-changing field of data analytics, especially in industries such as business, research, and technology where data is frequently copious yet prone to errors, gaps, or inconsistencies. The authors of Ref. [5] found that traditional data analysis approaches struggle to handle the complexity of real-world data because they are limited to using exact, deterministic algorithms and well-defined, organized data. In this case, the adaptability and lack of accuracy associated with soft computing approaches really show, allowing us to glean useful insights from data that would have been indecipherable using more conventional methods.

According to the authors of Ref. [6], soft computing has several practical and significant uses in data analytics. For example, by analyzing purchase behaviors, social media interactions, and customer feedback, soft computing helps firms in market analysis discover patterns of consumer behavior that might not be immediately obvious. As a result, advertising and new product development becomes more efficient. When dealing with large datasets that contain complicated, sometimes nonlinear interactions, as is the case in fields like genetics and climate modeling, soft computing plays a crucial role in scientific research [7]. Similarly, neural networks and fuzzy logic are used in predictive analytics, especially for consumer behavior prediction, to help companies see patterns and likelihoods rather than absolute accuracy in predicting what customers will do next based on past data [8]. This strategy has an impact on decisions in a variety of fields, including targeted marketing and inventory management. These examples show how important soft computing is in the modern data-centric world, as it is used to manage massive amounts of data and extract useful insights from data that is too complicated for conventional analytics. An in-depth analysis of soft computing and its pivotal function in data analytics is the goal of this chapter. In order to provide readers with a solid theoretical grounding as well as practical insights, the objectives are structured to lead them on a journey from basic principles to sophisticated applications. An overview of soft computing's foundational ideas, including fuzzy logic, neural networks, evolutionary algorithms, and the philosophies that underpin them, will be provided in the first section of the chapter. To grasp the distinctive benefits of soft computing for handling complicated, real-world data and how it differs from conventional computational methods, this groundwork is essential.

Demonstrating the use of soft computing in data analytics is one of the primary goals. Included in this will be in-depth analyses of how these methods handle and interpret incomplete, ambiguous, and massive information, revealing patterns that conventional approaches might overlook. In this chapter, we will examine the technical details of several soft computing approaches and their usage in a variety of contexts, from scientific research to business intelligence and market analysis, all with the goal of demonstrating how soft computing helps to get value from data. Just as crucial as comprehending the applications is acknowledging the difficulties. Computational complexity and the requirement for massive datasets are two of the main challenges that have been identified for the purpose of incorporating soft computing into data analytics. Furthermore, it will delve into the possible hazards and ethical concerns linked to these technologies.

Relating abstract ideas to real-world scenarios is a primary goal. You may make the learning experience more accessible and based on reality by offering real-world examples and case studies where soft computing has been effectively utilized. Lastly, the chapter will speculate on what's to come by talking about new developments in data analytics and soft computing. There will be some guesswork about what the area may become in light of new technologies and shifting data landscapes, as well as some speculation about possible new applications and future advances. By accomplishing these goals, the chapter will serve to educate readers while simultaneously piquing their interest in learning more about the fascinating and ever-changing area of soft computing as it pertains to data analytics.

In this work, we explore the cutting-edge role of soft computing techniques in revolutionizing data analytics, offering substantial contributions to the domain. By innovatively applying fuzzy logic, neural networks, and evolutionary algorithms, we address complex, uncertain, and imprecise data analytics challenges. We introduce a comprehensive, integrative framework for deploying these techniques effectively, bolstered by empirical evaluations and illustrative case studies that confirm their practical value. Furthermore, we delve into future research avenues, particularly the integration with big data and Internet of Things (IoT), to underscore the potential for further advancements. Simultaneously, we meticulously examine the ethical and computational hurdles inherent in these applications, providing insights on navigating these challenges. This multifaceted exploration not only highlights the transformative impact of soft computing on data analytics but also sets a foundational roadmap for future exploration and application in this rapidly evolving field.

The organization of our chapter is meticulously structured to guide readers through the intricate landscape of soft computing techniques in data analytics. It begins with an introduction to the fundamental concepts of soft computing, followed by an in-depth analysis of various methods, including fuzzy logic, neural networks, and evolutionary algorithms. Subsequent sections provide detailed case studies and empirical evaluations, showcasing the practical applications and effectiveness of these techniques. The chapter also explores future research directions, potential advancements, and ethical and computational challenges. The conclusion synthesizes the insights gained and reaffirms the transformative potential of soft computing in data analytics.

10.2 FOUNDATIONS OF SOFT COMPUTING

10.2.1 Definition, history, and key principles

In contrast to the conventional, rigid algorithms that characterize hard computing, soft computing represents a new paradigm shift in the field of computing [9]. When traditional computing's strict binary logic and exact analytical approaches fail to tackle complicated real-world situations, this approach is designed to step in. Soft computing revolves around its incredible adaptability, which allows it to tackle complex issues with solutions that are bearable and practical but not accurate. Since its inception in the 1960s and its official acknowledgment in the 1990s—thanks in large part to Lotfi A. Zadeh's fuzzy logic—soft computing has developed to mimic the way the human mind processes information, which is very skilled at dealing with ambiguity and imprecision [10]. A crucial portion of soft computing was established by Zadeh's fuzzy sets in 1965, which allowed for an approach based on approximation reasoning and partial truths. In a world where complicated issues defy exact mathematical answers, this approach is crucial. The advent of genetic algorithms in the 1970s and 1980s, which were influenced by natural systems, and neural networks in the 1980s, after Zadeh's work, both contributed to the field's enrichment. The incorporation of these varied approaches is what defines the evolution of soft computing, as they each offer something special to the table when it comes to handling the subtleties of uncertainty and imprecision in complicated situations [11].

Several fundamental principles govern the use and efficacy of soft computing. Recognizing and navigating the fuzziness and ambiguity that frequently characterize real-life scenarios, it displays a level of uncertainty and imprecision tolerance. The emphasis on accurate solutions in hard computing stands in sharp contrast to this. To add insult to injury, when solving complicated problems, finding the optimal solution is frequently not feasible or perhaps unattainable; in such cases, soft computing algorithms will seek practical solutions that are good enough. When a strong, flexible solution is more important than a flawless one, this comes in handy [12].

Soft computing also relies heavily on its capacity to learn and adapt. Its methods are built to learn from data and get better with time; this is especially true of neural networks and some evolutionary algorithms. Systems must adapt to stay efficient in an environment that is always changing, so flexibility is vital. In addition, computer science, mathematics, engineering, and the cognitive sciences are all integral parts of soft computing [13]. This eclectic mix of disciplines results in a more robust and adaptable toolbox for problem-solving, enabling novel approaches to challenging problems.

10.2.2 Theoretical underpinnings and basic concepts of soft computing

The foundational principles of soft computing have been shaped by a complex web of fields, the most important of which are genetic algorithms, fuzzy logic, and neural networks. Traditional computing technologies generally fail to mimic human decision-making abilities in contexts characterized by high levels of ambiguity and imprecision [14].

The foundation of this system is fuzzy logic, an idea that Lotfi Zadeh helped to establish. Fuzzy sets, in which components have degrees of membership rather than an absolute presence or absence, are introduced, departing from the strict dichotomy of conventional binary logic. When dealing with the ambiguity and uncertainty that often arise in real-world situations, this sophisticated method of classification is crucial. Another foundation of soft computing is neural networks, which mimic the complex human brain [15]. The linked nodes of these ANNs, which resemble real neurons, are very good at data processing and learning. Pattern recognition and predictive modeling are two areas where their adaptability and learning capabilities shine; these fields allow them to extrapolate and make well-informed conclusions using the information they have gathered.

When it comes to solving problems involving complicated, multidimensional search spaces, genetic algorithms, which are based on genetic and natural selection principles, provide strong answers. These algorithms are great at solving optimization and search problems with near-optimal solutions by using methods including selection, mutation, and crossover. When more conventional, sequential methods of addressing problems don't work, this facet of soft computing shines [16].

Several key ideas support the study of soft computing and, taken as a whole, make it more useful and applicable. Important to this is approximation reasoning, which, in contrast to conventional computers' reliance on exact, binary logic, enables soft computing to flourish in contexts where information is imperfect or imprecise. When dealing with noisy, partial, and imperfect data, as is often the case in real-world applications, this feature of soft computing gives a degree of resilience that is critical.

One of the most important features of soft computing approaches is their adaptability and flexibility. This allows them to accommodate different contexts and datasets, which is especially important in fields like pattern recognition where systems need to be constantly evolving to detect new patterns or adapt to changes in old ones. Soft computing encourages novel ways to tackle complicated problems since it is multidisciplinary and draws from many different areas, including mathematics, biology, computer science, and cognitive sciences. By bringing together experts from different fields, soft computing is able to efficiently and creatively tackle a broad variety of problems [17].

10.3 SOFT COMPUTING IN DATA ANALYTICS

10.3.1 Application of soft computing techniques in data analytics

Improving data analytics skills is where soft computing really shines. Fuzzy logic, evolutionary algorithms, and neural networks are some of its methodologies. These give powerful tools for evaluating complicated datasets, which is particularly useful when standard statistical approaches fail. These methods are great at dealing with the usual features of real-world data (Figure 10.1), such as uncertainty, non-linearity, and imprecision.

Fuzzy Logic in Data Classification: When it comes to categorizing data, fuzzy logic—an essential part of soft computing—is king. According to [18], this method shines in situations where data cannot be clearly divided into two categories. In market research, for instance, customer tastes are notoriously difficult to classify. To account for the inherently subjective character of human decision-making and action-taking, fuzzy logic permits the fuzzy categorization of these preferences. Because of this, market segmentation may be more precise and advanced, which helps companies learn more about their customers and develop better marketing and customer interaction strategies.

Fuzzy Logic in Data Classification	**Neural Networks for Pattern Recognition**	**Genetic Algorithms for Optimization Problems**
• Handles non-binary data categorization. • Useful in nuanced market analysis. • Classifies ambiguous consumer preferences. • Improves market segmentation accuracy. • Aids in developing targeted marketing strategies.	• Simulates brain-like data processing. • Identifies complex patterns in datasets. • Predicts trends in financial markets. • Analyses consumer behaviors patterns. • Detects anomalies in transaction data.	• Inspired by natural selection processes. • Searches for optimal solutions in large datasets. • Applied in logistics and supply chain management. • Optimizes routes and resource allocations. • Adaptable to changing conditions and parameters.

Figure 10.1 Soft computing techniques in data analytics.

Neural Networks for Pattern Recognition: When it comes to pattern detection in complicated datasets, neural networks—which are modeled after the way the human brain works—are invaluable. Financial analytics makes great use of their data-learning capabilities. To forecast stock market trends, for example, neural networks sift through mountains of financial data, both historical and real-time, in search of patterns that might point to where the market is headed [19]. Further, neural networks help with consumer analytics by interpreting complicated patterns of customer behavior and by detecting fraudulent actions through the identification of out-of-the-ordinary patterns in transaction data [20].

Genetic Algorithms for Optimization Problems: When searching through massive datasets for effective solutions becomes necessary, another aspect of soft computing, known as genetic algorithms (GAs), comes into play. Optimization of supply chains and logistics are two areas where these algorithms, which imitate natural selection, excel. Addressing issues such as transportation cost minimization, delivery time reduction, and inventory optimization, GAs can sift through mountains of data to find the best routes and allocate resources. They excel at optimizing processes in dynamic contexts with regularly changing circumstances and parameters because they can adapt solutions across iterations [21, 22].

10.3.2 Examples of soft computing applications

The adaptability and superior data-analysis capabilities of soft computing approaches are proving to be invaluable in a number of different sectors. The use of neural networks has had a notably revolutionary impact on healthcare. These networks outperform conventional approaches to predicting patient outcomes and illness diagnosis using medical imaging. When it comes to patient care and treatment outcomes, precise diagnoses and forecasts are life-changing, making this development all the more important in the medical industry. Modern healthcare analytics cannot function without soft computing due to its exceptional capacity to manage intricate, real-world medical data.

Fuzzy logic has become an invaluable resource for environmental scientists seeking to understand intricate systems. Particularly in cases where data is ambiguous or lacking, its use in calculating the effects of climate change and comprehending the behavior of ecosystems is priceless. Environmental scientists can build more effective conservation policies and tactics by using fuzzy logic to gain a more nuanced knowledge of ecological dynamics and environmental concerns (Figure 10.2) [23–25].

One area where soft computing has been used innovatively is in the financial sector, namely with GAs. In order to optimize portfolios and conduct

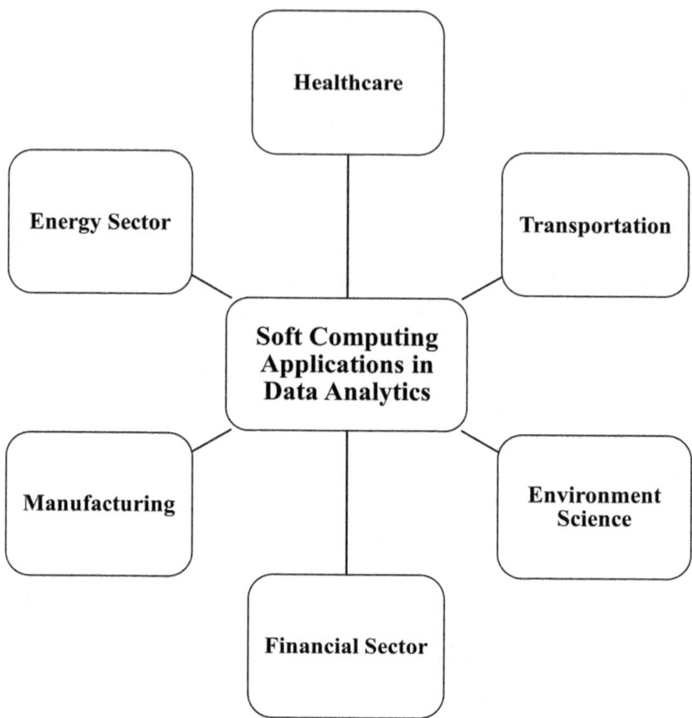

Figure 10.2 Soft computing applications.

risk assessments, these algorithms are used to build complex models. These models provide more durable and adaptable financial solutions for the intrinsically uncertain and turbulent financial markets. According to [26–28], they improve the overall profitability and efficiency of financial operations by assisting with market difficulties and making educated investment decisions.

Additionally, these aren't the only domains investigating the potential of soft computing approaches; other domains include manufacturing, where they may optimize production processes, and retail, where they can improve the consumer experience via trend analysis and tailored suggestions. The integration of soft computing into energy sectors for better resource management and transportation systems for better traffic flow optimization further demonstrates its widespread use [29, 30].

The wide range of applications of soft computing approaches across numerous sectors in today's data-driven world demonstrate their relevance. These methods are a boon in the face of the modern world's complicated and multifaceted problems because they provide more flexible, accurate, and dependable answers across a wide range of disciplines, from healthcare to environmental research, economics to manufacturing.

10.4 TECHNIQUES AND TOOLS IN SOFT COMPUTING FOR DATA ANALYTICS

10.4.1 Soft computing techniques

When faced with situations involving imprecise or confusing data, fuzzy logic systems shine at handling ambiguity and uncertainty. Their decision-making process is more natural and human-like, making them very effective in clustering, classification, and pattern recognition tasks. The membership functions in fuzzy logic are defined by expert knowledge, which might be subjective and doesn't necessarily reflect complicated data connections effectively [31–33].

Predictive analytics, picture and voice recognition, and natural language processing are just a few areas where neural networks shine because of their famous learning capabilities and flexibility. One major benefit is their capacity to learn from big data and get better with time. But a lot of computer power is needed for them, particularly for deep learning models. Furthermore, it can be difficult to understand the outcomes produced by neural networks since, like "black boxes," their decision-making process is not always open and obvious [34–38].

For difficult optimization issues, such as feature selection and model optimization, GAs provide strong answers. By taking an evolutionary stance, they are able to thoroughly investigate all potential options, which frequently results in the best conceivable results. On the other hand, GAs could take a long time to converge to a solution and might be computationally expensive. Another potential issue is becoming stuck in local optima, especially when dealing with extremely complicated problem areas [39–41].

Support In high-dimensional domains, support vector machines (SVMs) perform exceptionally well in regression and classification tasks. Their precision in bioinformatics, picture classification, and text classification has made them famous. When dealing with extremely large datasets, SVMs' computational inefficiency becomes apparent because of the lengthy training period. This is the main constraint of SVMs. Additionally, SVM kernel function choice is crucial and can dramatically affect model performance [42–44].

Data analytics optimization challenges are well-suited to evolutionary computing methods such as differential evolution (DE) and evolutionary strategies (ES). Parameter adjustment in ML models is where they really shine. A major drawback of these algorithms is their potential sluggish convergence and the high number of function evaluations needed to obtain an optimum solution. Also, the possibility of convergence to local optima exists, just as it does with gas [45–47].

10.4.2 Tools and software

Plenty of software and tools, each with its own set of features and capabilities, make it easy to apply soft computing approaches to data analytics.

Fuzzy logic and neural network techniques are widely implemented in MATLAB, a high-level language and interactive environment. It is well-liked by engineers and researchers because of the extensive range of tools it provides for creating and evaluating soft computing solutions. The robust computational capabilities, wide toolbox for various soft computing approaches, and intuitive user interface of MATLAB are its greatest assets. The price tag is high, and some people might not be able to afford it because of its proprietary nature [48].

Python libraries offer a flexible and user-friendly environment for doing soft computing. Neural network libraries provide a versatile and user-friendly framework for deep learning applications; examples include TensorFlow and Keras. Another well-liked package for many machine learning algorithms is scikit-learn, which is renowned for being both efficient and easy to use. As an evolutionary algorithm specialist, DEAP equips users with the means to put into action a wide range of genetic and evolutionary tactics. While more targeted programs, such as MATLAB, offer more specialized features, Python's open-source nature and large community support make it a desirable option. [49, 50].

Statistics and visual computing find a home in the R programming language and environment. It provides powerful tools for data analysis and visualization with packages like "e1071," which is for SVMs, and "nnet," which is for neural networks. The statistical analysis and visualization capabilities of R make it a popular option among researchers and academic institutions. Those just starting out in the world of computer programming may find that R's learning curve is steeper than Python's [51].

Data mining challenges may be tackled with the help of WEKA's array of machine learning algorithms. Both theoretical and practical uses benefit greatly from its support for various soft computing approaches. WEKA is well-known for its user-friendliness, which allows even those without substantial programming knowledge to utilize it. It may, however, fall short of the scalability and flexibility offered by languages like R or Python [52].

10.5 TRENDS AND FUTURE DIRECTIONS IN SOFT COMPUTING FOR DATA ANALYTICS

10.5.1 Current trends in soft computing integration

The integration of soft computing in data analytics is evolving rapidly, with several key trends shaping its future:

Deep Learning and Neural Networks: Deep learning, a subfield of neural networks, is seeing a huge upswing right now. This expansion is fundamentally altering fields like predictive analytics, natural language processing, picture and audio recognition, and natural language understanding. The networks are getting smarter and better at doing complicated jobs. These systems can now evaluate massive volumes of data with an unprecedented degree of complexity because of developments in deep learning algorithms and computer capacity. This is causing breakthroughs in a variety of fields [53].

Hybrid Models: Hybrid models, which combine several soft computing approaches, are becoming increasingly popular. This type of model takes advantage of the best features of both neural networks and fuzzy logic by combining the two into neuro-fuzzy systems. This integration enhances the models' capacity to learn from data in the face of uncertainty and imprecision. It is becoming clear that hybrid models outperform single-method approaches, especially in complicated situations [54, 55].

Explainable AI (XAI): Transparency and interpretability in results are becoming more important as the complexity of soft computing models, especially those in deep learning, increases. XAI's primary goal is to facilitate human specialists' comprehension of these models' inner workings and choices. In key applications such as healthcare, banking, and autonomous systems, where comprehending the reasoning behind choices is just as important as the conclusions themselves, this tendency is vital for acquiring confidence and wider adoption of AI systems[56, 57].

Quantum Computing: Soft computing is about to undergo revolutionary changes due to the advent of quantum computing. Thanks to quantum computing's very rapid processing rates, soft computing techniques will be able to perform much better. The processing of massive datasets and sophisticated computations is becoming increasingly widespread in numerous industries, making this breakthrough particularly essential in those areas. Potentially ushering in a new era of data analytics, quantum-enhanced soft computing has the ability to solve problems more efficiently and effectively than ever before [22].

10.5.2 Future potential and areas for research

The future landscape of soft computing in data analytics is shaping up to be both exciting and transformative, with several key developments on the horizon:

Advancements in Algorithmic Efficiency: A constant endeavor is being made to improve the efficacy of soft computing algorithms, particularly GAs and neural networks. These algorithms are now slow, but they should become much quicker and more efficient with time and technology. This enhancement is vital for more sustainably managing data sets of ever-increasing complexity and scale. With improved efficiency, these algorithms may handle more complicated studies with the same or even lower processing costs, expanding their usefulness in many other areas.

Integration with IoT and Big Data: IoT is a rapidly expanding field that generates enormous amounts of data. For effective analysis of this massive data set, soft computing is set to play a key role. Better, more autonomous IoT devices and systems are anticipated outcomes of this integration. When it comes to driving advancements in areas like smart cities, industrial automation, and customized healthcare, the interplay between IoT and soft computing will play a pivotal role in extracting valuable insights from massive datasets [58].

Focus on Real-Time Analytics: Industries such as financial trading, healthcare monitoring, and driverless cars are driving the need for real-time data processing. To keep up with these expectations, real-time analytics systems that rely on soft computing will have to improve in speed and efficiency. Systems that need to make decisions quickly rely on real-time data processing and analysis, which can only be made possible with the help of developments in soft computing [22, 59, 60].

Ethical AI and Bias Mitigation: The need for building ethical and bias-free techniques will grow as AI and soft computing are more integrated into decision-making across many sectors. Particularly in delicate domains like financing, criminal justice, and employment, it is essential to guarantee equity and fairness in AI-driven judgments. To promote the development of trustworthy and equitable AI systems, future advances in soft computing will probably incorporate methods to detect, reduce, and eradicate biases [61].

10.5.3 Predictions about the evolution of these technologies

As we look toward the future, the trajectory of soft computing in data analytics points toward several promising developments:

Increased Adoption Across Industries: The widespread use of soft computing techniques is anticipated to skyrocket in the next few years. Their potential application in healthcare lies in the improvement of diagnostic accuracy and the enhancement of patient care. They could improve financial risk analysis and market forecasting. Soft computing's increasing importance in data processing, both in terms of accuracy and efficiency, will propel its broad adoption [62].

Advancements in Learning Algorithms: We foresee a future where learning algorithms are more intelligent and can successfully learn from fewer datasets. In settings where data is limited or costly to get, this will be very helpful. Improved forecasts and analyses, especially in dynamic and complicated environments, will be the result of these developments, making soft computing approaches more reliable and useful for important decision-making [63].

Cross-disciplinary Applications: The use of soft computing is anticipated to go beyond conventional domains, encouraging creative answers in a wide range of subjects. More precise climate modeling and ecological management might be possible with its usage in environmental research. It has the potential to shed light on social dynamics and human behavior in the social sciences. As an example of its flexibility and adaptability, soft computing might even pave the way for new kinds of artistic expression and interpretation in the arts[64].

Focus on User-Friendly Tools: The creation of easier-to-use platforms and tools for deploying soft computing is expected to receive increased attention. Because of this, even those without strong technical experience will be able to take advantage of these powerful strategies. More individuals will be able to apply soft computing approaches to a wide variety of issues and scenarios if they are more easily available, which should lead to a spike in creativity and application.

10.6 CHALLENGES AND CONSIDERATIONS IN SOFT COMPUTING FOR DATA ANALYTICS

10.6.1 Challenges in implementing soft computing

To be effective, soft computing's use in data analytics must take into account a number of obstacles. Complex computational and resource-intensive soft computing methods, especially neural networks, provide a significant obstacle. In particular, smaller companies or projects with fewer resources may find these solutions prohibitive due to the high processing power they require. Particularly in settings with limited access to powerful computational resources, this restriction makes scaling and accessibility difficult.

The reliance on accessible and high-quality data is an additional formidable obstacle. There is a strong correlation between the amount and quality of the data used to train soft computing algorithms. Poor data quality, including biases, incompleteness, or noise, can severely hinder these models' performance, resulting in incorrect and wrong predictions. Because the input determines the accuracy and dependability of the model, it is crucial to have access to high-quality, representative data. This frequently calls for labor- and resource-intensive data gathering and preparation.

Another concern with soft computing models, particularly deep learning-based ones, is their interpretability and transparency. "Black box" situations arise when these models get too complicated and are no longer easy to understand. In contexts where knowing how decisions are made is vital, this opaqueness can be a huge problem. This makes model validation and debugging more challenging and has an impact on user trust and adoption, which are both essential for ensuring the stability and ethical use of AI systems.

Finally, there are unique difficulties associated with incorporating domain knowledge into models of soft computing. It is critical to successfully incorporate domain expertise into the models for various applications. Nevertheless, this integration may be intricate, and models could miss key details of the domain of the problem if it isn't included. In order to create reliable and situationally appropriate models, it is crucial to bring together domain specialists and data scientists.

10.6.2 Ethical considerations and potential risks

When it comes to using soft computing, ethical issues are key since they promote the equitable and responsible application of technology. Concerns about potential prejudice and unfairness rank high. Just like any other AI system, soft computing models might inherit or amplify the biases in the training data. Discrimination or unjust treatment may result, particularly in delicate fields like financing, recruiting, or law enforcement. To reduce this risk and make automated judgments that are fair, it is vital to train models using varied and impartial datasets.

Because soft computing frequently entails processing massive volumes of data, some of which may contain sensitive personal information, privacy considerations are of the utmost importance. Ensuring compliance with requirements such as the General Data Protection Regulation (GDPR) and safeguarding this data are of the utmost importance. In order to keep user confidence and stay in accordance with the law, it is important to create strong data protection mechanisms, anonymize data where feasible, and make sure that data usage is transparent.

Relying too much on automated decision-making systems is another ethical concern. Reducing human oversight—essential for mistake detection, making judgments with context, and accounting for subtleties that the model could miss—is a risk of relying too much on soft computing. Instead of completely replacing human decision-making, these technologies should be utilized to supplement it. One way to reduce the dangers of relying too much on automation is to put systems in place to allow for human scrutiny and involvement in crucial decision-making processes.

It is crucial to address ethical considerations including privacy, prejudice, and the risk of relying too much on automation when applying soft computing. Ensuring the long-term sustainability and efficacy of soft computing solutions, as well as the ethical and responsible use of technology, necessitates these issues.

10.6.3 Strategies to overcome these challenges

To effectively address the challenges associated with implementing soft computing in data analytics, several strategic approaches can be adopted:

Enhancing Computational Efficiency: Research and development focused on algorithm optimization is one of the main methodologies. We can make soft computing approaches more accessible and practicable for more applications, including ones with limited computer resources, by making them more computationally efficient. This entails creating algorithms with lower storage and processing requirements, lowering the hurdles to entry for smaller enterprises and organizations.

Improving Data Quality: To ensure that soft computing models use high-quality data, it is essential to employ strong data pretreatment and validation methods. Improved system performance and less error-prone modeling are the results of using high-quality data. Methods for removing mistakes, biases, and inconsistencies from data through cleaning, normalization, and transformation are all part of this category.

Advancing Explainable AI (XAI): Improving the interpretability of soft computing models is a top priority, especially increasingly complicated ones like deep learning models. More openness and reliability in these systems will result from XAI advancements. This entails developing methods and resources that make AI models more responsible and transparent by letting users comprehend and interpret their decision-making process.

Incorporating Domain Expertise: To make sure the models reflect the intricacies and subtleties of the actual world, it is a good idea to work with domain experts while building them. The models become more applicable and efficient in real-world scenarios as a result of this partnership's assistance in including crucial domain-specific information.

Addressing Bias and Ethical Issues: It is critical to regularly check models for biases. One way to reduce the ethical dangers of AI models is to use datasets that are reflective of society at large. Responsible and equitable AI development also requires adherence to ethical norms and legislation. This involves developing models with ethics in mind and making sure they don't reinforce preexisting social prejudices.

Balancing Automation with Human Oversight: It is critical to strike a balance between human oversight and fully automated decision-making processes. Human supervision is essential for contextual decision-making, error checking, and dealing with complicated situations that the model may not be able to handle, even though automation can improve consistency and efficiency. The advantages of automation may be achieved without sacrificing responsibility and dependability if this equilibrium is preserved.

10.7 CASE STUDIES IN SOFT COMPUTING FOR DATA ANALYTICS

10.7.1 Healthcare: Predicting patient outcomes

Predicting patient outcomes after surgery is one area where soft computing techniques have been successfully used in healthcare data analytics, demonstrating their potential to improve healthcare delivery. One prominent healthcare facility is used as an example of how neural networks, an important kind of soft computing, were applied for this specific goal. The neural networks were trained using data collected from patients in the past, which included demographics, medical history, and the procedures individuals had [65].

It was astonishing how this application turned out. The model created using artificial neural networks achieved a high rate of accuracy in forecasting post-surgery problems and recovery times. Healthcare practitioners were able to improve patient management techniques and create tailored care plans thanks to the accuracy of their predictions. Consequently, patient outcomes saw a marked improvement. Additionally, the hospital was able to better allocate its resources as a result of this soft computing implementation in healthcare. In order to improve the quality of healthcare services overall, hospitals may maximize their resource use by precisely forecasting post-surgery outcomes. This will guarantee that patients receive the most appropriate care at the right time. Soft computing has the potential to revolutionize healthcare by improving patient care and streamlining operations, as shown in this case study.

10.7.2 Finance: Credit scoring model

The use of soft computing has greatly improved the methods used to assess credit risk in the financial industry. One such example is a bank that used a system based on soft computing to determine a customer's creditworthiness. Because financial data and consumer profiles are inherently imprecise and ambiguous, this method relied heavily on fuzzy logic, a crucial soft computing concept [66].

A large number of qualitative and quantitative aspects impacting a customer's creditworthiness were processed using fuzzy logic, which can handle imprecise and misleading information. Aside from more obvious considerations like salary and work history, this category also includes intangibles like financial stability and spending patterns. Credit risk could be more thoroughly and intricately assessed when these varied elements were incorporated into the model. For the bank, this application had a game-changing result. The credit scoring model developed using soft computing techniques produced a more intricate and precise evaluation of credit risk. The decrease in false positives in credit denials—a problem that frequently arises with conventional credit scoring methods—was one of the major advantages. Since the algorithm could now distinguish between applicants with high and low risk more precisely, loan approval rates increased.

The bank was also able to expand its pool of potential borrowers thanks to the credit rating system that was based on soft computing. It opened the door for those whose complicated financial profiles might cause conventional models to ignore them. The growth of the client base did not, however, entail a corresponding rise in the likelihood of default. The bank might instead keep or lower its risk profile while serving more customers with its credit offerings.

This case study shows how soft computing may revolutionize conventional financial processes. Soft computing has the potential to improve operational efficiency and provide more egalitarian financial services by facilitating more precise, detailed, and inclusive credit risk assessments.

10.7.3 Retail: Customer segmentation

A retail organization that used these approaches to enhance its client segmentation process is a compelling example of a soft computing application in the retail sector. The creation of a hybrid model fusing neural networks with GAs was a crucial part of this strategic step in the field of soft computing. Complex patterns in consumers' buying habits and histories might be better understood with this novel method [67].

Learning from the massive datasets of consumer transactions and behaviors was the responsibility of the hybrid model's neural networks. Their proficiency in analyzing and drawing conclusions from massive datasets allowed them to achieve remarkable results in spotting hidden trends and patterns. Yet, in order to maximize segmentation accuracy, GAs were employed. Following a strategy similar to natural selection, they conducted iterative searches until they found the optimal method of client segmentation. This application's result had a significant influence. Thanks to the hybrid approach, we were able to separate our customers into several groups, each with their own particular tastes and buying habits. Thanks to this segmentation, the retail firm was able to better target certain categories of customers with its marketing methods. Because of the enhanced ability to reach specific audiences through tailored advertising, the business saw a meteoric rise in revenue. Customers were also far more satisfied as a result of receiving suggestions and offers that were tailored to their specific interests and requirements.

Soft computing's ability to revolutionize company strategy is demonstrated in this case study. Better marketing and overall company success may be in store for the retail giant if it harnesses the power of neural networks and GAs to better understand consumer behavior. Because of this strategy's effectiveness, we can see how soft computing may improve retail innovation and efficiency, which in turn benefits customers and companies.

10.7.4 Manufacturing: Quality control

One manufacturing company's experience demonstrates how the use of soft computing—and neural networks in particular—has greatly improved quality control procedures in the manufacturing business. A major change in the company's approach to quality control occurred when neural networks were used to analyze production data in real time [68, 69, 70].

Processing the massive amounts of complicated data produced by the production process was the primary goal of the neural network application in this case. These networks were programmed to spot irregularities and trends that may point to manufacturing line errors or poor quality. The organization was able to quickly detect and resolve any developing quality issues since the neural networks processed data in real time and offered rapid insights. A wide range of positive effects resulted from using this soft computing method. In the first place, the product quality was much improved. There were less waste and related expenses since fewer faulty items were generated because of the capacity to identify and fix problems quickly. This had a beneficial effect on the environment and helped save money.

The technique was also useful in improving predictive maintenance procedures. The business might avoid reactive maintenance by anticipating and fixing possible problems with the production process. As a result of this change to predictive maintenance, downtime was reduced, and the manufacturing line worked more effectively. Additionally, these improvements to product quality and production efficiency had a positive impact on consumer satisfaction. With fewer hiccups in the production process, the firm was able to dependably fulfill orders and provide customers with higher-quality products.

10.8 CONCLUSION

Soft computing and its crucial function in data analytics have been thoroughly examined in this chapter. First, we defined soft computing, then traced its historical growth and outlined its essential concepts, emphasizing its emphasis on approximate reasoning, uncertainty, and imprecision. We delved into the theoretical foundations and practical applications of soft computing's core techniques—fuzzy logic, neural networks, and evolutionary algorithms—to help you grasp their relevance in solving difficult computational issues. Soft computing has proven its worth in data analytics by showcasing its adaptability and capability. It has improved data analytics through a number of means, particularly in cases where conventional approaches fail because of complexity or unpredictability. Healthcare, retail, environmental research, manufacturing, and traffic management were among the industries that provided case studies illustrating the real-world uses of soft computing in data analytics. These practical instances demonstrated how soft computing has revolutionized data-driven problem-solving by offering more sophisticated, efficient, and effective answers. Soft computing is a promising field, according to recent developments and predictions for the future. Soft computing is well-positioned to further transform data analytics thanks to its emphasis on real-time analytics, integration with big data and the Internet of Things, and continuous improvements in algorithmic efficiency. In order to ensure the responsible and successful use of soft computing approaches, we addressed obstacles and ethical issues, such as computational complexity, data quality, model interpretability, and privacy concerns. We also provided ways to solve these challenges. Finally, in the dynamic domain of data analytics, soft computing is the bedrock. It is a priceless asset because of its capacity to deal with uncertainty, learn from facts, and adjust to novel situations. Technological progress and the ever-increasing complexity of real-world data will surely propel the further incorporation of soft computing in data analytics going forward. Data analytics, with the help of soft computing, will be able to comprehend and interact with the environment around us in increasingly complex ways in the future.

REFERENCES

[1] Dinesh G, Kumari PL, Lokhande M, et al. *SOFT COMPUTING*. GCS PUBLISHERS, 2022.

[2] Vigo R, Zeigler DE, Wimsatt J. Uncharted aspects of human intelligence in knowledge-based “intelligent” systems. *Philosophies* 2022; 7: 46.

[3] Rafatirad S, Homayoun H, Chen Z, et al. Machine learning for computer scientists and data analysts.

[4] Spector AZ, Norvig P, Wiggins C, & Wing, J. M. Data science in context: Foundations, challenges, opportunities.

[5] Ezugwu AE, Ikotun AM, Oyelade OO, et al. A comprehensive survey of clustering algorithms: State-of-the-art machine learning applications, taxonomy, challenges, and future research prospects. *Eng Appl ArtifIntell* 2022; 110: 104743.

[6] Kumar A, Jaiswal A. Systematic literature review of sentiment analysis on Twitter using soft computing techniques. *Concurr Comput* 2020; 32: e5107.

[7] de Pablo JJ, Jackson NE, Webb MA, et al. New frontiers for the materials genome initiative. *NPJ Comput Mater* 2019; 5: 41.

[8] Broby D. The use of predictive analytics in finance. *J Fin Data Sci* 2022; 8. 145–161.

[9] Kaushal M, Khehra BS, Sharma A. Soft Computing based object detection and tracking approaches: State-of-the-Art survey. *Appl Soft Comput* 2018; 70: 423–464.

[10] Zadeh LA. Fuzzy logic. In: *Granular, Fuzzy, and Soft Computing*. Springer, 2023, pp. 19–49.

[11] Konar A. *Artificial intelligence and soft computing: behavioral and cognitive modeling of the human brain*. CRC press, 2018.

[12] Atif M, Latif S, Ahmad R, et al. Soft computing techniques for dependable cyber-physical systems. *IEEE Access* 2019; 7: 72030–72049.
[13] Mallick PK, Balas VE, Bhoi AK, et al. *Cognitive informatics and soft computing*. Springer, 2021.
[14] Joseph S. The Impact of Artificial Intelligence on Modern Computer Software.
[15] Khajehzadeh M, Keawsawasvong S, Nehdi ML. Effective hybrid soft computing approach for optimum design of shallow foundations. *Sustainability* 2022; 14: 1847.
[16] Masood M, Nawaz M, Malik KM, et al. Deepfakes generation and detection: State-of-the-art, open challenges, countermeasures, and way forward. *Appl Intell* 2023; 53: 3974–4026.
[17] Charitopoulos A, Rangoussi M, Koulouriotis D. On the use of soft computing methods in educational data mining and learning analytics research: A review of years 2010–2018. *Int J ArtifIntell Educ* 2020; 30: 371–430.
[18] Škrjanc I, Iglesias JA, Sanchis A, et al. Evolving fuzzy and neuro-fuzzy approaches in clustering, regression, identification, and classification: A survey. *Inf Sci (N Y)* 2019; 490: 344–368.
[19] Su J, Jiang C, Jin X, et al. Large Language Models for Forecasting and Anomaly Detection: A Systematic Literature Review. arXiv preprint arXiv:240210350.
[20] Khan W, Haroon M. A Pilot Study and Survey on Methods for Anomaly Detection in Online Social Networks. In: *Human-Centric Smart Computing: Proceedings of ICHCSC 2022*. Springer, 2022, pp. 119–128.
[21] Sohail A. Genetic algorithms in the fields of artificial intelligence and data sciences. *Annal Data Sci* 2023; 10: 1007–1018.
[22] Kundra H, Khan W, Malik M, et al. Quantum-inspired firefly algorithm integrated with cuckoo search for optimal path planning. *Int J Modern Phys C* 2022; 33: 2250018.
[23] Wani AK, Rahayu F, Ben Amor I, et al. Environmental resilience through artificial intelligence: innovations in monitoring and management. *Environ Sci Pollut Res* 2024; 1–17.
[24] Johnson D, van Riper C, Stewart W, et al. Elucidating social-ecological perceptions of a protected area system in Interior Alaska: a fuzzy cognitive mapping approach. *Ecol Soc* 2022; 27: 34.
[25] Wajid, M. S., & Wajid, M. A. (2021). The importance of indeterminate and unknown factors in nourishing crime: a case study of South Africa using neutrosophy. *Neutrosophic Sets and Syst*, 41(2021): 15.
[26] Batool T, Abbas S, Alhwaiti Y, et al. Intelligent model of ecosystem for smart cities using artificial neural networks. *Intelligent Automation & Soft Computing*; 30.
[27] Zhang M, Wang L, Guo W, et al. Many-objective evolutionary algorithm based on dominance degree. *Appl Soft Comput* 2021; 113: 107869.
[28] Luo XJ, Oyedele LO. Forecasting building energy consumption: Adaptive long-short term memory neural networks driven by genetic algorithm. *Adv Eng Inform* 2021; 50: 101357.
[29] Abdullah SM, Periyasamy M, Kamaludeen NA, et al. Optimizing Traffic Flow in Smart Cities: Soft GRU-Based Recurrent Neural Networks for Enhanced Congestion Prediction Using Deep Learning. *Sustainability* 2023; 15: 5949.
[30] Ahmad T, Zhang D. Using the internet of things in smart energy systems and networks. *Sustain Cities Soc* 2021; 68: 102783.

[31] Sukhov A, Sihvonen A, Netz J, et al. How experts screen ideas: The complex interplay of intuition, analysis and sensemaking. *J Prod Innov Manag* 2021; 38: 248–270.
[32] Khosravi M, Fereidunian A. Fuzzy Realizations of Adaptive Autonomy in Smart Grid. In: *Applications of Fuzzy Logic in Planning and Operation of Smart Grids*. Springer, 2021, pp. 105–151.
[33] Nady A, Hassan DK, Assem A. A Proposed Framework for Automated Evaluation of Architectural Spatial Configurations Using Fuzzy Logic Approach. *JES J Eng Sci* 2023; 51: 19–48.
[34] Wajid, M. S., Terashima-Marin, H., Najafirad, P., Pablos, S. E. C., & Wajid, M. A. (2024). DTwin-TEC: An AI-based TEC district digital twin and emulating security events by leveraging knowledge graph. *Journal of Open Innovation: Technology, Market, and Complexity*, 10(2), 100297.
[35] Goldberg Y. *Neural network methods for natural language processing*. Springer Nature, 2022.
[36] Alam M, Samad MD, Vidyaratne L, et al. Survey on deep neural networks in speech and vision systems. *Neurocomputing* 2020; 417: 302–321.
[37] Khan W, Haroon M. An efficient framework for anomaly detection in attributed social networks. *International Journal of Information Technology Epub ahead of print* 2022. DOI: 10.1007/s41870-022-01044-2.
[38] Khan W, Haroon M. An unsupervised deep learning ensemble model for anomaly detection in static attributed social networks. *International Journal of Cognitive Computing in Engineering* 2022; 3: 153–160.
[39] Too J, Abdullah AR. A new and fast rival genetic algorithm for feature selection. *J Supercomput* 2021; 77: 2844–2874.
[40] Rostami M, Berahmand K, Forouzandeh S. A novel community detection based genetic algorithm for feature selection. *J Big Data* 2021; 8: 1–27.
[41] Sarkheyli A, Zain AM, Sharif S. Robust optimization of ANFIS based on a new modified GA. *Neurocomputing* 2015; 166: 357–366.
[42] Ardeshir N, Sanford C, Hsu DJ. Support vector machines and linear regression coincide with very high-dimensional features. *Adv Neural Inf Proces Syst* 2021; 34: 4907–4918.
[43] Hussain SF. A novel robust kernel for classifying high-dimensional data using Support Vector Machines. *Expert Syst Appl* 2019; 131: 116–131.
[44] Hsu D, Muthukumar V, Xu J. On the proliferation of support vectors in high dimensions. In: *International Conference on Artificial Intelligence and Statistics*. PMLR, 2021, pp. 91–99.
[45] Wajid, MA, & Zafar, A (2022). Neutrosophic Image Segmentation: An Approach for the Treatment of Uncertainty in Multimodal Information Systems. *Int J Neutrosophic Sci*, 19(1).
[46] Darwish A, Hassanien AE, Das S. A survey of swarm and evolutionary computing approaches for deep learning. *ArtifIntell Rev* 2020; 53: 1767–1812.
[47] Mohamed AW. A novel differential evolution algorithm for solving constrained engineering optimization problems. *J Intell Manuf* 2018; 29: 659–692.
[48] Poornima PR, Poornima L, Padmashree SRDS. *Brain Cancer Detection Using Artificial Neural Network*
[49] Zaccone G, Karim MR. *Deep Learning with TensorFlow: Explore neural networks and build intelligent systems with Python*. Packt Publishing Ltd, 2018.

[50] Khan W, Haroon M, Khan AN, et al. DVAEGMM: Dual variational autoencoder with gaussian mixture model for anomaly detection on attributed networks. *IEEE Access* 2022; 10: 91160–91176.
[51] Peng RD. *R programming for data science.* Leanpub Victoria, BC, Canada, 2016.
[52] Rajesh P, Karthikeyan M. A comparative study of data mining algorithms for decision tree approaches using weka tool. *Adv Nat Appl Sci* 2017; 11: 230–243.
[53] Khan W. An exhaustive review on state-of-the-art techniques for anomaly detection on attributed networks. *Turkish Journal of Computer and Mathematics Education (TURCOMAT)* 2021; 12: 6707–6722.
[54] Rahman MM, Shafiullah M, Rahman SM, et al. Soft computing applications in air quality modeling: Past, present, and future. *Sustainability* 2020; 12: 4045.
[55] Zgurovsky MZ, Sineglazov VM, Olena IC. *Artificial Intelligence Systems Based on Hybrid Neural Networks.* Springer, 2020.
[56] Minh D, Wang HX, Li YF, et al. Explainable artificial intelligence: a comprehensive review. *ArtifIntell Rev* 2022; 1–66.
[57] Wajid, MA, & Zafar, A (2022). A multimodal approach of information access and retrieval using neutrosophic sets. In *Emerging Trends in IoT and Computing Technologies* (pp. 382–388). Routledge.
[58] Khan W, Ishrat M, Haleem M, et al. An Extensive Study and Review on Dark Web Threats and Detection Techniques. In: *Advances in Cyberology and the Advent of the Next-Gen Information Revolution.* IGI Global, 2023, pp. 202–219.
[59] Chen W, Milosevic Z, Rabhi FA, et al. Real-Time Analytics: Concepts, Architectures and ML/AI Considerations. *IEEE Access.*
[60] Tsaramirsis G, Kantaros A, Al-Darraji I, et al. A modern approach towards an industry 4.0 model: From driving technologies to management. *J Sens* 2022; 2022: 1–18.
[61] Schwartz R, Vassilev A, Greene K, Perine L, Burt A, & Hall P *Towards a standard for identifying and managing bias in artificial intelligence.* (Vol. 3, p. 00). US Department of Commerce, National Institute of Standards and Technology.
[62] Devi M, Singh S, Tiwari S, Chandra Patel S & Ayan M T A survey of soft computing approaches in biomedical imaging. *J Healthc Eng*; 2021 (1), 1563844.
[63] Zhong S, Zhang K, Bagheri M, et al. Machine learning: new ideas and tools in environmental science and engineering. *Environ Sci Technol* 2021; 55: 12741–12754.
[64] Charitopoulos A, Rangoussi M, Koulouriotis D. On the use of soft computing methods in educational data mining and learning analytics research: A review of years 2010–2018. *Int J ArtifIntell Educ* 2020; 30: 371–430.
[65] Shaikh TA, Dar TR, Sofi S. A data-centric artificial intelligent and extended reality technology in smart healthcare systems. *Soc Netw Anal Min* 2022; 12: 122.
[66] Singh S, Singh S, Koohang A, et al. Soft computing in business: exploring current research and outlining future research directions. *Ind Manag Data Syst* 2023; 123: 2079–2127.

[67] Krishna GJ, Ravi V. Evolutionary computing applied to customer relationship management: A survey. *Eng Appl ArtifIntell* 2016; 56: 30–59.
[68] Pal S. Advancements in AI-Enhanced Just-In-Time Inventory: Elevating Demand Forecasting Accuracy. *Int J Res Appl Sci Eng Technol* 2023; 11: 282–289.
[69]. Bhushan, B., Sharma, S. K., Nand, P., Shankar, A., & Obaid, A. J. (Eds.). (2024). Emerging Trends for Securing Cyber Physical Systems and the Internet of Things.
[70] Pandey, S., & Bhushan, B. (2024). Recent Lightweight cryptography (LWC) based security advances for resource-constrained IoT networks. Wirel Netw, 30(4), 2987–3026.

Chapter 11

AI-infused respiratory diagnostics

A new era in healthcare

Afa, Mahfooz Alam, and Marghoob Enam

11.1 INTRODUCTION

The exchange of carbon dioxide and oxygen between the body and its surroundings is the main job of the respiratory system. Any component of this system may be impacted by respiratory diseases, which will limit its capacity to carry out these essential tasks. In addition to being brought on by a variety of underlying diseases, respiratory symptoms are frequently brought on by viral infections, such as the coronavirus or other newly developing viruses. Common symptoms of these respiratory viruses include coughing, fever, congestion in the nose, and breathing difficulties. On the other hand, timely identification of the viral type can be vital, particularly in situations like the COVID-19 pandemic. The COVID-19 pandemic was caused by a number of circumstances, including delayed diagnosis and misreading of COVID-19 symptoms as common flu-like symptoms [1]. AI is a key component of contemporary radiology and offers many benefits, including increased diagnostic precision, optimized workflow, and customized patient care. These developments portend a bright future for bettering patient outcomes. Examples of these innovations include tools for picture partitioning, categorization, computer-aided diagnosis, and novel diagnostic and prognostic tools powered by radiomics and predictive analytics. Nevertheless, issues with data security, privacy, and the "black box" nature of AI models still need to be resolved. The future looks bright despite these obstacles, as new algorithms and architectures expand the possibilities for medical picture processing. Building a bridge between academia and industry, interdisciplinary research requires the critical collaboration of radiologists and AI developers [2].

Interest in using artificial intelligence (AI) to solve biomedical and healthcare-related problems is on the rise due to the ever-increasing amount of biomedical data collected by hospital monitors, wearable technology, and the Internet of Things (IoT), as well as advancements in computational hardware and AI [3]. Traditional mathematical techniques may not be sufficient to address uncertainty issues in the domains of robotics, engineering,

 DOI: 10.1201/9781003606055-11

economics, medicine, and the environment. Numerous theories have been developed to address these shortcomings, including fuzzy set theory [4], intuitionistic fuzzy set theory [5], soft set theory [6], neutrosophic soft set theory [26], spherical linear Diophantine fuzzy sets [7], and artificial neural networks [8]. The complex dynamics of respiratory systems may be difficult for traditional mathematical methods to adequately capture since they frequently depend on linear models. The respiratory system is made up of many different tissues, organs, and feedback processes.

This chapter covers the typical side effects of current treatments as well as the significance of successful AI algorithms with accuracy of lung disorders. We observed how AI systems correctly detect these diseases because it could be argued that it is difficult to determine the exact type of lung disease because they all have similar symptoms. So the approach used to help predict the necessary therapy to cure the ailment is generalized *neutrosophic soft mapping*. With the help of this approach, we can find the exact diseases. Also, we postulated that a pulmonologist working in tandem with explainable AI (XAI) would do better when it comes to diagnosing and interpreting pulmonary function tests (PFTs) than a pulmonologist working alone.

By customizing therapy rather than applying a single strategy to all patients, AI is revolutionizing the healthcare industry. We also discussed the challenges and solutions that AI encounters despite its progress. The study concludes by contemplating the potential future impact of AI on respiratory healthcare, exploring AI algorithms, methodologies, and challenges in disease identification, with a focus on a successful respiratory system analysis AI algorithm. AI is being used to monitor respiratory health in real time, providing a novel method of managing diseases. This chapter also examines case studies that highlight real-world AI applications in the medical field. The difficulties that come with integrating AI are carefully explored, as are moral issues.

This study presents the importance of AI in the respiratory system in the era of healthcare. The complete structure of the chapter is shown in Figure 11.1. Section 11.2 delved into how AI algorithms are employed for disease identification, highlighting both their challenges and benefits. The discussion also encompassed successful AI algorithms and their advantages, with a specific focus on the side effects associated with existing treatments. In Section 11.3, the focus shifted to how AI can monitor respiratory health, facilitating customized treatment plans and early identification of respiratory issues. Section 11.4 explored the role of AI in personalized treatment approaches, such as tumor scanning and decoding subtle signals from the body, among other applications. Section 11.5 centered on case studies and practical applications of AI in healthcare. Section 11.6 addressed the challenges and ethical considerations pertinent to AI implementation in healthcare. In Section 11.7, the conversation revolved around future directions

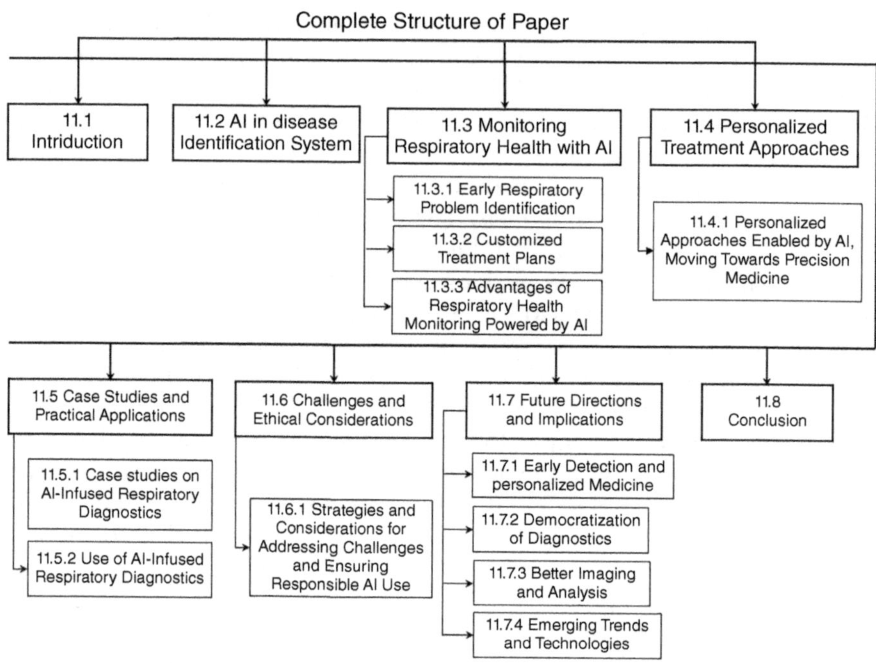

Figure 11.1 Complete structure of the chapter.

and implications, including the integration of various technologies to enhance the efficiency of existing AI systems. Lastly, Section 11.8 discussed real-world case studies showcasing the use of AI in disease identification, as well as the potential of employing generalized neutrosophic soft mapping for identifying lung diseases, suggesting it as a promising avenue for future exploration.

11.2 AI IN DISEASE IDENTIFICATION

AI is essential to the identification of diseases since it uses sophisticated algorithms to analyze medical data and provide early detection and precise diagnosis. The following methods have historically been employed by medical professionals to identify lung diseases:

1. *Spirometry*: This test determines whether a patient's lungs are adequately supplying oxygen to their body and performing their breathing function. When someone breathes into a spirometer, the amount of air they exhale is measured. It is the most widely used test to evaluate lung function.

2. *X-Ray*: A chest X-ray is useful for identifying a number of lung conditions, including emphysema, one of the main causes of COPD (Chronic obstructive pulmonary disease).
3. *CT Scan*: In severe cases, a CT scan may be useful when traditional methods are unable to diagnose a disease. It may reveal whether the patient needs any form of surgery.
4. *Bronchoscopy*: To visually inspect the lungs and obtain lung tissue or mucus samples for additional examination, a thin, flexible tube equipped with a camera (called a bronchoscope) is inserted into the airways. Lung illnesses, infections, and cancers of the lungs can all be diagnosed with bronchoscopy assistance.
5. Tests called *PFTs* measure things like gas exchange, lung capacity, and airflow to determine how effectively the lungs are working. These tests, which include the previously stated spirometry, can aid in the diagnosis of a number of lung conditions, including pulmonary fibrosis, chronic bronchitis, and asthma.
6. *Arterial Blood Gas Analysis*: This test determines whether breathing causes the blood's oxygen content to rise.

However, these methods now in use are limited by the following Patients who have had heart surgery recently or who have underlying cardiac issues are unable to do spirometry. And after the treatment *breathlessness, nausea, and dizziness* are common side effects that patients may have following these examinations. Patients who undergo CT scans or X-rays run the danger of dying from radiation exposure [9]. Therefore, as shown in Table 11.1, 3D CNN (Convolutional Neural Network), SVM (Support Vector Machines), RNN (Recurrent Neural Network), and NLP (Natural Language Processing) which was trained with 500 examples and then tuned via gradient descent to obtain an 89.2% Dice coefficient. In order to confirm accuracy, virtual structures generated by the AI system were compared to the anatomical structures of patients who underwent lobectomy or segmentectomy at the Peking University Cancer Hospital in retrospective cohorts (n = 113) and prospective cohorts (n = 139) [29]. The potential for AI and pulmonologists to work together to diagnose lung disease has not received much attention in research. We postulated that a pulmonologist working in tandem with explainable AI (XAI) would do better when it comes to diagnosing and interpreting PFTs than a pulmonologist working alone [10], as shown in Table 11.2. The use of deep learning algorithms to forecast COPD patients' exacerbations is examined in this work. It opens the door for proactive and customized management approaches by showcasing the viability and accuracy of AI-based prediction models [11]. The cost-effectiveness of using AI-powered monitoring systems in COPD care is evaluated by this study. It implies that by lowering hospital admission rates and enhancing patient outcomes, AI-based technology may eventually prove to be financially advantageous.

Table 11.1 AI algorithms and methodologies employed in disease identification

Disease	*AI algorithm/ methodology*	*Description*	*Benefits*	*Challenges*
Cancer (various types)	Convolutional Neural Networks (CNNs)	Analyze medical images (e.g., X-rays, mammograms) to identify suspicious patterns associated with cancer.	High accuracy in identifying tumors and other abnormalities, early detection, and personalized treatment recommendations.	Requires large datasets for training, the potential for bias in algorithms, and explainability limitations.
Skin diseases	Support Vector Machines (SVMs)	Analyze images of skin lesions to distinguish between different types of skin diseases.	Efficient classification of skin lesions, non-invasive and readily accessible diagnosis.	Limited ability to handle complex cases, dependence on high-quality images.
Cardiovascular diseases	Recurrent Neural Networks (RNNs)	Analyze electrocardiogram (ECG) data to detect arrhythmias and other heart abnormalities.	Continuous monitoring of heart rhythm, and early detection of cardiac events.	Requires specialized data collection equipment, and a limited understanding of underlying mechanisms.
Neurodegenerative diseases	Deep Learning	Analyze brain scans (e.g., MRI, PET) to identify early signs of Alzheimer's disease, Parkinson's disease, and other neurodegenerative conditions.	Early diagnosis and potential for intervention before significant cognitive decline.	High computational cost, reliance on expensive equipment, and ethical considerations regarding data privacy.
Infectious diseases	Natural Language Processing (NLP)	Analyze medical records, social media data, and other textual sources to detect and track outbreaks of infectious diseases.	Real-time monitoring of disease spread, improved public health response.	Can be influenced by misinformation and biases in data sources.
General disease identification	Machine Learning Ensemble Methods	Combine multiple AI algorithms to improve overall accuracy and robustness.	Can handle complex datasets with diverse features, less susceptible to bias.	Increased training time and computational resources required.

Table 11.2. Successful AI algorithm in the respiratory system

Disease	*AI technology*	*Example*	*Benefits*
Tuberculosis	Deep learning	CAD4TB	98% accuracy in TB detection
Asthma	AI-based audio analysis	WheezoMeter	Convenient and non-invasive screening
COPD	AI-driven prediction	COPDWatch	88% accuracy in predicting exacerbations
Sleep apnea	AI-powered sleep analysis	Nox Health	91% accuracy in sleep apnea detection
Respiratory diseases	AI-driven phenotyping	Aperiomics	Personalized treatment based on distinct disease subtypes

11.3 MONITORING RESPIRATORY HEALTH WITH AI

AI-powered real-time respiratory health monitoring is an exciting and quickly developing topic with enormous potential for preventative and early intervention in healthcare. As will be covered in Sections 11.3.1 and 11.3.2, AI algorithms are particularly good at identifying respiratory issues early on. Additionally, by offering personalized treatment plans, AI equips medical personnel with useful resources.

11.3.1 Early respiratory problem identification

AI algorithms are capable of identifying minute variations in breathing patterns, cough noises, and vital signs by analyzing data from wearable sensors, microphones, and cameras [12]. AI-powered systems have the capability to remotely monitor patients with chronic respiratory problems. This enables healthcare providers to keep a close eye on the patient's health and take immediate action if there is any decline in it.

11.3.2 Customized treatment plans

AI is capable of recommending tailored treatment plans and drug modifications based on an examination of a patient's medical background, present state of health, and external circumstances. Patient outcomes may be enhanced by more focused and efficient interventions brought about by this data-driven strategy.

11.3.3 Advantages of respiratory health monitoring powered by AI

AI-powered respiratory health monitoring has many advantages. There are several benefits of using AI in respiratory health monitoring such as:

- *Better Early Detection and Intervention*: AI's capacity to identify minute alterations in respiratory health can result in more timely diagnosis and treatments, averting problems and enhancing patient outcomes.
- *Improved Remote Patient Management*: AI-powered continuous monitoring enables remote care, cutting down on hospital stays and increasing access to healthcare, particularly for patients living in distant places.
- *Personalized Medicine*: AI-driven insights can enhance medication usage, customize treatment regimens, and *ultimately* raise patients' quality of life who suffer from respiratory disorders.
- *Artificial Intelligence as our Health Guardian*: Personalized and Ongoing Monitoring Demonstrated Imagine a time in the future when proactive health care replaces reactive care. Through ongoing and individualized monitoring, AI serves as your watchful health guardian, whispering insights and gently guiding you toward optimal well-being [13].
- *Collects and Evaluates data*: Describe how data is fed into AI models for ongoing monitoring via a variety of sensors, wearables, and electronic health records (EHRs). Finds trends and irregularities Showcase how AI systems can spot minute variations in behavior, vital signs, or prescription regimens that could be signals of impending health problems.
- *Customizes the Treatment*: Demonstrate how AI adapts interventions and suggestions to each patient's specific medical background, way of life, and preferences.
- *Foresees Difficulties*: Show how AI systems can predict possible health hazards and initiate timely treatments to prevent negative consequences.
- AI-driven respiratory monitoring systems are capable of ongoing learning and adaptation in response to patient and healthcare professional feedback. These systems can improve their accuracy and efficacy in detecting respiratory problems, maximizing treatment regimens, and projecting future health trajectories by evaluating result data and iteratively improving algorithms.
- *Patient Education and Engagement*: By offering patients individualized feedback and suggestions for taking care of their respiratory health, AI-powered respiratory monitoring devices can act as instructional resources. AI solutions can encourage adherence to treatment

plans and enhance long-term health outcomes by providing patients with the information and tools they need to self-monitor and manage their conditions.

11.4 PERSONALIZED TREATMENT APPROACHES

The battle against respiratory conditions like asthma, COPD (Chronic obstructive pulmonary disease), and pneumonia is often a complex one, demanding personalized strategies for optimal outcomes. Enter the realm of AI, poised to revolutionize treatment plans with its unique abilities. Sometimes doctors do not rely on one-size-fits-all approaches for respiratory conditions. Instead, AI can analyze: *Medical history*: Past diagnoses, treatments, and response to medications, *Biomarkers*: Genetic markers, blood tests, and lung function data, *Lifestyle factors*: Diet, exercise, and environmental triggers. AI algorithms can: *Predict disease progression*: Identify individuals at high risk of exacerbations or complications, *Tailor treatment plans*: Recommend specific medications, dosages, and treatment durations based on your unique needs, *Optimize medication regimens*: Identify potential drug interactions or side effects, and adjust accordingly, and *Personalize rehabilitation programs*: Design exercise routines and breathing techniques tailored to your physical abilities and respiratory capacity.

11.4.1 Personalized approaches enabled by AI, moving toward precision medicine

The power of AI is transforming healthcare from a one-size fits-all approach to a personalized journey. Imagine a future where a treatment plan is a unique fingerprint, meticulously crafted by AI to target your specific needs and maximize chances of success. Some examples are shown below:

1. *Cancer Care Gets Personal (AI Scans Tumors)*: Deep learning algorithms analyze tumor images as shown in Figure 11.2 [14] to identify genetic mutations and predict responses to specific therapies. This empowers oncologists to choose the most effective drugs, minimizing side effects and maximizing success rates. AI could accurately predict how breast cancer patients would respond to chemotherapy, helping personalize treatment plans and improve outcomes [15].
2. *Mental Health Finds its Match (AI Reads Between the Lines)*: Natural language processing (NLP) analyzes text and voice patterns to detect early signs of depression, anxiety, or other mental health conditions. This allows for early intervention and personalized therapy plans. AI could identify patients at high risk of suicide based on their medical records, enabling proactive interventions and potentially saving lives [16].

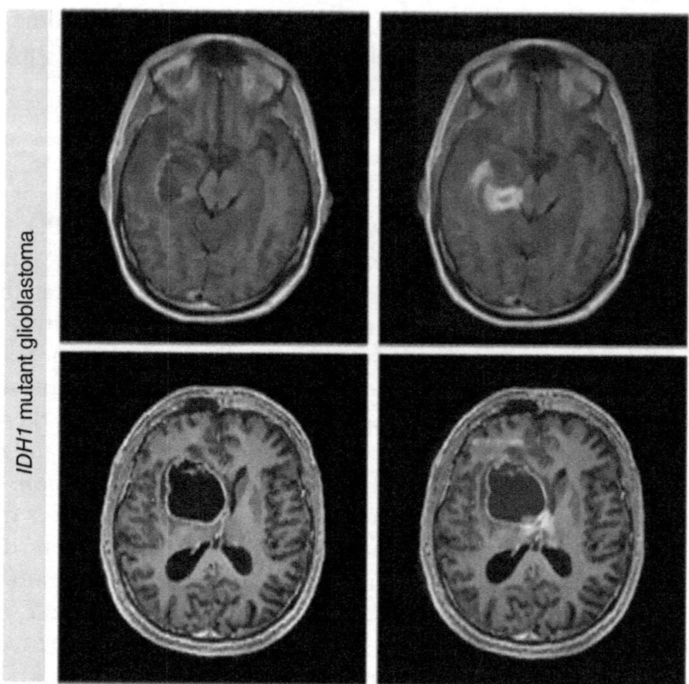

Figure 11.2 AI analyzing tumor images.

3. *AI Decodes the Body's Whispers*: Wearable sensors and AI algorithms track vital signs, blood sugar levels, sleep patterns etc. in real-time, providing insights into an individual's unique disease course [17]. Wang et al. present patients and healthcare professionals to adjust medication, lifestyle, and treatment plans proactively.
4. *AI-powered Early Detection of Pneumonia*: Pneumonia, a lung infection, often presents with subtle symptoms in its early stages, hindering timely diagnosis and intervention. Deep learning algorithms analyze chest X-rays or CT scans with remarkable accuracy, identifying early signs of pneumonia even before symptoms become evident demonstrating that an AI model could detect pneumonia on chest X-rays with an accuracy comparable to radiologists, potentially leading to faster diagnosis and treatment [18].
5. *Personalized Asthma Management with AI*: Asthma management often involves trial and error due to individual variations in response to medications and triggers. AI algorithms analyze patient data such as medical history, environmental factors, and sensor data from wearable devices to predict asthma exacerbations and personalize treatment plans. AI-powered mobile app helped asthmatic patients self-manage their condition, leading to fewer exacerbations and improved quality of life [19].

6. *AI-Assisted Lung Cancer Screening*: Detecting lung cancer early through low-dose CT scans often involves analyzing a large volume of images, risking missed lesions and unnecessary biopsies. AI algorithms assist radiologists in analyzing CT scans, highlighting suspicious lesions with high accuracy, improving detection rates, and reducing false positives. The remarkable performance of AI models in detecting lung nodules on CT scans, paved the way for improved lung cancer screening programs [20].

These effective applications highlight the observable advantages of using AI in a variety of fields. Table 11.3 shows examples of successful implementation and outcomes of AI:

7. *Genomic Analysis*: AI algorithms are capable of analyzing enormous volumes of genomic data in order to pinpoint genetic variants linked to medication metabolism, treatment response, and illness susceptibility. Healthcare professionals can administer drugs that are most likely to be beneficial and limit bad responses by knowing an individual's genetic composition.

Table 11.3 Examples of successful implementation and outcomes

Application	*Challenge*	*AI solution*	*Successful implementation*
Sepsis Early Detection	Slow and inaccurate traditional methods	AI analysis of EHRs, vital signs, and lab data	PRESERVE study (2020)
Tailored Cancer Treatment	Selecting optimal therapy	AI analysis of tumor data, medical history, and other factors	TAILORx study (2020)
Remote Patient Monitoring	Inconvenient traditional methods	AI-powered wearables track health data	CardiAI study (2021)
Chest X-ray Classification	-	A Large Scale Benchmark for Chest X-ray Classification	AI model achieved accuracy comparable to radiologists in detecting pneumonia.
Asthma Management System	-	A Randomized Controlled Trial of a Smartphone-Based Personalized Asthma Management System	AI-powered app reduced asthma exacerbations and improved the quality of life for patients.
Lung Analysis	-	A Database for Lung Nodule Analysis in 2016: Results and Insights from the LUNA16 Challenge	AI models achieved near-radiologist-level accuracy in detecting lung nodules.

8. *Clinical Decision Support*: To help healthcare professionals make individualized treatment decisions, AI-powered clinical decision support systems combine patient data, medical expertise, and evidence-based guidelines. To suggest the best courses of action for each patient, these systems examine the features of the patient, their medical background, the results of diagnostic tests, and available treatments.

11.5 CASE STUDIES AND PRACTICAL APPLICATIONS

Huge morbidity and mortality are caused by respiratory disorders, which represent a huge global health burden. Improved patient outcomes and efficacious therapy depend on an early and precise diagnosis. As a potent tool for transforming respiratory diagnostics, AI provides quicker, more precise, and less intrusive solutions.

11.5.1 Case studies on AI-infused respiratory diagnostics

1. *COVID-19 Diagnosis using CT and Chest X-rays*: Over 90% accuracy in COVID-19 identification by analyzing chest X-rays and CT scans using a deep learning algorithm. During the pandemic's peak, hospitals used AI-powered technology to expedite resource allocation and triage [21].
2. *Chest Radiograph Pneumonia Detection*: AI system that can autonomously identify pneumonia in chest X-rays with a level of accuracy that is on par with radiologists. In remote clinics with limited access to specialized physicians, an AI tool is being piloted [22].
3. *Using Cough Sound Analysis to Diagnose Tuberculosis*: AI model that can identify tuberculosis with a high degree of sensitivity and specificity by analyzing cough sounds. This non-invasive method may be especially helpful in situations with low resources when sputum testing presents difficulties [23].
4. *Low-Dose CT for Lung Cancer Screening*: A study showed how well AI algorithms can identify early-stage lung cancer from low-dose computed tomography (CT) scans. In detecting worrisome nodules and lowering false-positive rates, the AI system outperformed conventional screening techniques thanks to its high sensitivity and specificity. Patients with lung cancer may benefit from this technique in terms of early identification and treatment results [31].

AI-enabled respiratory diagnostics have great potential to enhance patient outcomes and the delivery of healthcare. Sustained research and development endeavors are necessary in order to surmount current obstacles and guarantee the fair distribution of these cutting-edge technologies.

11.5.2 Use of AI-infused respiratory diagnostics

The use of AI in respiratory diagnostics has enormous potential to enhance healthcare's precision, effectiveness, and accessibility. Case studies provide insightful information about the potential and problems that this integration presents in the actual world. Here, with an emphasis on particular case studies, we will examine important conclusions from pertinent research papers:

- *Enhanced precision in diagnosis*: To diagnose pneumonia, a study used deep learning algorithms to evaluate chest X-ray pictures. The AI system outperformed radiologists with an accuracy of 92.2% as opposed to 86.9%. This shows that AI has the potential to increase diagnostic precision, particularly in environments with restricted resources [27].
- *Early Detection and Intervention*: An AI model was created to identify early-stage lung cancer using CT images. With a 90.5% sensitivity and an 83.1% specificity, the model may have allowed for earlier intervention and better patient outcomes [28].
- *Increasing the Efficiency of Workflow*: In an emergency room, an AI-powered triage system for COVID-19 patients was put into place by a study conducted [24]. By streamlining patient flow, the technology lowered average wait times by 25%. This illustrates how AI may enhance workflow and enhance the provision of healthcare.
- *Accessibility and Scalability*: A model to diagnose tuberculosis from chest X-rays. The program may be easily implemented in distant places with limited access to healthcare, and it reached accuracy comparable to that of chest physicians. This demonstrates how AI could increase access to high-quality diagnostics [25].
- *Integration with EHR*: To detect respiratory infections early, a healthcare provider deployed an AI-powered system to scan unstructured data in EHRs. The system recognized patterns suggestive of respiratory illness by automatically retrieving and evaluating clinical notes, radiological reports, and laboratory findings, allowing for prompt diagnosis and treatment [30] [32].

11.6 CHALLENGES AND ETHICAL CONSIDERATIONS

While technological advancements hold great promise for enhancing healthcare, However, integrating technology into healthcare presents both challenges and ethical considerations that require careful evaluation like:

- *Limited and Skewed Datasets*: Intensive volumes of high-quality data labeled with precise diagnoses are needed for training AI models. However, data collection and annotation are difficult since respiratory disorders frequently appear with varied presentations and overlapping symptoms. Data biases (such as the underrepresentation of particular demographic groups) can also cause AI algorithms to make unfair or erroneous decisions.

- *Data Security and Provenance*: Concerns regarding patient privacy and data security are raised while sharing sensitive medical data for AI training. For AI-powered healthcare solutions to be transparent and trustworthy, strong ethical frameworks and data governance practices are essential.
- *Explainability and Transparency of the Model*: Some AI models are "black boxes," making it challenging to comprehend how they reach their judgments. In urgent situations, in particular, this lack of interpretability may make it more difficult for medical practitioners to accept and trust AI-based diagnoses.
- *Workflow Integration*: Adoption of AI tools depends on their smooth integration into current diagnostic workflows. Efficient clinical use requires addressing older system compatibility issues and guaranteeing user-friendly interfaces.
- *Misdiagnosis and Over-Reliance on AI*: Although AI has its uses, relying too much on its interpretations without taking clinical context into account can result in missing or incorrect diagnoses. It is essential to ensure that AI is used responsibly and under proper medical supervision.
- *Absence of Legal Frameworks*: Concerns of responsibility and liability in the event of errors are raised by the ongoing evolution of clear regulatory rules for AI-based medical devices. It is imperative to establish strong regulatory frameworks for the responsible development and application of AI in healthcare.
- *Equity and Accessibility*: Inequitable access to cutting-edge AI-powered diagnostics may make already-existing healthcare inequities worse. It is imperative to tackle issues of affordability, infrastructure constraints, and disparities in digital literacy to guarantee that these technologies are accessible to all.
- *Job Displacement and Ethical Issues*: There are worries over job displacement in the healthcare industry due to AI's potential to automate some operations. It's critical to think about the moral ramifications of adopting AI and to create plans for worker adaptation and reskilling.
- *Patient Consent and Autonomy*: In order to protect patients' privacy and autonomy, it is imperative to get their informed consent before using AI technologies in their treatment. Patients should be thoroughly informed about the possible hazards and benefits of AI-powered healthcare interventions, as well as how their data will be used and who will have access to it.
- *Algorithm Bias and Fairness*: Disparities in diagnosis, treatment, and outcome might result from AI algorithms that reinforce or magnify preexisting biases in the healthcare industry. In order to mitigate algorithmic bias and guarantee fairness in AI systems, the training data

must be carefully chosen, and algorithm performance must be continuously monitored and assessed across a range of patient groups.

- *Education and Training for Healthcare Professionals*: To effectively use and interpret AI-driven diagnostic tools and decision support systems, healthcare professionals require education and training. To optimize the advantages and reduce the hazards of AI adoption, healthcare practitioners must be well trained to incorporate AI into clinical practice.
- *Ethical Use of Patient Data*: When gathering, storing, and utilizing patient data for AI applications, healthcare institutions are required to adhere to ethical norms and guidelines. This entails protecting the privacy, security, and confidentiality of data as well as lowering the possibility of data breaches or misuse.
- *Education and Training for Healthcare Professionals*: To effectively use and interpret AI-driven diagnostic tools and decision support systems, healthcare professionals require education and training. To optimize the advantages and reduce the hazards of AI adoption, healthcare practitioners must be well-trained to incorporate AI into clinical practice.
- *Ethical Use of Patient Data*: When gathering, storing, and utilizing patient data for AI applications, healthcare institutions are required to adhere to ethical norms and guidelines. This entails protecting the privacy, security, and confidentiality of data as well as lowering the possibility of data breaches or misuse.
- *Patient Self-Determination and Knowledgeable Decision-Making*: Promoting patient autonomy and trust in healthcare systems requires enabling individuals to actively engage in their healthcare decisions and comprehend the role of AI technology in diagnosis and treatment. Making educated decisions can be facilitated by giving patients clear access to information regarding AI algorithms and their limits.

11.6.1 Strategies and considerations for addressing challenges and ensuring responsible AI use

Diversity and quality of data: Make use of inclusive datasets that are representative of the populations that the AI will be interacting with. Recognize and correct data biases to prevent them from being reflected in AI results.

- *Algorithmic Fairness*: Use algorithms that take into account the interests of all groups equally and prevent discriminating results. Counterfactual fairness is a technique that allows one to compare expected outcomes for individuals with varying qualities.
- *Comprehending and Elucidating*: Create comprehensible AI models that demonstrate the thought process behind their choices. Techniques for Explainable AI (XAI), such as LIME or SHAP, can contribute to

the transparency of black-box models, facilitating the identification of bias and fostering confidence.

- *Safeguarding Security and Privacy*: Differential privacy: To enable AI training while safeguarding individual privacy, apply differential privacy approaches to introduce controlled noise to data. TensorFlow Privacy and other libraries make implementation easier.
- *Safe Data Handling*: To protect sensitive data from unwanted access, leakage, and modification, use strong security measures including encryption, access controls, and intrusion detection systems.
- *Openness and Authority*: Prioritize user privacy in design and development, give users control over their data (e.g., through opt-out methods), and be transparent about data collection and usage.
- *Governance and Human Oversight*: Clearly define who is responsible for what and make sure that humans are involved in making important AI choices to minimize any unfavorable outcomes. Think of systems with people reviewing AI judgments before implementation, or human-in-the-loop systems.
- *Decisions That Can Be Explained*: Encourage openness in the decision-making processes used by AI. By dissecting intricate models and elucidating their logic, tools such as SHAP can build human confidence and facilitate the evaluation of possible biases or mistakes.
- *Participation and Communication with the Public*: Engage a variety of stakeholders in conversations regarding the societal ramifications of AI. Organize town halls, open forums, and citizen science projects to address issues openly and influence the development of responsible AI.
- *Interdisciplinary Collaboration*: To address the ethical, social, and legal ramifications of AI, promote collaboration between engineers, ethicists, policymakers, social scientists, and other stakeholders. Multidisciplinary methods can result in more comprehensive solutions that are better matched with the priorities and values of society.
- *Human-Centered Design*: Give human-centered design concepts top priority while creating AI technologies. In order to guarantee that AI systems are understandable, practical, and consistent with human values, it is necessary to comprehend user requirements, preferences, and concerns at every stage of the design process and to take feedback into account.
- *Responsible AI Education and Training*: Offer courses and training on responsible AI practices to developers, legislators, and other stakeholders. This entails bringing ethical issues, bias reduction strategies, and responsible AI system design and deployment best practices to the public's attention.
- *Collaborating and Engaging the Community*: Communicate with the communities that AI technologies affect to learn about their requirements, inclinations, and worries. Include members of the community

in co-design projects and decision-making processes to make sure AI systems take into account a variety of viewpoints and social demands.
- *Long-Term Impact Assessment*: Take into account how AI technology will affect society, the economy, and the environment in the long run. To determine the possible effects of widespread AI adoption on employment, inequality, privacy, and other important areas, conduct impact assessments. Make use of this knowledge while developing policies and making decisions.

11.7 FUTURE DIRECTIONS AND IMPLICATIONS

There are some future direction and implications given below:

11.7.1 Early detection and personalized medicine

Advanced Pattern Recognition: AI algorithms have the ability to examine intricate datasets containing genetic data, medical history, environmental factors, and real-time respiratory sounds. This allows for the identification of people who are at a high risk of developing respiratory diseases even before symptoms manifest. Machine learning algorithms have the potential to forecast how current respiratory disorders will develop, enabling individualized treatment regimens and specially designed treatments to avert consequences. AI-powered tests that make use of biosensors and image analysis may be able to quickly and accurately identify certain respiratory viruses and pathogens, enabling more timely and focused treatment as well as the control of outbreaks.

11.7.2 Democratization of diagnostics

Non-invasive Home Monitoring: Wearable technology and smartphone apps with AI algorithms built-in could allow for continuous home monitoring of respiratory health, giving people the ability to monitor their own health and identify possible problems before they become serious. These mobile, AI-powered diagnostic instruments can be used in resource-constrained or remote locations to give prompt, accurate diagnosis and tailored treatment recommendations. AI has the potential to help medical practitioners diagnose and treat respiratory disorders more remotely, thereby decreasing the need for in-person consultations and increasing access to expert treatment.

11.7.3 Better imaging and analysis

AI systems could perform extremely precise analysis of CT scans, X-rays, and other medical imaging, helping to diagnose respiratory conditions like pulmonary fibrosis, pneumonia, and lung cancer. During minimally invasive operations, AI-powered bronchoscopes and other imaging equipment may provide real-time guidance, increasing accuracy and lowering risks. AI might build individualized 3D models of patients' lungs to realistically test therapy alternatives and mimic the course of the disease, resulting in better therapeutic judgments [33].

11.7.4 Emerging trends and technologies

The state of development and progress in diverse domains, including technology, is constantly changing. Some technologies are used in healthcare such as:

- *Artificial Intelligence*: Diagnosis and Prognosis: Compared to conventional techniques, AI algorithms are able to identify respiratory disorders including asthma, COPD, and even lung cancer earlier and with more accuracy by analyzing lung scans, CT images, and other data. Better results and quicker interventions may result from this.
- *Personalized Medicine*: AI can forecast a patient's risk of respiratory diseases based on their unique genetic and health data, and then adjust treatment regimens accordingly. More effective and efficient care may result from this.
- *Real-Time Monitoring*: Wearables and sensors driven by AI can continuously monitor a patient's respiratory function, warning medical professionals of possible issues before they worsen. This can enhance quality of life and help avoid exacerbations.
- *Telemedicine and Remote Monitoring*: Home-based care: With the help of devices like pulse oximeters and smart inhalers, patients may keep an eye on their own respiratory health at home, which lowers the need for hospital visits and improves access to care, particularly in rural regions.
- *Telecare Consultations*: Patients can receive quick and convenient care from respiratory specialists virtually, saving on trip expenses and pathogen exposure.
- *Remote Monitoring and Management of Long-Term Respiratory Disorders*: Telemedicine enables remote monitoring and management of long-term respiratory conditions such as asthma and COPD, enabling real-time data-driven modifications to medication and treatment regimens.
- *Gene Therapy*: To address the underlying genetic causes of certain respiratory ailments, researchers are researching gene therapies, which may provide cures rather than merely symptom management. This is an example of precision medicine and regenerative medicine in action.
- *Stem Cell Therapy*: Patients with serious lung conditions including pulmonary fibrosis may find hope in using stem cells to repair damaged lung tissue.

- *Tailored Therapies*: As our knowledge of the genetic and molecular causes of respiratory disorders has grown, researchers are creating more specialized medications that work better for specific patients.
- *Additional Emerging Technologies*: Improved imaging methods: Devices such as low-dose CT scans and hyperpolarized MRIs are producing increasingly precise and detailed images of the lungs, which aid in diagnosis and therapy planning.
- *3D Printing*: although inhalers made with 3D printing can be personalized for each patient, lung models made with this technology can be used for surgery planning and education.
- *Nanomedicine*: Drugs can now be delivered to injured lung tissue directly through nanoparticles, increasing therapy efficacy and minimizing side effects.
- *Virtual Reality (VR) Therapy*: Patients with respiratory disorders like asthma and chronic obstructive pulmonary disease (COPD) may benefit from VR therapy. VR simulations can assist patients in practicing breathing exercises in realistic settings, enhancing lung function and lowering anxiety associated with respiratory problems.
- *Artificial Intelligence in Medication Discovery*: By evaluating enormous volumes of biological data, AI algorithms are being utilized to speed up the medication discovery process for respiratory illnesses. This method is more effective in identifying possible drug candidates, which could result in the creation of innovative therapies for diseases including cystic fibrosis and pulmonary hypertension.
- *Bioinformatics and Genomics*: These fields are developing at a rapid pace, which is helping scientists better understand the genetics of respiratory illnesses. Through the examination of genomic data, researchers may pinpoint genetic changes linked to ailments like idiopathic pulmonary fibrosis and create customized treatments based on the unique genetic characteristics of each patient.
- *Digital Biomarkers*: Real-time monitoring of respiratory symptoms and disease progression is being done with digital biomarkers, which include wearable technology and smartphone apps. The frequency of coughing, lung function, and medication adherence are just a few of the factors that these digital gadgets can measure. This data is crucial for individualized treatment planning and remote patient monitoring.
- *Exosome Therapy*: Studies are being conducted on exosomes, which are tiny vesicles secreted by cells, as a possible treatment for lung conditions. In preclinical research, exosome-based treatments have demonstrated potential for treating diseases like pulmonary fibrosis and acute respiratory distress syndrome, providing a regenerative method of healing damaged lung tissue.
- *Immunotherapy*: Being investigated as a potential lung cancer treatment, immunotherapy uses the body's immune system to target and eliminate cancer cells. Among the immunotherapy techniques being

researched for individuals with lung cancer are immune checkpoint inhibitors and chimeric antigen receptor T-cell therapy, which may provide more potent and long-lasting treatment results.

- *Microbiome Modulation*: The health and disease of the lungs are significantly influenced by the respiratory microbiome, which is made up of bacteria, viruses, and fungi in the respiratory tract. To improve respiratory outcomes in illnesses like chronic bronchitis and asthma, researchers are looking into microbiome modulation approaches like bacteriophage therapy and probiotics to restore microbial equilibrium.
- *Robot-Assisted Surgery*: As lung cancer and other respiratory disorders are treated during thoracic surgery, robotics technology is being utilized more and more. Compared to traditional surgical approaches, robot-assisted operations offer more precision and dexterity, which results in shorter recovery times and better patient outcomes.
- *Regenerative Medicine*: Techniques in this field, such as tissue engineering and cell therapy, show promise in the treatment of respiratory disorders by mending damaged lung tissue and restoring lung function. Particularly, stem cell-based treatments are being researched for their capacity to repair lung tissue that has been harmed by diseases like emphysema and pulmonary fibrosis.
- *Environmental Monitoring*: The identification and mitigation of environmental factors, such as air pollution and allergen exposure, that contribute to respiratory disorders is made possible by advancements in sensor technology and environmental monitoring systems. Individualized treatment strategies for patients with respiratory disorders and public health actions can be informed by real-time air quality monitoring and personalized exposure tracking.

11.8 CONCLUSION

The advent of AI marks a significant technological breakthrough in the healthcare sector, particularly in the real-time monitoring of respiratory health. AI serves as an innovative approach to disease management. The chapter also delves into real-world case studies showcasing practical applications of AI in the medical domain. It thoroughly examines the challenges associated with AI integration, including ethical considerations. The study concludes by pondering the potential future impact of AI on respiratory healthcare. The exploration encompasses AI algorithms, methodologies, and challenges employed in disease identification, highlighting a successful AI algorithm in analyzing the respiratory system. Despite the immense potential of technological advancements to improve healthcare, the integration of technology poses challenges and ethical dilemmas that necessitate careful scrutiny. Issues surrounding data security and privacy emerge as primary concerns in the healthcare-tech interface. The ongoing evolution of technology holds promise for real-time benefits in healthcare. In the future, there is a need to transition toward advanced technology, especially given the heightened severity of lung problems post-COVID and due to smoking, Leveraging advanced generalized neutrosophic soft mapping for identifying lung diseases emerges as a potential avenue for future exploration.

REFERENCES

[1] Belkacem, A. N., Ouhbi, S., Lakas, A., Benkhelifa, E., & Chen, C. (2021). End-to-end AI-based point-of-care diagnosis system for classifying respiratory illnesses and early detection of COVID-19: a theoretical framework. *Frontiers in Medicine*, *8*, 585578.

[2] Najjar, R. (2023). Redefining radiology: a review of artificial intelligence integration in medical imaging. *Diagnostics*, *13*(17), 2760.

[3] Sevakula, R. K., Au-Yeung, W. M., Singh, J. P., Heist, E. K., Isselbacher, E. M., & Armoundas, A. A. (2020). State-of-the-Art Machine Learning Techniques Aiming to Improve Patient Outcomes Pertaining to the Cardiovascular System. *Journal of the American Heart Association*, *9*(4), e013924. https://doi.org/10.1161/JAHA.119.013924

[4] Zadeh LA (1965) Fuzzy sets. *Inf Control* 8(3):338–353

[5] Atanassov K (1986) Intuitionistic fuzzy sets. *Fuzzy Set Syst* 20, 87–96. https://doi.org/10.1016/S0165-0114(86)80034-3

[6] Molodtsov, D. (1999). Soft set theory—first results. *Computers & Mathematics with Applications*, *37*(4–5), 19–31.

[7] Riaz, M., & Hashmi, M. R. (2019). Linear Diophantine fuzzy set and its applications towards multi-attribute decision-making problems. *Journal of Intelligent & Fuzzy Systems*, *37*(4), 5417–5439.

[8] Perchiazzi, G., Giuliani, R., Ruggiero, L., Fiore, T., & Hedenstierna, G. (2003). Estimating respiratory system compliance during mechanical ventilation using artificial neural networks. *Anesthesia and Analgesia*, *97*(4), 1143–1148. https://doi.org/10.1213/01.ANE.0000077905.92474.82

[9] Sharma, S., Pandey, S., & Shah, D. (2023). Enhancing Medical Diagnosis with AI: A Focus on Respiratory Disease Detection. *Indian Journal of Community Medicine*, *48*(5), 709–714.

[10] Das, N., Happaerts, S., Gyselinck, I., Staes, M., Derom, E., Brusselle, G., ... & Janssens, W. (2023). Collaboration between explainable artificial intelligence and pulmonologists improves the accuracy of pulmonary function test interpretation. *European Respiratory Journal*, *61*(5).

[11] Zeng, S., Arjomandi, M., Tong, Y., Liao, Z. C., & Luo, G. (2022). Developing a machine learning model to predict severe chronic obstructive pulmonary disease exacerbations: retrospective cohort study. *Journal of Medical Internet Research*, *24*(1), e28953.

[12] Zhang, C., Zhang, L., Tian, Y., Bao, B., & Li, D. (2023). A Machine-Learning-Algorithm-Assisted Intelligent System for Real-Time Wireless Respiratory Monitoring. *Applied Sciences*, *13*(6), 3885.

[13] Li, X., Zhou, H. P., Zhou, Z. J., Du, N., Zhong, E. H., Zhai, K., ... & Zhou, L. (2021). Artificial intelligence-powered remote monitoring of patients with

chronic obstructive pulmonary disease. *Chinese Medical Journal, 134*(13), 1546–1548.

[14] Jaber, N. (2022). Can artificial intelligence help see cancer in new, and better, ways?. *National Cancer Institute*, *22*(3).

[15] Byra, M., Dobruch-Sobczak, K., Piotrzkowska-Wroblewska, H., Klimonda, Z., & Litniewski, J. (2022). Prediction of response to neoadjuvant chemotherapy in breast cancer with recurrent neural networks and raw ultrasound signals. *Physics in Medicine and Biology*, *67*(18). https://doi.org/10.1088/1361-6560/ac8c82

[16] Lee, J., & Pak, T. Y. (2022). Machine learning prediction of suicidal ideation, planning, and attempt among Korean adults: A population-based study. *SSM-Population Health*, 19, 101231

[17] Wang, Z., Yang, Z., & Dong, T. (2017). A review of wearable technologies for elderly care that can accurately track indoor position, recognize physical activities and monitor vital signs in real time. *Sensors*, 17(2), 341.

[18] Wajid, M. S., & Wajid, M. A. (2021). The importance of indeterminate and unknown factors in nourishing crime: a case study of South Africa using neutrosophy. *Neutrosophic Sets and Systems*, 41(2021), 15.

[19] Fedele, D. A., Thomas, J. G., McConville, A., McQuaid, E. L., Voorhees, S., Janicke, D. M., ... & Gurka, M. J. (2021). Using mobile Health to improve asthma self-Management in early Adolescence: a pilot randomized controlled trial. *Journal of Adolescent Health*, *69*(6), 1032–1040.

[20] Balagurunathan, Y., Beers, A., Mcnitt-Gray, M., Hadjiiski, L., Napel, S., Goldgof, D., ... & Farahani, K. (2021). Lung nodule malignancy prediction in sequential ct scans: Summary of isbi 2018 challenge. *IEEE Transactions on Medical Imaging*, *40*(12), 3748–3761.

[21] Kanne, J. P. (2020). Chest CT findings in 2019 novel coronavirus (2019-nCoV) infections from Wuhan, China: key points for the radiologist. *Radiology*, *295*(1), 16–17.

[22] Majkowska, A., Mittal, S., Steiner, D. F., Reicher, J. J., McKinney, S. M., Duggan, G. E., ... & Shetty, S. (2020). Chest radiograph interpretation with deep learning models: assessment with radiologist-adjudicated reference standards and population-adjusted evaluation. *Radiology*, *294*(2), 421–431.

[23] Alqudaihi, K. S., Aslam, N., Khan, I. U., Almuhaideb, A. M., Alsunaidi, S. J., Ibrahim, N. M. A. R., ... & Alshahrani, M. S. (2021). Cough sound detection and diagnosis using artificial intelligence techniques: challenges and opportunities. *IEEE Access*, 9, 102327–102344.

[24] Arnaud, E., Elbattah, M., Ammirati, C., Dequen, G., & Ghazali, D. A. (2022). Use of artificial intelligence to manage patient flow in emergency department during the Covid-19 pandemic: a prospective, single-center study. *International Journal of Environmental Research and Public Health*, 19(15), 9667.

[25] Acharya, V., Dhiman, G., Prakasha, K., Bahadur, P., Choraria, A., Prabhu, S., ... & Kautish, S. (2022). AI-assisted tuberculosis detection and classification from chest X-rays using a deep learning normalization-free network model. *Computational Intelligence and Neuroscience*, 2022.

[26] Wajid, M. A., & Zafar, A. (2022). A multimodal approach of information access and retrieval using neutrosophic sets. In *Emerging Trends in IoT and Computing Technologies* (pp. 382–388). Routledge.

[27] Rajpurkar, P., Irvin, J., Ball, R. L., Zhu, K., Yang, B., Mehta, H., ... & Lungren, M. P. (2018). Deep learning for chest radiograph diagnosis: A retrospective comparison of the CheXNeXt algorithm to practicing radiologists. *PLoS medicine*, 15(11), e1002686.
[28] Wajid, M. S., Terashima-Marin, H., Najafirad, P., Pablos, S. E. C., & Wajid, M. A. (2024). DTwin-TEC: An AI-based TEC district digital twin and emulating security events by leveraging knowledge graph. *Journal of Open Innovation: Technology, Market, and Complexity*, 10(2), 100297.
[29] Li, X., Zhang, S., Luo, X., Gao, G., Luo, X., Wang, S., ... & Wu, N. (2023). Accuracy and efficiency of an artificial intelligence-based pulmonary broncho-vascular three-dimensional reconstruction system supporting thoracic surgery: Retrospective and prospective validation study. *EBioMedicine*, 87.
[30] Maarseveen, T. D., Meinderink, T., Reinders, M. J., Knitza, J., Huizinga, T. W., Kleyer, A., ... & Knevel, R. (2020). Machine learning electronic health record identification of patients with rheumatoid arthritis: algorithm pipeline development and validation study. *JMIR Medical Informatics*, 8(11), e23930.
[31] Ardila, D., Kiraly, A. P., Bharadwaj, S., Choi, B., Reicher, J. J., Peng, L., ... & Shetty, S. (2019). End-to-end lung cancer screening with three-dimensional deep learning on low-dose chest computed tomography. *Nature Medicine*, 25(6), 954–961.
[32] Bhushan, B., Sharma, S. K., Nand, P., Shankar, A., & Obaid, A. J. (Eds.). (2024). Emerging Trends for Securing Cyber Physical Systems and the Internet of Things.
[33] Pandey, S., & Bhushan, B. (2024). Recent Lightweight cryptography (LWC) based security advances for resource-constrained IoT networks. *Wireless Networks*, 30(4), 2987–3026.

Chapter 12

Intersection of neutrosophy and machine learning

Applications in agriculture

Vikram Singh and Gurcharan Dass

12.1 INTRODUCTION

Florentin Smarandache's proposal of Neutrosophy expands on Philosophy by addressing ambiguity, contradiction, and indeterminacy. Neutrosophic logic, a logical extension that allows for items to be simultaneously true, false, and indeterminate, is introduced in this framework [1]. At its core lies the neutrosophic set, comprising three subsets: definitely true (T), definitely false (F), and indeterminate (I), where elements can belong to one, two, or all three subsets concurrently. The sets' membership function is defined by three functions:

- the truth (μT),
- the indeterminacy (μI), and
- the falsity (μF) functions, assigning degrees of membership to T, I and F, respectively.

Neutrosophic numbers, represented as $N = (x, y, z)$, encapsulate membership or truth (x), indeterminacy (y), and non-membership or falsity (z), complying with the constraint $x + y + z \leq 1$.[1] These ideas open the door for mathematical models that can effectively handle the uncertainty involved in decision-making, particularly in domains such as agricultural resource management. Neutrosophic sets go beyond traditional set theory and offer a novel way to deal with ambiguity and inconsistencies inside a set. Within the universe U, each element x in a neutrosophic set N is linked to three membership functions:

- membership (truth) function $(T_N(x))$,
- uncertainty (indeterminacy) function $(I_N(x))$, and
- non-membership (falsity) function $(F_N(x))$.

 DOI: 10.1201/9781003606055-12

These functions gauge x's association with the set in terms of truth, indeterminacy, and falsity, respectively.

Mathematically represented as (12.1) below:

$$N = \left\{\left\langle x, \left(T_N(x), I_N(x), F_N(x)\right)\right\rangle : x \in U; T_N, I_N, F_N \in [0,1]\right\} \ldots \quad (12.1)$$

This set structure allows $T_N(x)$, $I_N(x)$, and $F_N(x)$ to span the closed interval $[0,1]$, emphasizing their membership values' range. Neutrosophic sets, in contrast to classical or fuzzy sets, offer a more nuanced portrayal of uncertainty by allowing elements that vary in degrees of truth, indeterminacy, and falsity to be simultaneous, partial, or non-membership. Neutrosophic sets also have no restrictions on the total membership values, which can be any value between 0 and 3 $\left(0 \leq T_N(x) + I_N(x) + F_N(x) \leq 3\right)$, enabling a broader representation of conflicting and uncertain data [3, 4], and [5].

The present chapter contributes significantly to two key agricultural domains: crop selection based on soil variability and weather forecast modeling. In the domain of Soil Variability and Crop Selection Modeling, a novel approach employing a neutrosophic kNN classifier predicts suitable crops based on specific soil properties using a small-scale dataset. This method accurately predicts the optimal crop for given soil conditions and showcases its potential in diagnosing agricultural diseases and pests. The "weather forecast modeling" offers a systematic framework for addressing uncertainties inherent in weather forecasting by incorporating neutrosophic mathematics, allowing for a comprehensive evaluation of weather forecasts. By providing farmers with valuable insights into weather patterns, this modeling technique assists them in resource allocation and informed decision-making regarding agricultural planning.

This chapter opens with an extensive abstract and comprises seven more sections. Section 12.1 "Introduction" lays the foundations of the discussion that furthers in the subsequent sections. Also, this section mentions the major contributions of the chapter toward the theme of the book. Section 12.2 discusses the neutrosophic rendering of three machine learning algorithms, namely, neutrosophic kNN, neutrosophic SVM, and neutrosophic multilinear regression. Also, the related mathematics has been presented in respect of the aforesaid machine learning techniques. Section 12.3 "Related Work" discusses an extensive review of topical research work published in reputed publications. Titled as "Neutrosophic Logic and Machine Learning in Agriculture," Section 12.4 presents general agricultural applications of neutrosophic variant of machine learning models. Sections 12.5 and 12.6 discuss two major contributions of this chapter namely, neutrosophic machine learning (NML) models of crop selection based on soil variability and weather forecasting. Section 12.7 concludes the chapter in terms of major "Conclusion".

12.2 NEUTROSOPHIC MACHINE LEARNING

Neurosophic machine learning expands on existing paradigms to address the ambiguity, incompleteness, and uncertainty present in complex datasets. This section has discussed neutrosophic variants of two classifiers, namely, kNN and SVM, as well as the neutrosophic linear regression model.

12.2.1 Neutrosophic kNN

Neutrosophic k-nearest neighbors (kNN) is a classification algorithm that operates based on similarities in neutrosophic feature spaces. Here are the steps in neutrosophic kNN algorithm:

(a) Choose a value of k (a positive integer).
(b) Calculate neutrosophic distances.

Calculate the neutrosophic distance (Euclidean distance or some other distance measure) between the test data instance and each instance in the training dataset. Presently, neutrosophic Euclidean distance (NED) has been computed using eq. (12.2) to (12.5):

$$\text{NED}_j = \sqrt{(D_{tj})^2 + (D_{ij})^2 + (D_{fj})^2 \ldots} \tag{12.2}$$

Where:

- NED_j is the neutrosophic Euclidean distance between j^{th} training data instance and the test data (unknown) instance.
- D_{tj}, D_{ij} and D_{fj} are Euclidean distance between truth, indeterminate, and false components of jth training data instance and the test data (unknown) instance.

Further:

$$D_{tj} = \sqrt{\Sigma\left((pH_{tj} - pH_{tu})^2 + (MC_{tj} - MC_{tu})^2 + (NL_{tj} - NL_{tu})^2\right)\ldots} \tag{12.3}$$

Where:

- pH_{tj} and pH_{tu} are the truth components of neutrosophic pH value of jth training instance and test data (unknown) instance, respectively.
- MC_{tj} and MC_{tu} are the truth components of neutrosophic "Moisture Content" of jth training instance and test data (unknown) instance, respectively.
- NL_{tj} and NL_{tu} are the truth component of neutrosophic "Nutrient Level" of jth training instance and test data (unknown) instance, respectively.

Similarly:

$$D_{ij} = \sqrt{\Sigma\left(\left(pH_{ij} - pH_{iu}\right)^2 + \left(MC_{ij} - MC_{iu}\right)^2 + \left(NL_{ij} - NL_{iu}\right)^2\right)} \ldots \quad (12.4)$$

$$D_{fj} = \sqrt{\Sigma\left(\left(pH_{fj} - pH_{fu}\right)^2 + \left(MC_{fj} - MC_{fu}\right)^2 + \left(NL_{fj} - NL_{fu}\right)^2\right)} \ldots \quad (12.5)$$

Performing these steps provides the predicted crop label for the given test instance based on its similarity to the nearest instances in the dataset.

(c) Sort the training dataset in ascending order of NED.
(d) Select the top k instances of the training dataset.

These instances are the "nearest neighbors" to the test data instance.

(e) Perform majority voting.

Find out the majority crop label among the selected k instances. This will be the predicted crop label for the test instance based on the neutrosophic kNN algorithm.

12.2.2 Neutrosophic SVM

The N-SVM algorithm, a variation of the standard SVM, uses the neutrosophic components to improve accuracy and reduce the impact of unusual data points during learning. Assuming the availability of a set of neutrosophic training data (x_i, y_i) where each x_i represents neutrosophic input feature space data and y_i shows its class in neutrosophic terms.

(a) Calculate three Neutrosophic components (t_i, i_i, f_i) for each data point:
t_i is the measure of data's distance from the positive class center.
i_i is the measure of data's distance from the average center.
f_i is the measure of data's distance from the negative class center.
(b) Define g_j as a combining function using these neutrosophic components:

$$g_j = t_i + i_i + f_i \ldots \quad (12.6)$$

(c) Optimize the hyperplane using the combining function g_j:
The goal is to minimize the weighted sum of these components. The optimization seeks the best hyperplane that separates classes well.
(d) Ensure that the optimized hyperplane satisfies the conditions specified in eq. (12.7) to correctly classify each data point:

$$(y_j.\left(\omega_j + b\right) > 1 - \zeta_j - g_j) \ldots \quad (12.7)$$

Where:

- y_j represents the neutrosophic class label or output associated with the data point j.
- ω_j denotes the neutrosophic weight vector assigned to the data point j.
- b represents the neutrosophic bias or intercept term in the model.
- ζ_j represents the neutrosophic slack variable associated with data point j, allowing for misclassification or outliers within the margin.

The equation (12.6) states that the product of the class label y_j and the dot product of the weight vector ω_j and the data point plus the bias term b should be greater than $\left(1-\zeta_j-g_j\right)$ for proper classification. This formulation ensures that correctly classified points lie beyond the margin and are not within the margin boundary (controlled by ζ_j).

12.2.3 Neutrosophic linear regression

Extending the notion of classical linear regression to neutrosophic datasets with only one input feature, the simple linear regression model may be represented as (12.8) below:

$$Y_i = \beta_0 + \beta_1.X_i \ldots \tag{12.8}$$

Where:

- Y_i denotes the neutrosophic predicted value of target variable for i^{th} training instance,
- X_i denotes the neutrosophic independent variable in i^{th} instance of training data,
- β_0 is the estimated (to be optimized) neutrosophic intercept, and
- β_1 is the estimated (to be optimized) neutrosophic slope or regression coefficient.

Further, for a multicriteria decision situation, the regression model (multilinear regression) takes the form of *(12.9)* below:

$$Y_i = \beta_0 + \sum_{1}^{m} \beta_j.X_{ji} \ldots \tag{12.9}$$

Where:

- Y_i is the neutrosophic predicted value of the target feature for i^{th} instance of data,
- X_{ji} denotes the jth neutrosophic input variable in i^{th} instance of training data,

- β_0 is the estimated (to be optimized) neutrosophic intercept,
- β_j represents the neutrosophic regression coefficient for j^{th} input feature, and
- m is the number of input features.

In neutrosophic linear regression, adjusting the intercept β_0 and regression coefficients $\beta_j \forall j = 1 \text{ to } m$ involves employing gradient descent to minimize the error between predicted and actual neutrosophic values. The formulas for updating these parameters iteratively using the gradient descent method are derived from the partial derivatives of the error function for each parameter. The update rules are based on the direction and magnitude indicated by the gradients to move toward the minimum error:

The update rule for the intercept β_0 and β_j's in each iteration is:

$$\beta_0^{t+1} = \beta_0^{t} + \gamma \frac{2}{N} \sum_{1}^{N} \left(Y_{A_i} - Y_{P_i} \right) \ldots \tag{12.10}$$

$$\beta_j^{t+1} = \beta_j^{t} + \gamma \frac{2}{N} \sum_{1}^{N} X_i . \left(Y_{A_i} - Y_{P_i} \right) \ldots \tag{12.11}$$

Where:

- β_0^{t+1} is the updated value of the intercept for the $\left(t + 1^{st}\right)$ iteration,
- β_j^{t+1} is the updated value of the intercept for the $\left(t + 1^{st}\right)$ iteration,
- β_0^{t} is the current value of the intercept in the tth iteration,
- β_j^{t} is the current value of the intercept in the tth iteration,
- γ is the learning rate (step size) controlling the rate of gradient descent,
- N is the total number of instances in the dataset,
- Y_{A_i} is the actual value of target feature for i^{th} data instance.
- Y_{P_i} is the predicted value of target feature for i^{th} data instance.
- X_i is the ith input feature vector

Here j varies from 1 to m: m is the count of input features and i varies from 1 to N: N is the count of training data instances.

12.3 RELATED WORK

In controlled contexts like greenhouses or open fields, [6] offer a Smart Agriculture Mechanism, which monitors ecosystem functions after irrigation and seeding. The study offers a novel approximation technique for neutrosophic data—neutrosophic Bezier surfaces—that makes matrix-based computer processing possible. The chapter uses neutrosophic if-then principles for adaptive visualization and creates a mathematical model for creating Bezier surfaces in matrix form. It demonstrates how neutrosophy can be used in cloud computing, smart farming, and the IoT to handle uncertain data in these systems. Nevertheless, there are still issues that need to be resolved, such as de-neutrosophication techniques for n-gonal neutrosophic numbers and geometric approaches for visualizing neutrosophic data, which can be explored further in the future. This study highlights the potential advantages of neutrosophic theory and lays the groundwork for future investigations in this area by outlining how it might be included in IoT and smart farming systems.

A computer vision-based technique for diagnosing illnesses in basil leaves—a critical step for conventional medical treatments—is presented by [7]. Despite a lot of studies with sensors and DNA/RNA analysis, computer vision is still not widely used in this field. To categorize basil leaves as healthy or unhealthy, the study suggests a two-phase approach that consists of a novel segmentation technique based on neutrosophic logic and feature extraction thereafter. Images from a variety of herbal gardens, comprising both healthy and diseased leaves of several types of basil, are included in the collection. The literature review emphasizes the limitations of manual disease detection techniques, stressing their subjective and time-consuming character. It examines current developments in computer vision with a focus on feature extraction and picture collection for the identification of crop diseases. To determine the best leaf classification model, the study assesses nine classifiers and finally shows encouraging outcomes. In conclusion, it highlights the new feature set and segmentation method that combines texture and leaf intensity, potentially leading to precise basil leaf classification.

The study reported by [8] presents a novel technique for the classification of breast tumors using neutrosophic score features, utilizing ultrasound pictures. The significance of detecting breast cancer is underlined, and the limits of ultrasonography due to picture ambiguity are highlighted. The suggested strategy uses a support vector machine (SVM) classifier in conjunction with a supervised feature selection technique to integrate texture and morphologic data based on neutrosophic similarity scores. A remarkable 99.1% classification accuracy on 112 instances highlights the importance of early identification and the function of computer-aided systems. Image segmentation, feature extraction, and SVM classification are the phases of the method that address the discriminatory ability of different features and their combinations. This work proposes a novel approach for classifying breast tumors that combines morphologic and textural information using scores for neutrosophic similarity. The feature space is narrowed by using a supervised feature selection method. With the help of a SVM classifier, the suggested approach shows strong discriminating abilities. With an astounding 99.1% classification accuracy, validation with 112 patients (58 malignant, 54 benign) shows promise and highlights the possibility of precise tumor classification in ultrasonography pictures.

In their investigation on maintaining data accuracy in big data analytics, authors in [9] pay particular attention to how to deal with errors, imprecisions, and inconsistencies that arise throughout the data preparation process. It emphasizes the value of single-valued neutrosophic numbers, which provide a reliable method of modeling complex data and a tool for conveying complex information. While data mining algorithms already in use handle incomplete data during pre-processing, this work clarifies the disregard for inaccurate and inconsistent data at the modeling stage. It gives a summary of the efforts made to move machine learning algorithms from environments with crisp numbers to ones with neutrosophic. To address faulty information, the study investigates the integration of machine learning techniques with single-valued neutrosophic numbers, showcasing their usefulness in resolving real-world problems. It also presents a taxonomy of neutrosophic learning algorithms, separating processed from unprocessed algorithms, and it sketches directions for further research. The goal of this taxonomy is to make it easier for academics to navigate this domain. Neutrosophic theory, machine learning, single-valued neutrosophic numbers, and applications such as neutrosophic C-means, neutrosophic k-NN, neutrosophic-SVM, and neutrosophic simple linear regression are among the notable topics discussed.

In [10], a complex model for predicting the risks of medication toxicity has been presented. The model operates in three main phases, evaluating four different harmful effects on a dataset of 553 pharmaceuticals that undergo liver biotransformation. The model first concentrates on feature selection, using rough set-based methods to determine which characteristics are the most discriminative. By removing superfluous features, this pruning procedure greatly reduces computing overhead and increases the effectiveness of following analyses. In the second stage, the model employs random under-sampling, and random over-sampling to address data imbalances. To improve the accuracy of the model, it is imperative to have a balanced dataset, and these techniques are essential for efficiently organizing the dataset. The third and final phase presents two novel categorization algorithms based on neutrosophic rule generation. By demonstrating great sensitivity across a wide range of harmful effects during experimental evaluations, these novel models seek to categorize unknown substances as toxic or non-toxic. The study's findings demonstrate encouraging outcomes and emphasize the model's potential for early drug toxicity prediction. The report makes future research recommendations for hybrid models that combine neutrosophic systems with genetic algorithms in ensemble-based classifiers. These initiatives seek to increase predictive power even further, resulting in more reliable and efficient drug toxicity prediction systems.

A neutrosophic classifier was presented by [11] as an addition to the well-known fuzzy classification algorithms. Although fuzzy logic has shown to be useful in managing ambiguous data, there are still some issues with it. The chapter suggests neutrosophic logic as a more flexible and capable framework for dealing with ambiguity, inconsistency, and uncertainty to overcome these drawbacks. It contrasts the neutrosophic classifier's performance with that of conventional fuzzy classifiers, paying particular attention to variables such as membership functions, rule counts, and outcome indeterminacy. In comparison to its fuzzy equivalent, the study shows that the neutrosophic classifier maximizes these parameters. The theoretical underpinnings of neutrosophic logic are discussed, emphasizing how, in contrast to fuzzy logic, it can provide a continuous spectrum between concepts and their opposites, allowing for more subtle degrees of neutrality. The Iris dataset is used in the chapter's actual implementation using MATLAB, which highlights the variations in results between fuzzy and neutrosophic classifiers. The experimental results demonstrate the effectiveness of the neutrosophic technique by reducing the computing load, training time, and classifier complexity by forming non-overlapping decision zones in the pattern space. The chapter's conclusion highlights the possibility of investigating neutrosophic logic in more intricate fields and makes the case that it has the potential for use in real-time applications, especially if it closely resembles human reasoning.

Authors in [12] have examined a unique method for multimodal categorization termed neutrosophic CNN-based image and text fusion. The study focuses on e-commerce platforms that show products using both photos and text descriptions in the age of vast internet data, where information arrives in several forms like text, photographs, and videos. This research manages uncertainty in multimodal data for information retrieval tasks by using neutrosophic fuzzy sets, whereas previous classification approaches typically dealt with a single modality. The study uses neutrosophic CNNs to build feature representations for image classification, combining text and picture data by embedding text over images. It solves the problem of conventional convolutional neural networks' sensitivity to noise during testing, which impairs the performance of the networks in terms of classification. The suggested neutrosophic CNN-based methodology shows promising results when compared to established multimodal fusion approaches such as early and late fusion, as well as individual sources.

The use of single-valued neutrosophic sets (SVNS) to address ambiguity, imprecision, and inconsistency in real-world data is covered in [13]. It suggests using a single-valued neutrosophic cross-entropy measure in multicriteria decision-making processes to handle ambiguous and inconsistent data. The method is new not only because it applies single-values neutrosophic set theory but also because it takes into account indeterminacy in addition to truth and falsity information when comparing alternatives to predetermined standards. This method works well because it captures the imprecise, unclear, and inconsistent information necessary for multicriteria decision-making studies, effectively reflecting the ambiguous character of subjective judgments. Compared to conventional methods, the suggested strategy is more practicable and usable in real-world decision-making settings because of its practicality. Prospective research entails delving into single-valued neutrosophic multicriteria group decision-making quandaries and implementing the suggested single-valued neutrosophic cross-entropy measure across diverse fields like information fusion systems, expert systems, and medical diagnosis.

Artificial intelligence has been viewed by authors in [14] as a vital tool for detecting coronavirus and controlling its transmission. The SVM algorithm is one of many AI and machine learning algorithms that have been used to diagnose diseases. However, because traditional SVMs cannot address cognitive difficulties like ambiguity and inconsistency that are inherent in human perception, they may produce incomplete answers. Neutrosophic SVMs, which take into account all possible situations during sample analysis and lessen the impact of outliers, have been investigated by researchers as a potential solution to this problem. This approach is expected to improve diagnostic accuracy. The review describes a study that used neutrosophic SVMs to diagnose symptoms of coronavirus. It describes the methodology used, which included creating a neutrosophic dataset, training N-SVMs on new data, and comparing the results with those of traditional SVM algorithms. The study extracted features from chest radiographs using the Gray-Level Co-occurrence Matrix. The results highlight the promise of N-SVMs in medical applications by indicating that they provide better illness detection accuracy when compared to traditional SVM methods.

A unique paradigm for implementing neutrosophy in deep learning models has been developed by the authors in [15]. To better understand the concept, a data point's class has been predicted as three membership functions rather than just predicting one class as the outcome. The two pieces that make up the suggested model are feature extraction and feature categorization. Every model can be used as a feature extractor because it has its own feature extraction block. Bidirectional Encoder Representations from Transformers (BERT), GloVe (Global Vectors for word representation), ALBERT (A Lite BERT), RoBERTa (Robustly optimized BERT method), MPNet, and stacked ensemble models were used in the experiments with BiLSTM. Feature categorization reduces the dimensionality of features and makes predictions about them. Membership functions of SVNS were defined using the properties of the intermediate layer and the experiment was conducted using SemEval 2017 Task 4 dataset for forecasting. Proposed models have been assessed against the most recent state-of-the-art systems and the top five teams in the task. The stacked ensemble model (proposed) has yielded a Recall value of 0.733.

In the study published in [16], neutrosophic logic has been used in the analytic network process and multiattribute utility theory to reduce subjectivity related to expert-driven decisions and generate a trustworthy ranking of hospital building assets based on their variable criticality levels and performance deficiencies. This is further combined with the cutting-edge application of machine learning techniques in this domain, including Decision Trees, k-Nearest Neighbors, and Naïve Bayes, to automate and make repeatable the process of selecting priorities, hence reducing the need for further expert opinions. The created model was used in Canadian healthcare institutions, and its superiority was amply displayed when its related predictive performance was verified by comparison with an earlier model. Thus, it is anticipated that the integrated framework that has been developed will help establish a systematic, objective, and automated system for hospital asset renewal prioritization. This, in turn, will help ensure that resources are allocated in a way that is effective, sensible, and well-informed.

A technique for detecting items inside utilizing a rotational ultrasonic array and neutrosophic logic is described in the study [17]. Because a neutrosophic set lacks the undetermined membership value found in normal fuzzy sets, it has been viewed as the next development of the fuzzy set. Based on the degree of truthiness, degree of indeterminacy, and degree of falsity for the reflected distance, the proposed approach is designed to reflect the position of the walls (obstacle distance) and to allow the traffic to travel freely (forward, right, or left). The trials' findings demonstrate the good performance of the suggested indoor object-detecting system, with an accuracy rate of $97.2 \pm 1\%$ (mean average precision).

A novel hybrid deep learning-based layer-fusion and neutrosophic-set technique for diagnosing skin lesions is proposed by the research presented in [18]. Using transfer learning on the International Skin Imaging Collaboration (ISIC) 2019 skin lesion datasets, the off-the-shelf networks are investigated for their ability to classify eight categories of skin lesions. The accuracy of the top two networks, GoogleNet and DarkNet, was 77.41% and 82.42%, respectively. The suggested approach operates in two steps: first, it increases each trained network's individual classification accuracy. By enhancing the extracted features' descriptive power by the application of a recommended feature fusion methodology, the accuracy is raised to 79.2% and 84.5%, respectively. To further enhance, the second stage investigates ways to combine these networks. Using fused DarkNet and GoogleNet feature maps, respectively, the error-correcting output codes (ECOC) paradigm is applied to build a set of well-trained true and false SVM classifiers. The coding matrices used by the ECOC are made to train each true classifier against its adversary in a one-on-one manner. As such, discrepancies in the classification scores of true and false classifiers give rise to an ambiguity zone that is measured by the indeterminacy set. This uncertainty is resolved by recent neutrosophic approaches, which tip the scales in favor of the right skin cancer class. Consequently, the categorization score rises to 85.74%, clearly surpassing the latest proposals. The SVNSs that have been proposed, together with the trained models, will be made publicly available to support pertinent research.

Neutrosophic logic, a fuzzy logic generalization, is employed by the authors in [19] to address the uncertainty issue in networked real-time deadlock-resolving systems. The suggested approach is set up to take into account various forms of knowledge and the relationships between each feature and tripartition. Depending on the degree of truthiness, indeterminacy, and falsity membership, the aspects of the transaction include the degree of slackness, the degree of validation factor, and the degree of deadline-missed transaction. The authors have created a set of tools and used benchmark datasets to run experiments. When this new method is applied, the detection rate increases and the rollback rate significantly decreases. Authors have claimed to resolve all database deadlocks through the proposed solution, which also markedly improved database performance by up to three orders of magnitude.

The research article [20] suggests a paradigm that makes use of neutrosophic logic to evaluate risk and guarantee system safety. The model collects expert subjective data using linear trapezoidal neutrosophic values. To make it easier to use neutrosophic logic IF-THEN rules, the gathered facts are quantified and de-neutrosophicated. An example involving the risk and safety analysis of a critical system with a high degree of uncertainty on board a ship is used to assess and validate the model. Notably, applying IF-THEN principles, a type of neutrosophic logic, to risk/safety assessment is a novel approach that can be quite successful in qualitatively modeling systems for decision-making.

[21] addresses the problem of enhancing the classification of stress by concentrating on particular speech characteristics. Using neutrosophic methods to estimate a probability vector, the strategy entails classifying stress levels. In the classification of stress that is contextually sensitive, machine learning algorithms have demonstrated encouraging outcomes. Furthermore, different speaker stress combinations, such as quick and intense speaking, can be quantified using the output stress probability vector. It is proposed that a stress mixture model could be useful for tasks such as emergency phone message prioritization or enhancing the efficiency of traditional speech processing systems. In summary, the study indicates that the Neutrosophic speech recognition algorithm's targeted characteristics for stress classification may accurately estimate the stress levels of speakers and offer important insights for improving voice recognition system performance.

The main objective of the research reported in [22, 23] is to evaluate the application of predictive Machine Learning algorithms in a neutrosophic framework empirically to determine the probable return on investment for potential investors who are thinking about investing in a particular business plan. The purpose of this validation is to enable investors to make well-informed judgments about financing Micro and Small Enterprises (MSEs) and close any gaps in the MSEs' growth trajectory that may exist in the Peruvian setting. This study uses a post-test-only design with a control group and combines descriptive and predictive approaches. The main analytical technique is the Neutrosophic Technique for Order Preference by Similarity to Ideal Solution (NEBS). The results provide statistical evidence for the hypotheses put out and show how effective NEBS is at enabling the use of machine learning in the finance sector of micro and small businesses in Peru. In particular, the findings show that the use of machine learning increases formal funding sources' utilization, decreases complaints, shortens the time it takes to evaluate requirements, and increases financing opportunities. To fully investigate the possibilities and ramifications of machine learning technology, further study in this field is necessary, given its developing nature and intrinsic complexity [24].

12.4 NEUTROSOPHIC LOGIC AND MACHINE LEARNING IN AGRICULTURE

When it comes to dealing with the uncertainties that exist in agricultural settings, neutrosophy can be an invaluable resource. Neutrosophy examines elements that are neutral or ambiguous in addition to true and false elements to address indeterminacy, inconsistencies, and uncertainties. By providing answers to the uncertainty and complexity that are inherent in this sector, the combination of machine learning and neutrosophic logic offers a promising frontier in agriculture that will ultimately improve farming techniques' profitability, sustainability, and productivity. The following scenarios explain how to use neutrosophic logic in agricultural settings:

Pest and crop disease identification: The use of neutrosophic logic in crop disease identification, which addresses the complex interplay between overlapping features in insect infestations and the frequently confusing and conflicting symptoms in crop diseases, transforms agricultural diagnostics. It provides a special area for navigating uncertainty because of its basic recognition of elements that exist between absolute truths and falsities. This approach promotes a more thorough understanding of intricate agricultural challenges by accounting for the subtleties of incomplete or ambiguous data. Neutrosophy's incorporation of unknown factors facilitates a flexible diagnostic approach, enabling researchers and farmers to incorporate disparate, sometimes contradictory, sources of knowledge. This flexibility embraces the complexities and uncertainties inherent in crop disease detection and pest control, improving diagnostic accuracy and decision-making in agriculture. In the end, it creates opportunities for more complex, practical agricultural solutions.

Weather forecasting: Farmers now have a genuine method to deal with the uncertainty and ambiguity that are frequently associated with meteorological forecasts thanks to the application of neutrosophic logic in interpreting ambiguous weather forecasts. Through the integration of machine learning techniques like data analysis and predictive modeling, neutrosophy enables a greater understanding of the inherent uncertainties in weather forecasts. This method recognizes that weather patterns are unpredictable and gives farmers the option to prepare for a range of possible outcomes instead of depending just on one, potentially inaccurate, forecast. Adaptive and resilient agricultural techniques that take into account the uncertainty and unpredictability inherent in weather forecasting are made possible by combining machine learning with neutrosophic concepts to enable a more sophisticated comprehension of uncertain weather data.

Crop selection and soil variability: When combined with machine learning capabilities, neutrosophic analysis is an essential tool for farmers to choose the best crop varieties that can adapt to changing weather patterns and different types of soil. This method explores the neutral features of agricultural traits, which helps identify adaptable plant types that can flourish in a range of environmental circumstances. Neutrosophic analysis is a technique that helps identify resilient and adaptable crop varieties by processing and analyzing large amounts of agricultural data using machine learning algorithms. By optimizing crop selection tactics that take weather and soil type variability into account, this amalgamation gives farmers the power to make well-informed decisions that eventually improve agricultural production and sustainability.

Agriculture resource management: Machine learning techniques are integrated to enhance Neutrosophy's ability to identify neutral or indeterminate features in supply chain dynamics, irrigation management, and resource demand estimation. This allows for the control of unpredictability in these domains. The dynamics of supply chains and resource demand are inherently uncertain and ambiguous, as recognized by neoclassical concepts. Neutrosophy facilitates flexible decision-making in response to the dynamic needs and constantly shifting conditions in these industries by utilizing machine learning algorithms to examine past data and trends. Decision-makers can use this synergy to gain insights from these uncertain factors and make quick modifications and strategic plans that are in line with supply chain management and resource utilization's dynamic nature.

Adaptation to climate change: Neutrosophic thinking, supported by advances in machine learning, is essential for strengthening agricultural resilience to uncertainties brought on by climate change. Through recognition of the unpredictability of future environmental conditions, Neutrosophy enables agricultural operations to adjust in advance. To predict future climate changes, machine learning algorithms examine large datasets that include climate variables, historical trends, and prediction models. This combination approach makes it easier to formulate adaptive solutions that take into account complex consequences and the unpredictable nature of changing climate trends. Agricultural systems can plan for a range of possible outcomes thanks to neutrosophic frameworks, which improve readiness and build the resilience required to handle the challenges of a changing climate in the agricultural environment.

For farmers and agricultural decision-makers, neutrosophy presents a significant paradigm change when combined with machine learning skills. Its skill in handling ambiguities, contradictions, and inconsistencies offers a flexible and sophisticated way to address the complexity of agricultural landscapes. The agricultural community acquires a framework that can handle contradicting and ambiguous facts by adopting neutrosophy. This strategy is strengthened by machine learning algorithms, which sort through enormous and varied amounts of agricultural data, spot trends, and conclude from many sources. This combination makes it possible to make well-informed decisions in the face of uncertainty, which in turn makes agriculture's strategies more flexible and resilient. This, in turn, increases the sector's resilience and efficacy in addressing a variety of complex issues.

12.5 SOIL VARIABILITY AND CROP SELECTION MODELING

To predict suitable crop(s) for soil with particular characteristics, a neutrosophic flavor of kNN classifier, a supervised machine learning engine, has been designed. The modus operandi of the neutrosophic kNN classifier has been demonstrated by considering a small neutrosophic labeled dataset comprising three soil characteristics (pH value, moisture content, and nutrient level) as independent input features and the "suitable crop" as the class label (target feature). The following discussion illustrates the working of the use of the neutrosophic kNN classifier for selecting a crop suitable to a particular soil with given characteristics. In this illustration, the procedure outlined in subsection "*12.2.1 Neutrosophic kNN*" has been used with a value of k=5.

(a) Table 12.1 shows the training dataset used for illustration purposes. The working of Neutrosophic kNN has been illustrated using a small sample of training data.

Table 12.1 Soil characteristics and crop dataset (neutrosophic)

pH Level (T, I, F)	*Moisture content (T, I, F)*	*Nutrient level (T, I, F)*	*Crop*
<0.8, 0.1, 0.1>	<0.6, 0.3, 0.1>	<0.7, 0.2, 0.1>	Rice
<0.7, 0.2, 0.1>	<0.5, 0.4, 0.2>	<0.5, 0.4, 0.1>	Maize
<0.75, 0.15, 0.1>	<0.4, 0.3, 0.3>	<0.6, 0.3, 0.1>	Wheat
<0.65, 0.3, 0.05>	<0.7, 0.2, 0.1>	<0.6, 0.3, 0.2>	Rice
<0.7, 0.25, 0.05>	<0.6, 0.2, 0.2>	<0.65, 0.25, 0.1>	Wheat
<0.6, 0.3, 0.1>	<0.6, 0.3, 0.2>	<0.5, 0.3, 0.2>	Maize
<0.6, 0.2, 0.2>	<0.45, 0.25, 0.3>	<0.55, 0.3, 0.15>	Maize
<0.7, 0.2, 0.1>	<0.65, 0.25, 0.1>	<0.65, 0.25, 0.2>	Rice

(b) Working of the Neutrosophic kNN classifier has been tested using unknown data instance of Table 12.2.

Table 12.2 Neutrosophic test data instance

pH Level (T, I, F)	*Moisture content (T, I, F)*	*Nutrient level (T, I, F)*	*Crop*
<0.65, 0.25, 0.1>	<0.7, 0.2, 0.1>	<0.6, 0.35, 0.05>	Rice

(c) Calculate the neutrosophic Euclidean distances (NEDs) of all eight data instances from the test data instance and arrange them in ascending order of NED (Table 12.3).

Table 12.3 Neutrosophic Euclidean distances

Instance#	*NED*	*Suitable Crop*
4	0.1732	Rice
5	0.2000	Wheat
8	0.2121	Rice
6	0.2645	Maize
1	0.3157	Rice
2	0.3316	Maize
7	0.3670	Maize
3	0.4062	Wheat

(d) The most frequent class label in the top five data instances (k nearest neighbors) is the "Rice." Accordingly, the neutrosophic kNN classifier predicts "Rice" as a suitable crop for the soil with neutrosophic characteristics of the test data instance. The performance of the classifier has, however, not been evaluated in the present chapter. The use of a larger dataset for training and testing is an open research thread in this area.

Further, neutrosophic kNN may also be used in agricultural settings to diagnose pests and crop diseases, for this situation is fraught with uncertainty because of shifting environmental conditions, vague symptoms, and irregular historical trends. It is difficult for traditional decision-making to deal with these uncertainties. To handle this complexity, the neutrosophic mathematical model permits the expression of truth, indeterminacy, and falsehood in each criterion, allowing for a thorough evaluation that takes uncertainties into account.

12.6 WEATHER FORECAST MODELING

Because weather forecasts are erratic or ambiguous, they frequently introduce uncertainty into agricultural practices. Neutrosophy provides an organized method for handling and interpreting such hazy predictions. The mathematical model under discussion makes use of three parameters, each represented by three neutrosophic values: accuracy, reliability, and consistency with real situations (truth, indeterminacy, and falsity). The process of giving numerical values and weights to these criteria makes it possible to evaluate weather forecasts thoroughly. Despite the inherent uncertainties in weather forecasting, farmers are empowered to make educated decisions thanks to the weighted scores, which aid in assessing and planning for a variety of possible weather scenarios. By modifying the weights or criteria values, decision-making systems for agricultural planning can be made more flexible to accommodate varying forecast conditions.

A machine learning model for weather forecasting relies heavily on several features to accurately predict the weather. These models frequently incorporate the following features:

- Ambient temperature: an important factor influencing weather patterns.
- Humidity: the amount of moisture in the air is a key factor in determining the weather.
- Pressure: atmospheric air pressure affects the current and future weather patterns.
- Wind: speed and direction of the wind cast a big influence on the weather.
- Precipitation: comprising sleet, rain, and snow, is an important factor in forecasting the likelihood and volume of precipitation.
- Cloud cover: the kind and amount of cloud cover can affect temperature and precipitation projections.
- Sunshine: the strength and duration of sunshine influence the temperature and other meteorological factors.

- Geographical features: local weather patterns can be influenced by geographic features of the area, such as elevation or proximity to significant bodies of water.
- Historical weather data also has a bearing on weather forecasts for the future.
- Seasonality and time of day also are important considerations in weather forecasting.

Accurate weather forecasting models are constructed with the aid of these characteristics as well as extra data such as atmospheric models, radar information, and satellite photos. Weather conditions are predicted for both short- and long-term forecasts using the aggregation and analysis of these features. There are two methods you can use when working with neutrosophic data in a multifeature dataset for precipitation prediction:

- Direct neutrosophic regression: Multilinear regression can be performed directly on the neutrosophic dataset without converting it into crisp data. This method involves handling neutrosophic values throughout the regression process, considering the truth, indeterminacy, and falsity memberships within the regression framework. It's a complex method that accounts for uncertainties inherent in the data but might require specialized algorithms tailored for neutrosophic regression.
- Transformation to crisp data: Alternatively, you can transform the neutrosophic data into crisp numerical values and then apply conventional multilinear regression. This approach involves extracting crisp values from neutrosophic sets, such as using weighted averages or other methods to convert neutrosophic values to numerical ones. After transforming, you can use regular multilinear regression techniques suited for numerical data.

The decision is based on the particular situation, the instruments at hand, and how crucial it is to preserve ambiguous information. A more appropriate approach would be direct neutrosophic regression if your goal is to preserve the subtleties of the neutrosophic data. However, converting to crisp data can be the better option if ease of use and familiarity with conventional regression procedures are top priorities. In this chapter, a direct neutrosophic regression model has been used to predict the precipitation given the ambient temperature, humidity, wind speed, and cloud cover.

12.6.1 Training dataset

Although, the weather datasets comprise numerous features, albeit, for illustration and brevity purposes, only the following four predictors (input features) and the dataset (comprising only five data instances of Table 12.4)

Table 12.4 Weather dataset (neutrosophic)

(T) <T, I, F>	(H) <T, I, F>	(WS) <T, I, F>	(CC) <T, I, F>	(P_A) <T, I, F>
{0.7,0.2,0.1}	{0.6,0.3,0.1}	{0.8,0.15,0.05}	{0.5,0.4,0.1}	{0.6,0.2,0.2}
{0.6,0.3,0.1}	{0.5,0.3,0.2}	{0.7,0.2,0.1}	{0.4,0.3,0.3}	{0.7,0.1,0.2}
{0.8,0.1,0.1}	{0.7,0.2,0.1}	{0.6,0.3,0.1}	{0.5,0.2,0.3}	{0.6,0.3,0.1}
{0.5,0.3,0.2}	{0.4,0.5,0.1}	{0.8,0.1,0.1}	{0.7,0.2,0.1}	{0.7,0.2,0.1}
{0.6,0.1,0.3}	{0.6,0.3,0.1}	{0.5,0.3,0.2}	{0.4,0.3,0.3}	{0.5,0.3,0.2}

have been considered. Using a bigger dataset for training purposes is the further research thread in this direction.

- Temperature (T) – predictor feature
- Humidity (H) – predictor feature
- Wind speed (WS) – predictor feature
- Cloud cover (CC) – predictor feature
- Precipitation (P_A) – target feature (actual value)

12.6.2 Regression model

For a multiple linear regression model to be deemed credible, certain assumptions must be fulfilled. Specifically, a linear connection between the predictors and the target variable is assumed. In the instant case, the mathematical form of eq. (12.12), derived from eq. (12.9) of Section 12.2.3, is assumed by the multilinear regression model:

$$P_p = \beta_0 + \beta_1.T + \beta_2.H + \beta_3.WS + \beta_4.CC\ldots \tag{12.12}$$

Where:

- P_p is the predicted (neutrosophic) value for precipitation.
- β_0 is the intercept (neutrosophic constant term).
- β_1, β_2, β_3, and β_4 are the neutrosophic coefficients for the predictors.
- T, H, WS, and CC are the neutrosophic values of the predictors.

The objective of the regression modeling is to minimize the difference between the predicted and actual values of precipitation. This is typically done by minimizing the sum of squared differences between the actual (observed) precipitation P_A and predicted precipitation level P_p. Often, the ordinary least squares method is used as the error function to estimate and optimize the coefficients by finding values that minimize the sum of squared residuals. Further, the common metrics for evaluating regression models include absolute error, mean squared error (MSE), root mean squared error (RMSE), etc. These metrics measure the goodness of fit and predictive performance of the model. In the underlying illustration, gradient descent optimization has been used to estimate neutrosophic parameters β_0, β_1, β_2, β_3, and β_4 that minimize the error between predicted and actual neutrosophic Precipitation values.

(a) Assume the initial values of neutrosophic regression parameters and learning rate γ :
β_0 = (0.4, 0.3, 0.3)
β_1 = (0.3, 0.4, 0.3)
β_2 = (0.2, 0.5, 0.3)
β_3 = (0.1, 0.6, 0.3)
β_4 = (0.1, 0.3, 0.6)
γ = (0.65, .0.25, 0.10)

(b) Estimate/predict and normalize the target feature—precipitation level for each instance of training data using the regression function of eq. *(12.12)*
$P_{p_1} = \{0.419, 0.346, 0.234\}$
$P_{p_2} = \{0.364, 0.378, 0.258\}$
$P_{p_3} = \{0.405, 0.332, 0.273\}$
$P_{p_4} = \{0.375, 0.379, 0.245\}$
$P_{p_5} = \{0.354, 0.340, 0.296\}$

(c) Compute the chosen model optimization metric (say RMSE) using P_A*'s* and P_p*'s*. In the context of neutrosophic multilinear regression, the formula of eq. (12.13) is used to compute the *RMSE* between the predicted neutrosophic precipitation values P_p*'s* and the actual neutrosophic precipitation values P_A*'s*.

$$\text{RMSE} = \sqrt{\frac{1}{N}\sum_{1}^{N}\left(\left(P_{P_i}^T - P_{A_i}^T\right)^2 + \left(P_{P_i}^I - P_{A_i}^I\right)^2 + \left(P_{P_i}^F - P_{A_i}^F\right)^2\right)\ldots} \quad (12.13)$$

Where:

- N is the total number of instances in the dataset.
- $P_{P_i}^T$, $P_{P_i}^I$ and $P_{P_i}^F$ are truth, indeterminacy, and falsity components, respectively, of the predicted neutrosophic precipitation for i^{th} neutrosophic instance.
- $P_{A_i}^T$, $P_{A_i}^I$, and $P_{A_i}^F$ are truth, indeterminacy, and falsity components, respectively, of the observed (actual) neutrosophic precipitation for i^{th} neutrosophic instance.

The formula calculates the squared differences between the components of predicted and actual neutrosophic precipitation values, sums them across all instances, takes the average, and then computes the square root to obtain the RMSE.

$$RMSE_1 = 0.3189 \text{ (first iteration } RMSE)$$

(d) Recast the values of neutrosophic regression parameters β_0, β_1, β_2, β_3, and β_4 using eq. *(12.10)* and *(12.11)* from *Section 12.2.3*.
(e) Repeat steps ***(b)*** to ***(d)*** above, till the stopping criteria is not met (say, model metric RMSE keeps descending).
(f) For the neutrosophic regression-based prediction model, use the values of regression parameters β_0, β_1, β_2, β_3, and β_4 corresponding to the global minimum value of the gradient.

12.7 CONCLUSION

Some applications and implications of NML in the field of agriculture have been discussed in this chapter. The introduction of NML into crop selection—especially when it comes to soil variability—marks a significant breakthrough and the beginning of a new age of crop variety recommendations that are dynamically adjusted to changing soil conditions. It makes precision agriculture possible, allowing farmers to maximize yield potential even in a variety of environmental settings. The discussion in this chapter emphasizes how important NML is for creating models of agricultural diseases and pests. Through the use of neutrosophic logic, these models can deftly negotiate the complex network of overlapping features and unclear symptoms, providing precise and nuanced diagnostic approaches that greatly improve crop health management. Increased use of disease mitigation techniques is encouraging for the sustainability and productivity of agriculture.

Further, NML's involvement in creating weather forecasting models specifically for agriculture has also been covered in this chapter. Because of NML's special capacity to understand uncertainty in weather forecasts, farmers are better able to plan and be prepared for a variety of climatic conditions. The combination of machine learning principles and neutrosophic logic, which provides flexible and adaptable solutions to the industry's many problems, has the potential to benefit the agriculture sector in several ways. As NML integration becomes more widespread in agriculture, it represents more than just technological advancement; rather, it is a critical step toward the development of efficient and sustainable farming methods. This chapter attempts to highlight the all-encompassing effect of NML in strengthening the agriculture sector's resilience against the constantly shifting environmental dynamics while also improving agricultural operations' precision and efficiency. The union of NML and agriculture represents a major change toward a future in agriculture that is more resilient, productive, and sustainable than it is merely a step forward.

Owing to the time and space constraints, the chapter could not delve into agricultural applications in the domains of pest and disease identification, agro-resource management, and climate change adaptation. The two models presented in this chapter could be trained and tested on larger and real datasets. Further research in this area may concentrate on creating NML models for precision farming, improving disease diagnosis techniques, and enhancing the precision of weather prediction and crop selection models. Furthermore, research on the long-term sustainability of NML integration in farming systems could also be explored.

NOTE

1 This relation holds for normalized neutrosophic sets and further for the sets when all the three components, namely, truth (x), indeterminacy (y), and non-membership or falsity (z), are dependent upon each other. In general, for single valued neutrosophic logic, the sum of the components is: $0 \le x + y + z \le 3$ when all three components are independent; $0 \le x + y + z \le 2$ when two components are dependent and the third one is independent from them; $0 \le x + y + z \le 1$ when all three components are dependent [2].

REFERENCES

[1] Smarandache, F. (1995). Neutrosophy/eutrosophic Probability, *Set, and Logic* http://doi.org/10.5281/zenodo.57726

[2] Smarandache, F. (2013). Introduction to Neutrosophic Measure, Neutrosophic Integral, and Neutrosophic Probability. *Sitech* https://arxiv.org/ftp/arxiv/papers/1311/1311.7139.pdf

[3] Smarandache, F. (1998). A unifying field in logics. *Neutrosophy: Neutrosophic probability, set and logic, American Research Press, Rehoboth.* https://arxiv.org/ftp/math/papers/0101/0101228.pdf

[4] Wajid, M. S., & Wajid, M. A. (2021). The importance of indeterminate and unknown factors in nourishing crime: a case study of South Africa using neutrosophy. *Neutrosophic Sets and Systems*, 41(2021), 15.

[5] Smarandache, F. (2019). Neutrosophic Set is a Generalization of Intuitionistic Fuzzy Set, Inconsistent Intuitionistic Fuzzy Set (Picture Fuzzy Set, Ternary Fuzzy Set), Pythagorean Fuzzy Set, q-Rung Orthopair Fuzzy Set, Spherical Fuzzy Set, and n-HyperSpherical Fuzzy Set, while Neutrosophication is a Generalization of Regret Theory, Grey System Theory, and Three-Ways Decision (revisited). *Journal of New Theory* Retrieved from https://digitalrepository.unm.edu/math_fsp/21

[6] Topal, S., Tas, F., Broumi, S., & Kirecci, O. A. (2020). Applications of Neutrosophic Logic of Smart Agriculture via Internet of Things. *International Journal of Neutrosophic Science (IJNS)*, 12(2), 105–115. http://doi.org/10.54216/IJNS.120205

[7] Dhingra, G., Kumar, V., & Joshi, H.D. (2019). A novel computer vision based neutrosophic approach for leaf disease identification and classification. *Measurement* https://doi.org/10.1016/j.measurement.2018.12.027

[8] Amin, K. M., Shahin, A. I., & Guo, Y. (2016). A novel breast tumor classification algorithm using neutrosophic score features. *Measurement*, 81, 210–220. https://doi.org/10.1016/j.measurement.2015.12.013
[9] Elhassouny, A., Idbrahim, S., & Smarandache, F. (2019). Machine learning in Neutrosophic Environment: A Survey. *Neutrosophic Sets and Systems*, 28, 58. https://link.gale.com/apps/doc/A600451179/AONE?u=anon~76a0c009&sid=googleScholar&xid=14ca122a
[10] Basha, S. H., Tharwat, A., Abdalla, A., & Hassanien, A. E. (2019). Neutrosophic Rule-based Prediction System for Toxicity Effects Assessment of Biotransformed Hepatic Drugs. *Expert Systems with Applications*, 121, 142–157. https://doi.org/10.1016/j.eswa.2018.12.014
[11] Wajid, M. A., & Zafar, A. (2022). A multimodal approach of information access and retrieval using neutrosophic sets. In *Emerging Trends in IoT and Computing Technologies* (pp. 382–388). Routledge.
[12] Wajid, M. A., Zafar, A., Terashima-Marín, H., & Wajid, M. S. (2023). Neutrosophic-CNN-based image and te fusion for multimodal classification. *Journal of Intelligent & Fuzzy Systems*, 45(1), 1039–1055. https://doi.org/10.3233/JIFS-223752
[13] Ye, J. (2014). Single valued neutrosophic cross entropy for multi-criteria decision-making. *Applied Mathematical Modelling*, 38, 1170–1175. https://doi.org/10.1016/j.apm.2013.07.020
[14] Mohammed A., Maissam J., & Said B. (2023). Artificial Intelligence and Neutrosophic Machine learning in the Diagnosis and Detection of COVID 19. *Journal of Prospects for Applied Mathematics and Data Analysis*, 1(2), 17–27. https://doi.org/10.54216/PAMDA.010202
[15] Sharma, M., Kandasamy, I., & Vasantha, W. B. (2021). Comparison of neutrosophic approach to various deep learning models for sentiment analysis. *Knowledge-Based Systems*, 223, 107058. https://doi.org/10.1016/j.knosys.2021.107058
[16] Ahmed, R., Nasiri, F., & Zayed, T. (2021). A novel Neutrosophic-based machine learning approach for maintenance prioritization in healthcare facilities. *Journal of Building Engineering*, 42, 102480. https://doi.org/10.1016/j.jobe.2021.102480
[17] Darwish, S. M., Salah, M. A., & Elzoghabi, A. A. (2023). Identifying indoor objects using neutrosophic reasoning for mobility assisting visually impaired people. *Applied Sciences*, 13(4), 2150. https://doi.org/10.3390/app13042150
[18] Abdelhafeez, A, Mohamed, H.K., Maher, A. & Khalil, N.A. (2023). A novel approach toward skin cancer classification through fused deep features and neutrosophic environment. *Frontiers in Public Health* 11:1123581. doi: 10.3389/fpubh.2023.1123581
[19] Hassan, M. H., Darwish, S. M., & Elkaffas, S. M. (2022). An efficient deadlock handling model based on neutrosophic logic: Case study on real-time healthcare database systems. *IEEE Access*, 10, 76607–76621. https://doi.org/10.1109/ACCESS.2022.3192414
[20] Pai, S. P., & Prabhu Gaonkar, R. S. (2021). Safety modeling of marine systems using neutrosophic logic. *Proceedings of the Institution of Mechanical Engineers, Part M: Journal of Engineering for the Maritime Environment*, 235(1), 225–235. https://doi.org/10.1177/1475090220925733

[21] Nagarajan, D., Broumi, S., & Smarandache, F. (2023). Neutrosophic speech recognition algorithm for speech under stress by machine learning. *Neutrosophic Sets and Systems*, 55, 1.
[22] Juro-Barrios, J., Gamboa-Cruzado, J., Romero Baylon, A., & del Valle Jurado, C. (2023). Practical Validation in a Neutrosophic Environment of the NEBS Methodology for the Optimization of SME Financing through Machine Learning. *Journal of International Journal of Neutrosophic Science*, 20(3), 137–149. https://doi.org/10.54216/IJNS.200313
[23] Bhushan, B., Sharma, S. K., Nand, P., Shankar, A., & Obaid, A. J. (Eds.). (2024). Emerging Trends for Securing Cyber Physical Systems and the Internet of Things.
[24] Pandey, S., & Bhushan, B. (2024). Recent Lightweight cryptography (LWC) based security advances for resource-constrained IoT networks. *Wireless Networks*, 30(4), 2987–3026.

Index

Pages in *italics* refer to figures and pages in **bold** refer to tables.

For Product Safety Concerns and Information please contact our EU representative GPSR@taylorandfrancis.com Taylor & Francis Verlag GmbH, Kaufingerstraße 24, 80331 München, Germany

Batch number: 10397790

Printed by Printforce, the Netherlands